The Essential Guide to Passing the Structural Civil PE Exam Written in the form of Questions

175 CBT Questions Every PE Candidate Must Answer

Request latest Errata, or add yourself to our list for future information about this book by sending an email with the book title, or its ISBN, in the subject line to:
Errata@PEessentialguides.com
or
info@PEessentialguides.com

The Essential Guide to Passing the Structural Civil PE Exam Written in the form of Questions

175 CBT Questions Every PE Candidate Must Answer

Jacob Petro
PhD, PMP, CEng, PE

Petro Publications LLC
PE Essential Guides
Hillsboro Beach, Florida

Report Errors For this Book

We are grateful to every reader who notifies us of possible errors. Your feedback allows us to improve the quality and accuracy of our products.

Report errata by sending an email to Errata@PEessentialguides.com

The Essential Guide to Passing the Structural Civil PE Exam Written in the form of Questions – 175 CBT Questions Every PE Candidate Must Answer

Second Edition

For written permissions contact: Permissions@PEessentialguides.com

For general inquiries contact: Info@PEessentialguides.com or PEessentialguides@outlook.com

Imprint name: PE Essential Guides

Company owning this imprint: Petro Publications LLC. Established in Florida, 2023.

ISBN: 979-8-9891857-2-6

Disclaimer

The information provided in this book is intended solely for educational and illustrative purposes. It is important to note that the technical information, examples, and illustrations presented in this book should not be directly copied or replicated in real engineering reports or any official documentation.

While there may be resemblances between the examples in this book and real structures, users must exercise caution and conduct comprehensive verification of all information before implementing it in any practical setting. The author and all affiliated parties explicitly disclaim any responsibility or liability arising from the misuse, misinterpretation, or misapplication of the information contained in this book.

Furthermore, it is essential to understand that this book does not constitute legal advice, nor can it be considered as evidence or exhibit in any court of law. It is not intended to replace professional judgment, and readers are encouraged to consult qualified experts or seek legal counsel for any specific legal or technical matters.

By accessing and utilizing the information in this book, readers acknowledge that they do so at their own risk and agree to hold the author and all affiliated parties harmless from any claims, damages, or losses resulting from the use or reliance upon the information provided herein.

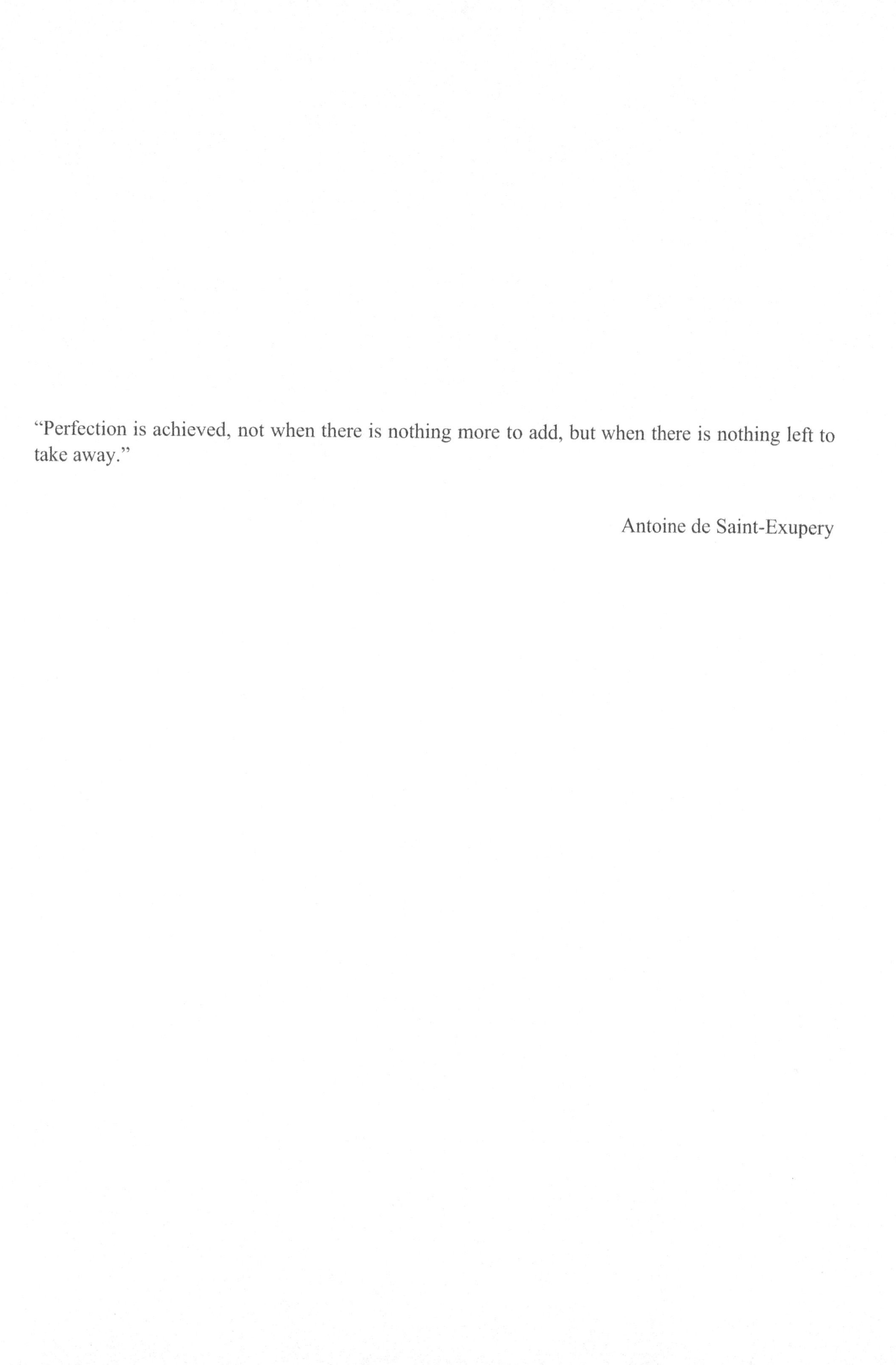

"Perfection is achieved, not when there is nothing more to add, but when there is nothing left to take away."

Antoine de Saint-Exupery

Preface

General information about the Book

This book is designed to help civil engineers pass the NCEES exam which is a prerequisite for obtaining the professional engineering PE license in the United States. This book is tailored to provide you with comprehensive knowledge, detailed examples, and step-by-step solutions with ample graphics that are directly related to the subjects covered by the NCEES exam.

In this book, you will find an extensive collection of civil engineering problems that are carefully selected to build your knowledge, skills, and ability to apply fundamental principles and advanced concepts in all fields of civil engineering. These problems are accompanied by detailed explanations, diagrams, and equations to help you understand the underlying principles and solve the problems efficiently and accurately.

Whether you are a recent graduate, a seasoned engineer, or a professional who wants to obtain the engineering license in the United States, this book will prepare you for the exam and equip you with the necessary tools to succeed.

The book is structured in a way such that it provides the reader with a comprehensive understanding of the core topics that are covered by the NCEES exam the morning session (the general session). The book covers topics in general civil engineering such as, but not limited to environmental engineering, water resources, transportation, surveying, planning and construction, geotechnical, structural, etc.

The book provides the reader with a full coverage and understanding for all NCEES exam topics and possible question scenarios during a real test situation. If certain topics or methods were not covered in this book, the book method of presentation will ultimately guide you on how to find the solution you seek on your own, know where to find it, and provide the solution on a timely manner that saves you time during the real exam.

The questions in this book are neither easy nor difficult. They are constructive and creative in nature. They have been authored in a way to have you remember the core concept of the engineering topic you seek. They are designed to challenge engineers to think critically and apply their knowledge to exam and real-world scenarios. These questions require a deeper level of analysis and understanding than simple recall of information. They may involve multiple steps or require the engineer to consider different perspectives or solutions, while they can be difficult and require significant amount of effort and research to solve them.

Reasons I wrote this Book

I decided to author this book because I have a strong passion for engineering. I have a deep interest and understanding of civil engineering, and I wanted to share this knowledge and passion with others.

I am also very enthusiastic about engineering, and this has allowed me to explore various concepts and develop unique perspectives.

By writing this book I hope I can inspire others to pursue and improve on their career in engineering, help them pass the NCEES exam, improve on their skills and advance their knowledge in this field and provide them with the tools they need to succeed.

Lastly, the energy and enthusiasm I have, and I brought into this work is infectious and I wanted to channel this energy into this fascinating project and share it with others. I strongly believe this book will be a valuable resource for anyone interested in learning more about civil engineering.

Acknowledgement and Dedication

I would like to thank the readers of this book, who I hope will find it informative, engaging, and thought-provoking. It is my sincere hope that this book will inspire others to pursue their own passion, and that it will serve as a valuable resource for all those interested in the field of engineering.

Table of Contents

Preface .... **v**

General information about the Book v
Reasons I wrote this Book v
Acknowledgement and Dedication vi

About the Author .... **ix**

New in Second Edition .... **ix**

Introduction **1**

General description of this Book 1
Book Structure 1

About the Exam **2**

General information 2
Dissecting the Exam 3

How to use this Book **3**

Which References to Own **4**

Map of Problems Presented **5**

Part I: General Breadth 5
Part II: Structural Depth 6
NCEES Matrix of General Breadth Knowledge Areas Covered in this Book 10
NCEES Structural Depth Knowledge Areas Covered in this Book 14

PART I: General Breadth Problems & Solutions **15**

General Civil Engineering 15

PART II: Structural Depth Problems & Solutions **99**

Structural Analysis 101
Cross Section Analysis 123
Loads and Load Analysis 139
Structural Design 163
Codes and Specifications 213

References **227**

References' Key Chapters **229**

General 229
Chapters per Reference 229
IBC International Building Code 230
ACI 318 Building Code for Structural Concrete 231
TMS 402/602 Building Code Requirements and Specifications for Masonry Structures 232
PCI Design Handbook 233
AISC Steel Construction Manual 234
NDS National Wood Design Specifications 235
ASCE 7 Minimum Design Loads for Buildings and Other structures 236
AASHTO LRFD Bridge Design Specification 238

About the Author

Dr. Petro is a professional engineer and a business leader with over 20 years of experience in leading and growing engineering companies. Throughout his career, he worked with some of the most prestigious engineering firms. With a vast background in design and construction, he earned a reputation for delivering innovative and cutting-edge projects throughout his career.

Dr. Petro is a civil engineer, he holds a Doctorate degree, he has earned a Professional Engineering (PE) license as well as Chartered Engineer (CEng) certification. Additionally, Dr. Petro has earned a Project Management Professional (PMP) certification, further demonstrating his expertise in managing complex projects. Over the years, he successfully led and managed teams of engineers, designers, and other professionals, overseeing complex projects from conception to completion.

Throughout his career, Dr. Petro designed and delivered numerous innovative and interesting projects that have contributed significantly to various industries he has worked in. His passion for engineering and business has driven him to publish several papers and articles in industry-leading journals and magazines. His work has been recognized as state-of-the-art and has been referenced by many industry professionals.

As an international civil engineer who worked across the globe, Dr. Petro brings an interesting perspective to the table. He has a deep understanding of how civil facilities and structures work and how to optimize them for maximum efficiency and safety. His ability to communicate complex engineering concepts to both technical and non-technical stakeholders has been key to his success.

New in Second Edition

The following have been added/updated in the Second Edition of this book:

- ✓ The addition of 40 more General Breadth questions.
- ✓ The addition of four more important Depth Questions that the writer thought they were important to achieve a comprehensive coverage through this publication.
- ✓ The update of the NCEES Handbook version used in this edition to version 1.2.

Introduction

General description of this Book

The Essential Guide to Passing the Structural Civil PE Exam Written in the form of Questions, is a guide designed in the form of questions. It is set to prepare engineers for the two sessions of the Civil PE exam – the breadth (the morning session) and the structural depth (the afternoon session).

This book has been meticulously crafted to provide civil/structural engineers with the knowledge and guidance necessary to succeed in the exam and achieve their career goals.

This book contains 175 carefully curated questions with detailed answers that aims a complete coverage for a range of topics relevant to the NCEES PE exam the general and the structural depth. These questions are accompanied by graphics and detailed step-by-step explanations to help engineers understand the concepts better.

The first 80 questions cover the general breadth of the exam and civil engineering, while the remainder of the questions are focused on the structural engineering depth.

The structural engineering depth (Part II of this book) is divided into five sections, each section covers a specific area of expertise. These sections are designed to cover all topics of the exam and guide structural engineers and deepen their understanding in their field and gives them the needed confidence to tackle the most complex exam questions.

The book contains comprehensive explanations of key concepts, step-by-step problem-solving techniques, and real-world examples that will also help engineers apply this knowledge to real-world scenarios.

Book Structure

Part I of this book contains 80 problems on the civil engineering general breadth. Those questions cover a wide range of topics, including transportation, general civil, structural, statics, construction, economics, water resources and environmental, geotechnical and more.

Part II of this book has the structural depth questions that are compliant with the NCEES structural exam, and the recent references and guides used in the CBT exam. This part of the book comprises of 95 detailed questions, each accompanied by a comprehensive answer.

Part II of this book is divided into five sections: (1) Structural Analysis, (2) Cross-Section Analysis, (3) Loads and Load Analysis, (4) Structural Design, and (5) Codes and Specifications.

The *Structural Analysis Section* covers topics such as deflection, indeterminate structures, and stability analysis, 3D trusses, moving loads and others.

The *Cross-Section Analysis Section* focuses on the behavior of cross-sections under axial, bending, shear, torsion, and combined loading conditions. This section provides examples on the different types of cross-sections, such as rectangular, circular, and I-shaped. It explains how to calculate their properties and how to maximizes inertias via rotation. It provides insights into the principles of mechanics of materials for composite structures, prestressed sections, and during asymmetric loading.

The *Loads and Load Analysis Section* covers the principles of structural loads, including dead loads, live loads, wind loads, seismic loads, snow loads, rain loads, etc. This section provided ample methods of how to calculate these loads and how to analyze them and incorporate them into structural analysis and design. It also covers the use of various codes when it comes to defining load requirements, such as the AASHTO LRFD, ASCE 7, ACI and AISC.

The *Structural Design Section* covers the design of structural elements such as beams, columns, foundations, and connections. This section of the book provides ample explanation on design requirements, such as strength, serviceability, and stability, and how to calculate the required sizes of members. This section also includes questions on the design of steel, concrete, masonry, and wood structures. This section alone has more than 30 problems with coverage of various design topics.

The *Codes and Specifications Section* contains problems that cover the codes and standards that are used in structural engineering, including the American Concrete Institute (ACI), American Institute of Steel Construction (AISC), the American Society of Civil Engineers (ASCE 7), the International Building Code (IBC), the AASTHO LRFD Bridge Design Specifications, the Building Code Requirements for Masonry Structures (TMS 402, ACI 530 and TMS 602) and others.

Although Part II of this book is divided into five distinct sections, it is important to recognize that these sections are not entirely separate from each other. In fact, and due to the nature of civil and structural engineering, these sections share common themes and ideas that overlap in various ways. However, to make the content more accessible and easier to navigate, also to make sure this book has the best NCEES exam requirements coverage, the second part was divided as such.

About the Exam

General information

The NCEES PE exam is a rigorous exam that is administered in two sessions, a morning session, and an afternoon session. The morning session is four hours long and focuses on broader, foundational concepts in the engineering field along with a wide range of engineering problems compared to the afternoon session. Topics during this session could vary and the selection is wide, and this could cover geotechnical, general civil, transportation, water resources, statics, material and mix designs, structural design, and analysis, and much more.

A good preparation resource would be the *NCEES Handbook* that you can download version 1.2 free of charge from their website. Make sure that you go over this handbook and make sure you familiarize yourself with all the topics it has.

The afternoon session is also four hours long and is generally more focused on specific areas of expertise, either structural engineering, transportation, geotechnical, construction, or water resources and environmental.

Both sessions consist of multiple-choice questions, and candidates are typically required to demonstrate their ability to analyze and solve complex engineering problems.

The exam is designed to test not only one's knowledge and technical skills, but also their ability to think critically, communicate effectively, and work under pressure.

Dissecting the Exam

The morning session consists of 39 questions and the afternoon session consists of 41 questions; the questions count in each session may vary but will never go above 80 questions in total.

During the morning session of a certain depth exam, you may find more questions targeting a particular relevant depth topic that you may need to access certain codes or other material apart from the *NCEES Handbook* to solve them. So be prepared to do so.

During this exam you will be given a total of 480 minutes to solve 80 complex questions. This means that you have an average allotted time of six minutes per question. It is very important to keep in mind that some questions during the real exam will take only one minute to solve while others could take you up to ten minutes to complete. To prepare for the exam, it is crucial to practice solving questions that have longer duration and that are more difficult, which is what this book is trying to provide.

Let me give you a reflection from my own experience sitting for the NCEES PE exam. During the exam, I was thoroughly prepared to confidently answer each question within the allotted time, and even expected to complete the exam at a faster pace. However, I regretfully failed to study two specific chapters of a certain manual that I had assumed would have a lower probability of being included in the exam. To my dismay, two difficult questions emerged from these topics, leaving me with no alternative but to study the relevant material from the codes and specifications provided. Consequently, I spent approximately 20-30 minutes ensuring that my answers were accurate. This unexpected event significantly impacted the remainder of my allotted exam time, and although I ultimately passed, it was an avoidable situation.

This experience is one of the reasons why I strived to provide a comprehensive coverage of all possible topics in the book. In case I had inadvertently failed to cover any subject, the book, and its solutions are presented in a manner that facilitate faster problem-solving.

How to use this Book

The questions presented in this book have been designed to be a mix of varying lengths. Some may only take a minute or two to answer, while others may require up to 10 or 15 minutes. This design has been done as such intentionally, as it reflects the format of the actual PE exam with more questions that are longer and more difficult than the exam.

By practicing with questions of varying lengths, you will be better prepared to manage your time during the exam. You will learn to quickly identify the easier questions and move through them efficiently, while also having the skills to tackle the more time-consuming questions effectively. It is important to note that practicing only short questions, or questions that are six minutes long, may not be enough to fully prepare you for the exam.

Furthermore, the variety of questions' lengths helps to keep your mind engaged and challenged. It can be easy to become bored or disengaged when faced with a series of similar questions, but by mixing up the lengths, you will be forced to stay focused and adapt to the different types of questions.

Therefore, it is important to practice longer and more difficult questions in addition to the shorter ones. This will help you develop your ability to think critically, analyze complex problems, and apply your knowledge to solve them. It will also help you build your endurance and focus, which is critical for success on this exam.

During your practice sessions solving questions provided in this book, it is okay to spend more time on difficult questions or even short ones. This will help you identify your weaknesses and areas where you need to improve. By practicing longer and more difficult questions, you will be well prepared to pass the exam and you will be better equipped to handle any challenges that may arise on exam day. Also don't forget that you are learning and improving your ability by using this book, which reinforces that you should take your time doing so.

As a final note, it is important to view this book as a textbook as it contains, not only straightforward questions and answers, but comprehensive explanations and guidance and wide coverage to the most complex exam problems with detailed elaborations of both questions and answers. In addition, this book provides proper referencing and in-depth analysis of the topics in hand while also exploring alternative solutions and offering valuable insights and materials that can be beneficial not only during an exam but also in one's professional career.

Which References to Own

The question at hand is whether owning references is necessary at this stage or not, and, if so, how many references are required, and which ones should you own and which ones you can spare.

The exam, and this book, is based on nine references, along with the NCEES PE handbook, all mentioned and detailed in the last section of this book, and they are all required to pass the exam. Owning all of them could cost you a fortune.

Although all references are provided during the CBT test, you don't want to find yourself in a situation where you are flipping through the references attempting to answer complex questions and running short on time. You should thoroughly familiarize yourself with all these references beforehand, including all their chapters and the location of relevant topics within them. This is the golden rule to pass the exam.

So, shall you own all the references to study and pass the exam?

You can either buy the references if you feel you would like to have an uninterrupted access or borrow them from a library or a colleague. Owning all references can be beneficial as it provides you with the freedom to study at your own pace and in the manner that suits you best.

On the other hand, it is important to know that you can access some of the references online if not all, should one prefer not to incur significant expenses. However, you need to acknowledge that accessing free material is not risk free. For instance, free material obtained online may be older in version, or a version that has been shared for the purpose of commentary. While it is acceptable to utilize such material and references for study purposes only, it is crucial to be cognizant of the associated risks and to comprehend any modifications that may have been made in recent versions. Ultimately, the objective is to acquaint yourself with these references, irrespective of whether they are recent or outdated, obtained for free or borrowed from a friend or a library.

Map of Problems Presented

Part I: General Breadth

Problem 1 *Watershed Rainfall Depth*	**Problem 2** *Precipitation Methods*
Problem 3 *Soil Moisture Content*	**Problem 4** *Seawater Canal*
Problem 5 *Water Channel*	**Problem 6** *Project Planning*
Problem 7 *Settlement in Clay*	**Problem 8** *Project Planning*
Problem 9 *Boreholes Locations*	**Problem 10** *Soil Shear Strength*
Problem 11 *2D Truss Deflection*	**Problem 12** *Horizontal Curve*
Problem 13 *Pressure Under Footing*	**Problem 14** *Construction Material*
Problem 15 *Construction Methods*	**Problem 16** *Shear Stress Cross Section*
Problem 17 *Soil Testing*	**Problem 18** *Water Discharge External Forces*
Problem 19 *Travel Time for Shallow Flow*	***Problem 20*** *Vertical Curve Design*
Problem 21 *Construction Material*	**Problem 22** *Net Present Value Analysis*
Problem 23 *Construction Activities*	**Problem 24** *Cross Section Analysis*
Problem 25 *Material Properties*	**Problem 26** *Material Properties*
Problem 27 *Retention Pond Sizing*	**Problem 28** *Environmental Chemical Reactions*
Problem 29 *Concrete Design/Wall Thickness Determination*	**Problem 30** *Soil Properties*
Problem 31 *Soil Classification System*	**Problem 32** *Geotechnical Engineering*
Problem 33 *Foundation Settlement*	**Problem 34** *Soil Classification System*
Problem 35 *Effective Stress Over Time*	**Problem 36** *Retaining Wall Safety Factor*
Problem 37 *Budgetary Cost Acceptable Range*	**Problem 38** *Head Losses*
Problem 39 *Concrete Mix Design*	**Problem 40** *2D Truss Reactions*
Problem 41 *Roadway Density*	**Problem 42** *Points on Vertical Curve*
Problem 43 *Frame Moment Diagram*	**Problem 44** *Benefit Cost Analysis*

Problem 45 *Shallow Flow*	**Problem 46** *Resources Histogram*
Problem 47 *Double Horizontal Curve*	**Problem 48** *Mass Haul Diagram*
Problem 49 *Peak Hour Factor*	**Problem 50** *Volume of Excavation*
Problem 51 *Skid Marks Distance*	**Problem 52** *Unconfined Aquifer*
Problem 53 *Transit Average Speed*	**Problem 54** *Retention Basin Design*
Problem 55 *Elevation of Water Surface in Reservoir*	**Problem 56** *Concrete Beam Deflection*
Problem 57 *Salvage Value*	**Problem 58** *Soil Erodibility*
Problem 59 *Soil Loss Prevention*	**Problem 60** *Optimum Moisture Content*
Problem 61 *Slope Stability/ Slope Safety Factor*	**Problem 62** *Permissible Noise Exposure*
Problem 63 *Consolidation Settlement*	**Problem 64** *Excavation Construction Safety*
Problem 65 *Distresses in Flexible Pavements*	**Problem 66** *Formwork Design*

Problem 67 *Dam Site Rock Quality*	**Problem 68** *Concrete Section Design*
Problem 69 *Combined Stresses*	**Problem 70** *Fastener Group in Shear*
Problem 71 *Steel Props Spacing*	**Problem 72** *Rate of Concrete Placement*
Problem 73 *85th Percentile Speed*	**Problem 74** *Safety Incidence Rate*
Problem 75 *Framework Prop Loading*	**Problem 76** *Bearing Capacity for a Square foundation*
Problem 77 *Construction Operation Over a Single Footing*	**Problem 78** *Estimated Cost and Profitability*
Problem 79 *Ground Anchor Capacity*	**Problem 80** *Pile Depth Calculation*

Part II: Structural Depth
Section 1: Structural Analysis

Problem 1.1 *Two Fixed Ends*	**Problem 1.2** *A Strand With a Cantilever Beam*
Problem 1.3 *3D Truss Reactions*	**Problem 1.4** *Structural Changes*
Problem 1.5 *Moment Distribution Method*	**Problem 1.6** *Moving Load on a Truss*

Problem 1.7 *Moving Load on a Continuous Beam*	**Problem 1.8** *3D Truss Deflection*
Problem 1.9 *Frame Deflection*	**Problem 1.10** *3D Frame Deflection and Torque*
Problem 1.11 *Frame Moment Diagram*	**Problem 1.12** *Moment Distribution*

Section 2: Cross Section Analysis

Problem 2.1 *C Section Internal Stresses*	**Problem 2.2** *Shear Center*
Problem 2.3 *Section Unsymmetrical Loading*	**Problem 2.4** *Shear Stress Calculation*
Problem 2.5 *Maximizing Moment of Inertia*	**Problem 2.6** *Composite Prestressed Beam Analysis*
Problem 2.7 *Moment of Inertia Calculation*	**Problem 2.8** *Shear Flow Determination*
Problem 2.9 *Shear Flow Profile*	**Problem 2.10** *Shear Stress Profile*

Section 3: Loads and Load Analysis

Problem 3.1 *Base Shear for a Tall Building*	**Problem 3.2** *Vertical Seismic Load Distribution*
Problem 3.3 *Seismic Design Requirements*	**Problem 3.4** *Seismic Load Directions*
Problem 3.5 *Vertical Seismic Load Distribution*	**Problem 3.6** *Horizontal Seismic Load Distribution*
Problem 3.7 *Rain Load Calculation*	**Problem 3.8** *Flood Elevation Design*
Problem 3.9 *Retaining Wall Applicable Loads*	**Problem 3.10** *Distribution of Pressure /Footing*
Problem 3.11 *Traffic Dynamic Load*	**Problem 3.12** *Lateral Distribution of Loads*
Problem 3.13 *Load Combination*	**Problem 3.14** *Snow Load Calculation*
Problem 3.15 *Live Load Reduction*	**Problem 3.16** *Live Load Maximization*
Problem 3.17 *Vehicular Collision Force*	**Problem 3.18** *Load Combination*
Problem 3.19 *Wind Pressure on Roof*	**Problem 3.20** *Wind Force on a Sign board*

Section 4: Structural Design

Problem 4.1 *Critical Bucking Load Nonsway Frames*	**Problem 4.2** *Critical Bucking Load Sway Frames*
Problem 4.3 *One Way Slab Thickness*	**Problem 4.4** *Two Way Slab Thickness*
Problem 4.5 *Single Footing Thickness*	**Problem 4.6** *Concrete Slab Moment Calculation*

Problem 4.7 *Flat Slab Moment*	**Problem 4.8** *Flat Slab Shear Analysis*
Problem 4.9 *Punching Shear caused by Circular Columns*	**Problem 4.10** *Concrete Section Design*
Problem 4.11 *Concrete Section Design*	**Problem 4.12** *Masonry Wall Bearing Analysis*
Problem 4.13 *Masonry Wall Bearing Design*	**Problem 4.14** *Masonry Beam Section Design*
Problem 4.15 *Masonry Wall Design*	**Problem 4.16** *Steel Section Column Design*
Problem 4.17 *Steel Shear Connection Analysis*	**Problem 4.18** *Steel Corbel Analysis*
Problem 4.19 *Steel Section with Web Opening*	**Problem 4.20** *Steel Section with Reinforced Opening*
Problem 4.21 *Steel Corbel Analysis*	**Problem 4.22** *Truss Connection Weld Design*
Problem 4.23 *Gantry Crane WT Section Design*	**Problem 4.24** *Steel WT Section Design*
Problem 4.25 *Steel W Section Design*	**Problem 4.26** *Steel Plate in Tension*
Problem 4.27 *Glulam Wooden Beam Design*	**Problem 4.28** *Sawn Lumber Column Design*
Problem 4.29 *Lumber Loss of Strength Due to Fire*	**Problem 4.30** *Adjustment Factors for Glulam*
Problem 4.31 *Wooden Truss Joint*	**Problem 4.32** *Effective Char Depth*

Section 5: Codes and Specifications

Problem 5.1 *Inspection of Construction Activities*	**Problem 5.2** *Reinforced Concrete Basement Wall*
Problem 5.3 *Concrete Mix Design*	**Problem 5.4** *Concrete Mix Design*
Problem 5.5 *Construction Safety*	**Problem 5.6** *Concrete Mix Design*
Problem 5.7 *Masonry Basement Wall*	**Problem 5.8** *Concrete Strength in Bridge Design*
Problem 5.9 *Abutment Design*	**Problem 5.10** *Concrete Testing*
Problem 5.11 *Highway Bridge Column Design*	**Problem 5.12** *Seismology Retaining Wall Safety*
Problem 5.13 *Masonry Wall Openings*	**Problem 5.14** *Masonry Wall Reinforcement Req's*
Problem 5.15 *Minimum Distance From Slopes*	**Problem 5.16** *Masonry Mix Design*

Problem 5.17 *Masonry Construction Inspection*	**Problem 5.18** *Steel Members Strength During Fire*
Problem 5.19 *Structural Steel Weld Inspection*	**Problem 5.20** *Shotcrete Mix Design*
Problem 5.21 *Masonry Wall Minimum Requirements*	

NCEES Matrix of General Breadth Knowledge Areas Covered in this Book

There are eight knowledge areas in the Civil PE Exam. You can see these knowledge areas and the number of items covered for each one on the following three pages in a matrix format.

The question numbers and their titles in this book in Part I are also mapped to these knowledge areas in the matrix. This helps if you would like to focus on specific breadth questions or you need more practice in a particular area, or you would like to check how much material this book covers for the breadth exam.

This matrix is included here to ensure that this book provides comprehensive coverage of all possible exam topics in the breadth Part – Part I. For instance, on the leftmost column, you'll find the number of expected exam questions, as reported on the NCEES official website. When you compare this to the number and variety of questions presented in this book for each knowledge area, you will see that there is sufficient coverage per knowledge area.

Knowledge Area	Items required	Relevant questions in this book
Project Planning *Expected Questions in the exam: 4-6*	A. Quantity take-off methods B. Cost estimating C. Project schedules D. Activity identification and sequencing	Problem 6: *Project Planning* Problem 8: *Project Planning* Problem 22: *Net Present Value Analysis* Problem 37: *Budgetary Cost Acceptable Range* Problem 44: *Benefit Cost Analysis* Problem 46: *Resources Histogram* Problem 57: *Salvage Value* Problem 78: *Estimated Cost and Profitability*
Means and Methods *Expected Questions in the exam: 3-5*	A. Construction loads B. Construction methods C. Temporary structures and facilities	Problem 15: *Construction Methods* Problem 23: *Construction Activities* Problem 66: *Formwork Design* Problem 71: *Steel Props Spacing* Problem 72: *Rate of Concrete Placement* Problem 75: *Framework Prop Loading* Problem 77: *Construction Operation Over a Single Footing*
Soil Mechanics *Expected Questions in the exam: 5-8*	A. Lateral earth pressure B. Soil consolidation C. Effective and total stresses D. Bearing capacity E. Foundation settlement F. Slope stability	Problem 7: *Settlement in Clay* Problem 13: *Pressure under Footing* Problem 33: *Foundation Settlement* Problem 35: *Effective Stress Over Time* Problem 36: *Retaining Wall Safety Factor* Problem 61: *Slope Stability/Slope Safety Factor* Problem 63: *Consolidation Settlement* Problem 76: *Bearing Capacity for a Square Foundation* Problem 79: *Ground Anchor Capacity* Problem 80: *Pile Depth Calculation*

Structural Mechanics *Expected Questions in the exam: 5-8* 5 + 12 + 5 = 22	A. Dead and live loads B. Trusses C. Bending, shear, axial D. Combined stresses E. Deflection F. Beams, columns, slab, footings G. Retaining walls	Problem 11: *2D Truss Deflection* Problem 16: *Shear Stress Cross Section* Problem 24: *Cross Section Analysis* Problem 25: *Material Properties* Problem 26: *Material Properties* Problem 29: *Concrete Design/Wall Thickness Determination* Problem 40: *2D Truss Reactions* Problem 43: *Frame Moment Diagram* Problem 56: *Concrete Beam Deflection* Problem 68: *Concrete Section Design* Problem 69: *Combined Stresses* Problem 70: *Fastener Group in Shear*
Hydraulics and Hydrology *Expected Questions in the exam: 6-8*	A. Open-channel flow B. Stormwater collection and drainage C. Storm characteristics D. Runoff analysis E. Detention/retention ponds F. Pressure conduits G. Energy and/or continuity equation	Problem 1: *Watershed Rainfall Depth* Problem 2: *Precipitation Methods* Problem 4: *Seawater Canal* Problem 5: *Water Channel* Problem 18: *Water Discharge External Forces* Problem 19: *Travel Time for Shallow Flow* Problem 27: *Retention Pond Sizing* Problem 38: *Head Losses* Problem 45: *Shallow Flow* Problem 52: *Unconfined Aquifer* Problem 54: *Retention Basin Design* Problem 55: *Elevation of Water Surface in Reservoir*
Geometrics *Expected Questions in the exam: 3-5*	A. Basic circular curve elements B. Basic vertical curve elements C. Traffic volume	Problem 12: *Horizontal Curve* Problem 20: *Vertical Curve* Problem 41: *Roadway Density* Problem 42: *Points on Vertical Curve*

		Problem 47: *Double Horizontal Curve* Problem 49: *Peak Hour Factor* Problem 51: *Skid Marks Distance* Problem 53: *Transit Average Speed* Problem 73: *85th Percentile Speed*
Materials *Expected Questions in the exam: 5-8*	A. Soil classification and boring log interpretation B. Soil properties C. Concrete D. Structural steel E. Material test methods and specifications conformance F. Compaction	Problem 3: *Soil Moisture Content* Problem 9: *Boreholes Locations* Problem 10: *Soil Shear Strength* Problem 14: *Construction Material* Problem 17: *Soil Testing* Problem 21: *Construction Material* Problem 28: *Environmental Chemical Reactions* Problem 30: *Soil Properties* Problem 31: *Soil Classification System* Problem 32: *Geotechnical Engineering* Problem 34: *Soil Classification System* Problem 39: *Concrete Mix Design* Problem 60: *Optimum Moisture Content* Problem 67: *Dam Site Rock Quality*
Site Development *Expected Questions in the exam: 4-6*	A. Excavation and embankment B. Construction site layout and control C. Temporary and permanent soil erosion and sediment control D. Impact of construction on adjacent facilities E. Safety	Problem 48: *Mass Haul Diagram* Problem 50: *Volume of Excavation* Problem 58: *Soil Erodibility* Problem 59: *Soil Loss Prevention* Problem 62: *Permissible Noise Exposure* Problem 64: *Excavation Construction Safety* Problem 65: *Distresses in Flexible Pavements* Problem 74: *Safety Incidence Rate*

NCEES Structural Depth Knowledge Areas Covered in this Book

There are three knowledge areas in the structural civil PE Exam as set per the requirements and specifications of the NCEES organization. Those three knowledge areas are broken down into six sub knowledge areas and they are all summarized in the table below.

All these knowledge areas are covered in the five sections of Part II of this book.

Knowledge Area	***Expected No. of Questions***
Analysis of Structures	**13-20**
A. Loads and load applications: 1. Dead loads 2. Live loads 3. Construction loads 4. Wind loads 5. Seismic loads 6. Moving loads 7. Snow, rain, ice 8. Impact loads 9. Earth pressure and surcharge 10. Load paths (lateral & vertical) 11. Load combinations 12. Tributary areas	4-6
B. Forces and load effects: 1. Diagrams (shear & moment) 2. Axial (tension and compression) 3. Shear 4. Flexure 5. Deflection 6. Special topics (torsion, buckling, fatigue, progressive collapse, thermal deformation, bearing, etc.)	9-14
Design and Details of Structures	**16-24**
A. Materials and material properties: 1. Concrete (plain, reinforced cast-in-place, prestressed, etc.) 2. Steel (structural, reinforcing, cold-formed, etc.) 3. Timber 4. Masonry (brick, veneer, CMU.)	4-6
B. Component design and detailing: 1. Horizontal members (beams, slabs, diaphragms, etc.) 2. Vertical members (columns, bearing walls, shear walls, etc.) 3. Systems (trusses, braces, frames, composite structures, etc.) 4. Connections (bearing, bolted, welded, anchored, etc.) 5. Foundations (retaining walls, footings, combined footings, slabs, etc.)	12-18
Code and Construction	**6-10**
A. Codes, standards, and guidance documents: 1. International Building Code (IBC) 2. American Concrete Institute (ACI 318, 530) 3. Precast/Prestressed Concrete Institute (PCI Design Handbook) 4. Steel Construction Manual (AISC) 5. National Design Specification for Wood Construction (NDS) 6. LRFD Bridge Design Specifications (AASHTO) 7. American Welding Society (AWS D1.1, D1.2 and D1.4) 8. OSHA 1910 General Industry and OSHA 1926 Construction Safety Standards	4-6
B. Temporary structures and other topics: 1. Special inspections 2. Submittals 3. Formwork 4. Falsework and Scaffolding 5. Shoring and reshoring 6. Concrete maturity and early strength evaluation 7. Bracing 8. Anchorage 9. OSHA regulations 10. Safety management	2-4

PART I

GENERAL BREADTH

General Civil Engineering

Problems & Solutions

PROBLEM 1 ***Watershed Rainfall Depth***
The following watershed has a total area of $0.31\,Acre$ and is plotted to scale on a $10\,ft \times 10\,ft$ grid.

Rain gauge stations 1, 2, 3 and 4 have been placed as shown and they measure the following rain depths:

Station 1 = $7.5\,in$

Station 2 = $5.5\,in$

Station 3 = $11.5\,in$

Station 4 = $7.5\,in$

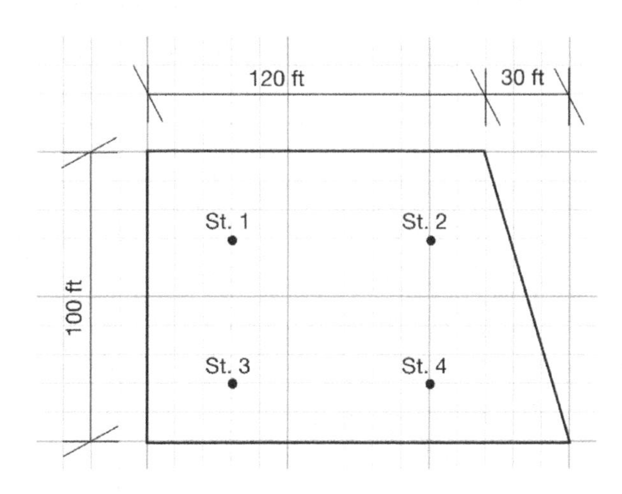

The average rainfall depth over the shown watershed using the Thiessen method is most nearly:

(A) $8.0\,in$

(B) $7.8\,in$

(C) $7.6\,in$

(D) $7.5\,in$

PROBLEM 2 ***Precipitation Methods***
The most accurate method for averaging precipitation over an area is:

(A) The mathematical averaging method.

(B) The Isohyetal method.

(C) The Thiessen method.

(D) None of the above, each of those methods has a specific use and accuracy is irrelevant in this case.

PROBLEM 3 ***Soil Moisture Content***
The optimum moisture content of a soil is close to its:

(A) Liquid limit

(B) Shrinkage limit

(C) Plastic limit

(D) Plasticity index

PROBLEM 4 ***Seawater Canal***
The below seawater return canal is made of concrete lining and is situated in an industrial area. Its purpose is to return seawater used to cool down equipment from factories in this industrial area back to the sea through an outfall.

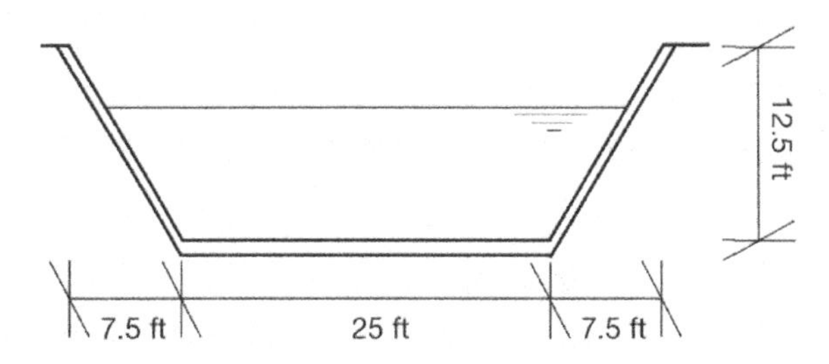

Assuming a continuous daily operation of this canal with a steady uniform flow, a design freeboard of $3.75\,ft$ and a slope of 2%, the seawater intake structure should be sized for a maximum intake (in MGD) of nearly:

(A) 9,000

(B) 7,750

(C) 5,300

(D) 9,200

PROBLEM 5 ***Water Channel***

The below is a cross section of a V shaped open channel that delivers water at $70^o\ F$ with a 1 ft free board. The mean velocity of the flow is 0.3 ft/sec.

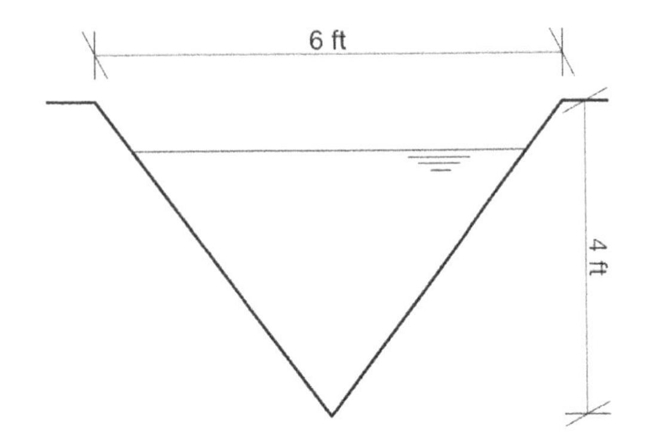

Reynolds number for this flow is most nearly:

(A) 0.32

(B) 32,000

(C) 0.26

(D) 26,000

PROBLEM 6 ***Project Planning***

The below table summarizes four activities which belong to a larger program:

Activity	Pre-decessor	Budget (USD)	Actual Cost (USD)	Progress
(A) Excavate pit	-	10,000	4,850	85%
(B) Compact & prep pit to receive precast units	**(A)**	2,000	-	0%
(C) Pour, cure, prep & transport precast units	-	18,000	8,500	60%
(D) Installation and backfilling	**(C)**	10,000	-	0%
	100%	**40,000**	**13,350**	

Activity	Wk1	Wk2	Wk3	Wk4	Wk5	Wk6	Wk7	Wk8
A	■	■						
B			■					
C		■	■	■	■	■	■	
D								■

At the end of week 3 the following statement best describes the project progress:

(A) Project is ahead of schedule and the estimate at completion is expected to be $27,668.

(B) Project is ahead of schedule and the estimate at completion is expected to be $34,050.

(C) Project is delayed as excavation works are behind schedule, based on cost performance, the project is expected to finish with a total cost of $27,668.

(D) Project is delayed as excavation works are behind schedule, based on cost performance, the project is expected to finish with a total cost of $34,050.

PROBLEM 7 ***Settlement in Clay***

Time settlement in saturated clays when loaded, due to the addition of a building for example, is attributed to the following:

(A) Expulsion of clay particles

(B) The increase in effective stress of clay

(C) The deformation of clay particles

(D) All the above

PROBLEM 8 ***Project Planning***
The below table represents project activities, durations in days, predecessors, and successors:

Activity	Duration	Predecessor	Successor
A	2	-	B, C
B	5	A	F, G
C	2	A	E, D
D	4	C	H
E	2	C	G
F	7	B	I
G	3	B, E	I
H	2	D	I
I	2	F, G, H	-

Assume work starts on a Monday, ignoring weekends and holidays, Total Float and Free Float for activity 'G' are:

(A) 5, 1

(B) 4, 4

(C) 4, 0

(D) 3, 3

PROBLEM 9 ***Boreholes Locations***
The below is a cross section of a 300 ft long, yet to be designed, retaining wall.

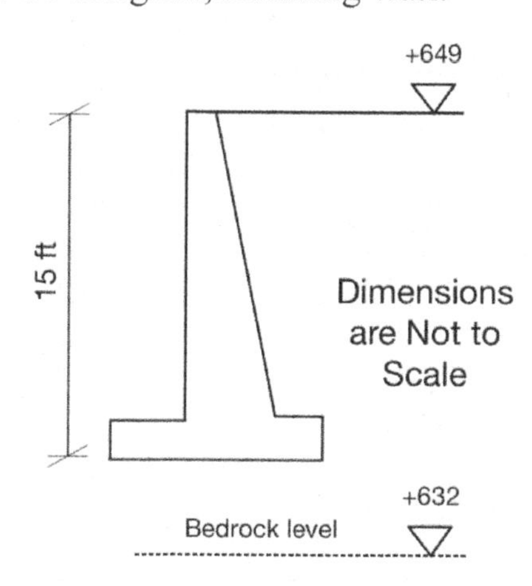

The minimum number, locations, and depth of the needed exploration boreholes to design this wall is/are:

(A) One borehole 6 ft deep in front of the wall below its footing level. Another one 21 ft (*) behind the wall.

(B) One borehole 16 ft (*) deep in front of the wall below its footing level. Another one 31 ft (*) behind the wall.

(C) One borehole 21 ft (*) deep and another borehole 35 ft (*) deep both behind the wall.

(D) One borehole 35 ft (*) deep behind the wall at mid span.

* Take datum level as (+649) for boreholes taken behind the wall.

PROBLEM 10 ***Soil Shear Strength***
The below is a particle distribution chart for two sands, Sand 1 (top) and Sand 2 (bottom).

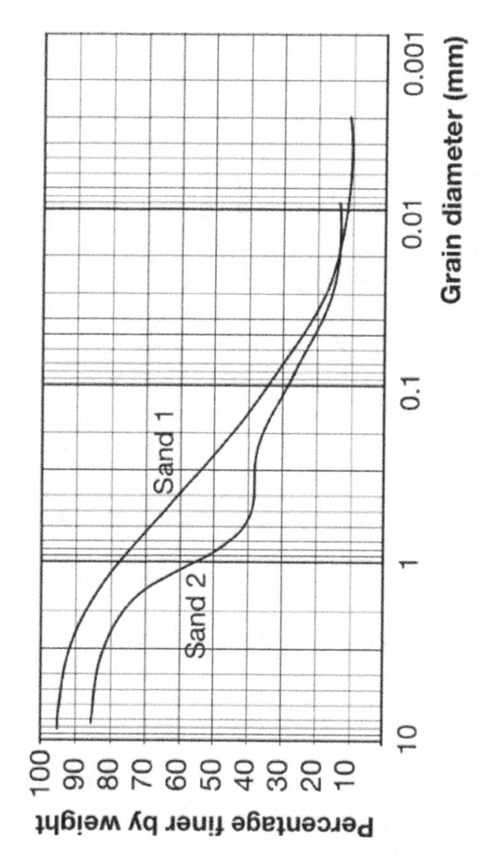

Based on this chart, the following statement is accurate during normal loading for the two sands:

(A) Sand 2 has a higher shear strength compared to Sand 1.

(B) Sand 1 has a higher shear strength compared to Sand 2.

(C) Both Sands have approximately the same shear strength.

(D) More information is required.

PROBLEM 11 ***2D Truss Deflection***

The below two-member truss has the following members' properties:

$L6 \times 6 \times 9/16$ Section

$E = 29{,}000\ ksi$

$A_g = 6.45\ in^4$

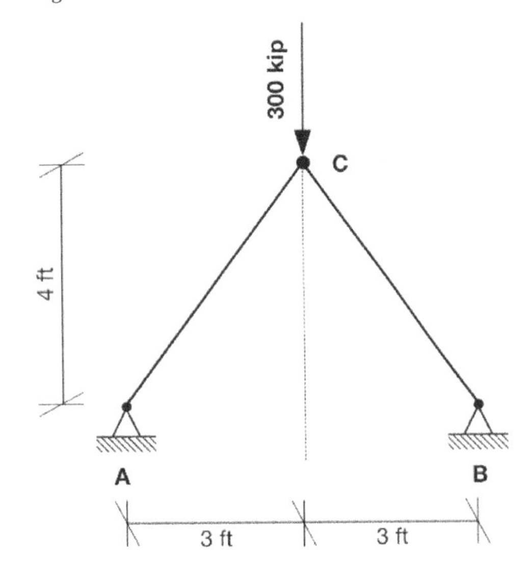

Vertical deflection at point 'C' is:

(A) $0.08\ in$

(B) $0.06\ in$

(C) $0.16\ in$

(D) $0.04\ in$

PROBLEM 12 ***Horizontal Curve***

The below is a plan view for an 11 ft wide road with a horizontal curve and an obstruction as shown. The setback from this obstruction to the edge of the road should not be less than 12 ft.

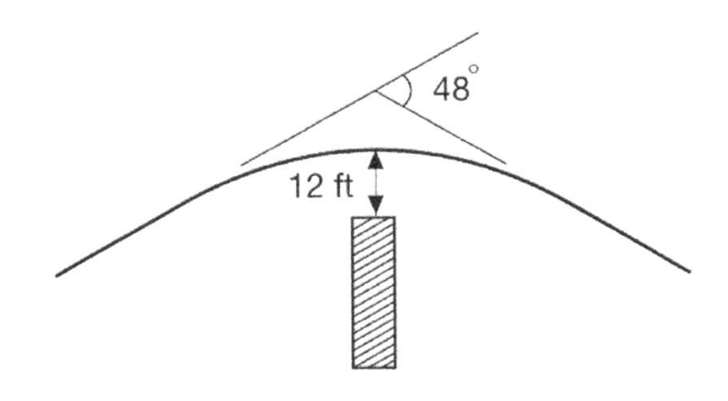

Based on this, the radius, and the length of the curve in ft should nearly be:

(A) $138\ and\ 116$

(B) $36\ and\ 31$

(C) $53\ and\ 44$

(D) $200\ and\ 170$

PROBLEM 13 ***Pressure Under Footing***

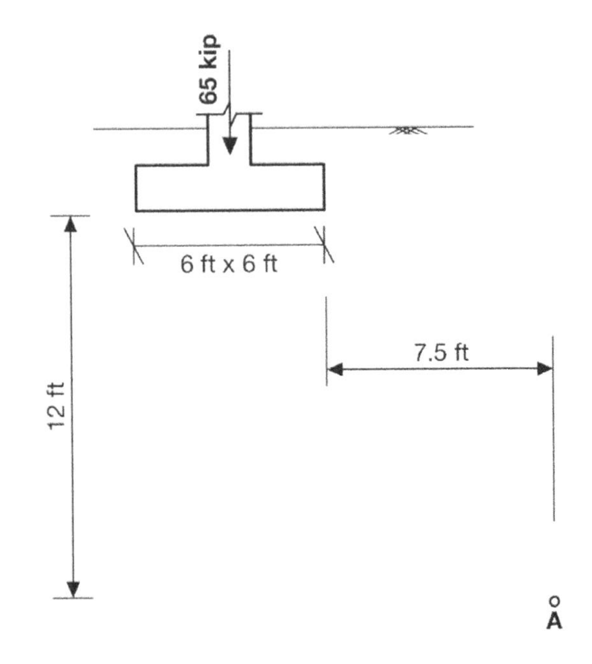

The increase in the vertical pressure at point 'A' due to loading the square footing using those two theories respectively:

- The Boussinesq's Theory
- The 2:1 theory

are as follows:

(A) 110 psf (Boussinesq) and zero (2:1 theory)

(B) 110 psf (Boussinesq) and 200 psf (2:1 theory)

(C) 270 psf (Boussinesq) and zero (2:1 theory)

(D) 270 psf (Boussinesq) and 200 psf (2:1 theory)

PROBLEM 14 ***Construction Material***

The best type of cement, or combination thereof, that is preferred for the use in prestressed bridge decks is:

(A) Type III cement

(B) Type IIIA cement

(C) Type K cement

(D) Type III mixed with fly ash

PROBLEM 15 ***Construction Methods***

Choose three from the below list of construction machinery used in asphalting:

☐ Motor grader

☐ Paver screed

☐ Pneumatic tire roller

☐ Milling Machine

☐ Double steel roller

☐ Single smooth drum roller

PROBLEM 16 ***Shear Stress Cross Section***

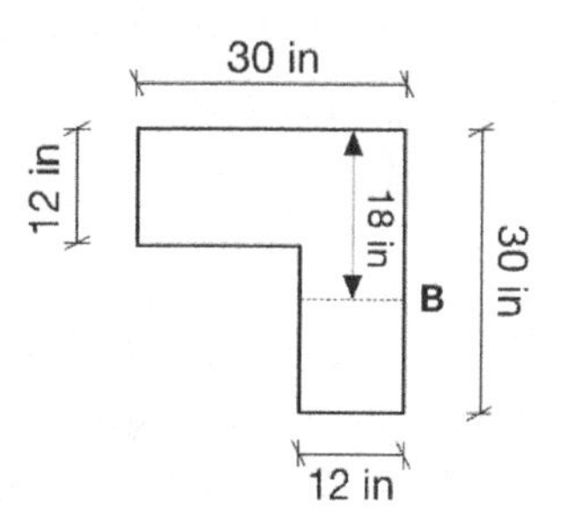

The above is an inverted concrete L-section that is subjected to a total ultimate shear of $V_u = 100\ kip$. Shear stress in ksi at horizontal line 'B' is most nearly:

(A) 0.24

(B) 0.37

(C) 4.44

(D) 0.19

PROBLEM 17 ***Soil Testing***

The following test is recommended to be used to determine the coefficient of permeability for materials with lower permeability such as silts and clays:

(A) The constant head permeameter test

(B) The Piezometric test

(C) The falling head permeameter test

(D) The flexible wall permeameter test

PROBLEM 18 ***Water Discharge External Forces***

An inclined nozzle at 'B' with a diameter of 1.5 in, discharges 1 cfs of water into an open tank.

The supply pipe 'A' has a diameter of 3.5 in, and the nozzle is held in place by hinge 'C' as shown below.

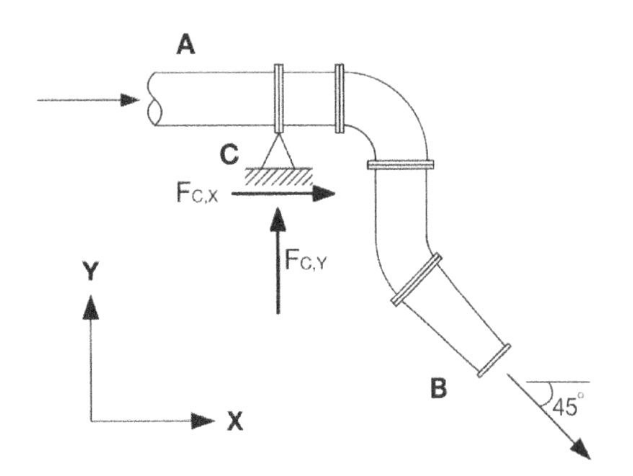

Ignoring the weight of water and the pipe arrangement, the horizontal and vertical reactions at support 'C' when pressure at 'A' is 100 psi:

(A) $F_{C,x} = 754.5\ lb \leftarrow$
$F_{C,y} = 236.75\ lb \downarrow$

(B) $F_{C,x} = 77.25\ lb \rightarrow$
$F_{C,y} = 113.0\ lb \downarrow$

(C) $F_{C,x} = 837.0\ lb \leftarrow$
$F_{C,y} = 176.7\ lb \downarrow$

(D) $F_{C,x} = 132.0\ lb \leftarrow$
$F_{C,y} = 22.5\ lb \downarrow$

PROBLEM 19 ***Travel Time for Shallow Flow***

The below paved parking lot drains into the channel at its left side as shown. The parking has a slope of 0.5%.

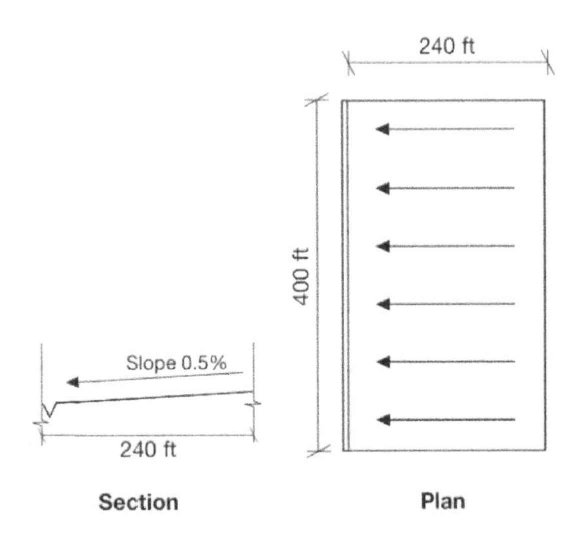

With a manning roughness of '0.011' and a 2-year 24-hour rainfall intensity of 2.0 in/hr. The travel time in minutes for a shallow flow over this plane to the channel is:

(A) 42.0 minutes

(B) 7.7 minutes

(C) 6.2 minutes

(D) 57.2 minutes

PROBLEM 20 ***Vertical Curve Design***

The below vertical curve has a 14 ft bridge passing over the curve's PVI as shown:

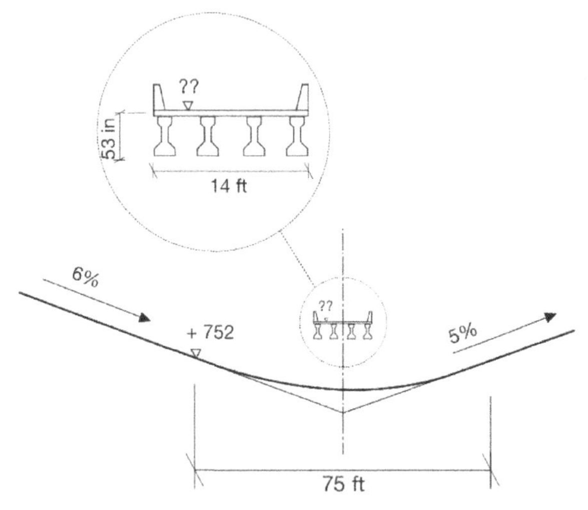

With a 16 ft clearance, and asphalt level at PVC of +752, bridge deck level should be designed at:

(A) + 750.9

(B) + 771.3

(C) + 771.0

(D) + 769.9

PROBLEM 21 ***Construction Material***

A road contractor is requested to stabilize the subbase layer before the construction work starts as it includes clay gravel soils within its particles.

Considering that the *Plasticity Index* (PI) for the affected layer is '40', the best improvement method is as follows:

(A) Cement stabilization by mixing 3% Portland cement with the subbase material.

(B) Cement stabilization by mixing 9% Portland cement with the subbase material.

(C) Apply a small percentage of lime, typically 0.5% to 3%, to the affected material with a process called lime modification.

(D) Lime stabilization by mixing nearly 3 – 5% of lime to the affected layer.

PROBLEM 22 ***Net Present Value Analysis***
One of the city's pumping stations require major mechanical rehabilitation for its pumps. The city has four options:

Option A: Replace old pumps with new pumps at a cost of $25,000 inclusive of labor. Yearly maintenance costs $3,000. The selected pumps to be replaced every 10 years.

Option B: Replace old pumps with new pumps at a cost of $27,500 inclusive of labor. Yearly maintenance costs $2,000. The selected pumps to be replaced every 10 years.

Option C: Replace old pumps with new pumps at a cost of $40,000 inclusive of labor. Yearly maintenance costs $1,500. The selected pumps to be replaced every 15 years.

Option D: Replace old pumps with new pumps at a cost of $47,500 inclusive of labor. Yearly maintenance costs $700. The selected pumps to be replaced every 15 years.

The remaining life of the civil structure is 30 years.

Average yearly inflation is 5% applied to capital and maintenance costs.

The city uses a discount rate of 7% for their capital projects.

Based on the above, the cheapest and more feasible option amongst the four above would be:

(A) Option A

(B) Option B

(C) Option C

(D) Option D

PROBLEM 23 ***Construction Activities***
A 12 in thick subgrade layer is being placed for an under-construction 11 ft wide road. Work should be undergoing at a rate of 6.0 $miles\ per\ day$ (*).

Consider an average roller/compacter speed of 3.5 mph, 5 $passes\ per\ roller$ to achieve required density and a driver efficiency of 50 $minutes\ per\ hour$.

How many 66 in wide rollers are required for this job site to keep up with the stated rate of production:

(A) 1

(B) 2

(C) 3

(D) 4

* Consider that this operation runs on a 10 hours shift.

PROBLEM 24 ***Cross Section Analysis***
The below section is subjected to a shear load.

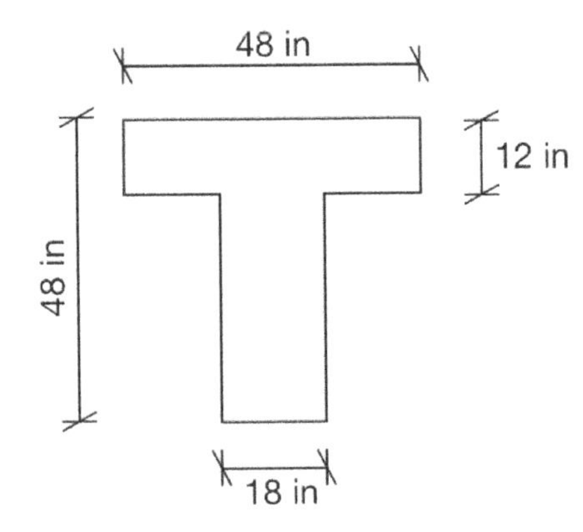

Shear flow profile for this cross-section is best represented by:

(A) Profile A

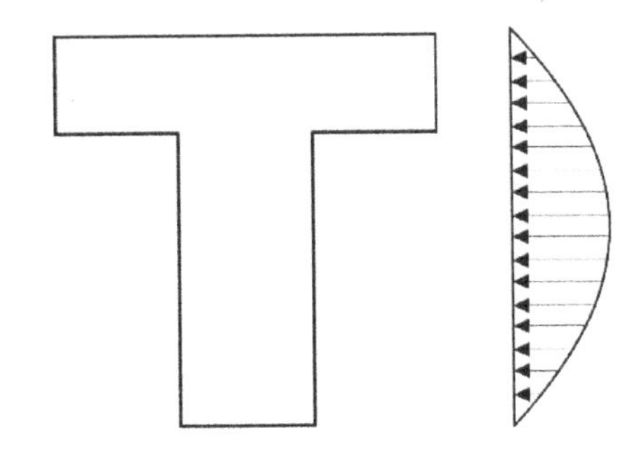

(B) Profile B

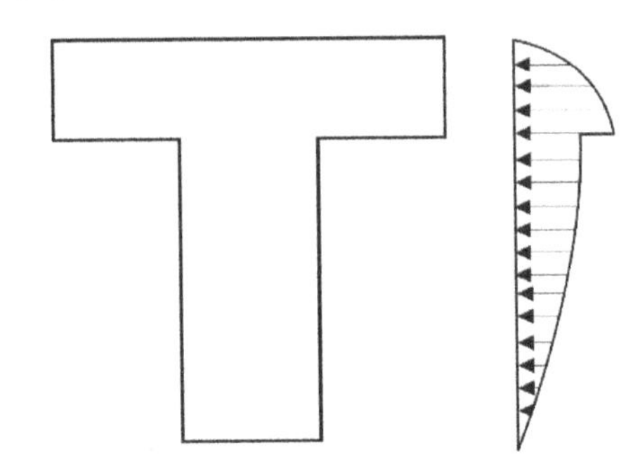

(C) Profile C

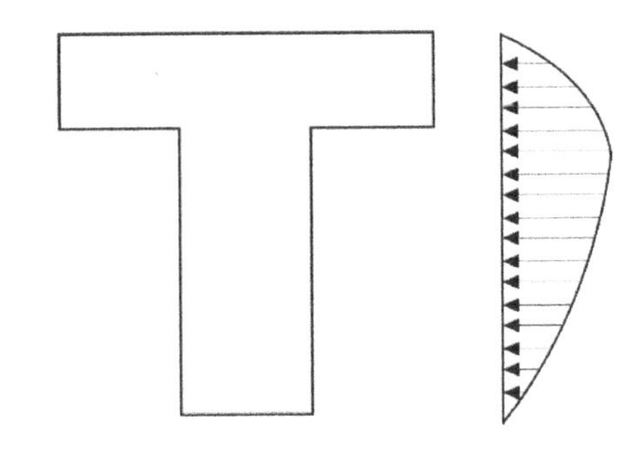

(D) Profile D

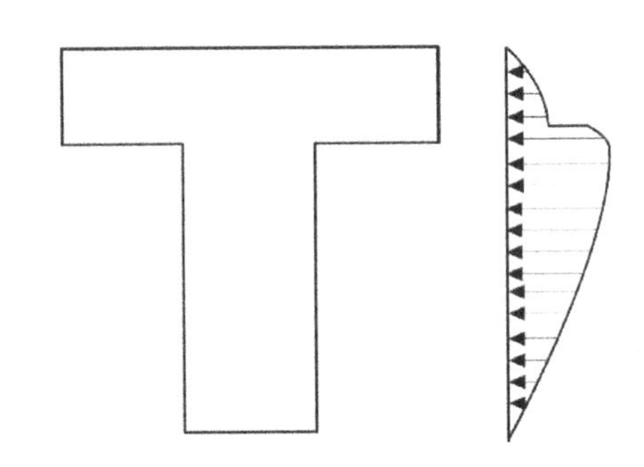

PROBLEM 25 ***Material Properties***
The below figure shows two different sections placed under pressure with force P and are parallel to each other. The two sections share a similar cross-sectional area (i.e., $A_1 = A_2$), and a similar length (L). Their moduli of elasticity differ as follows:
$E_1 = 2E_2$

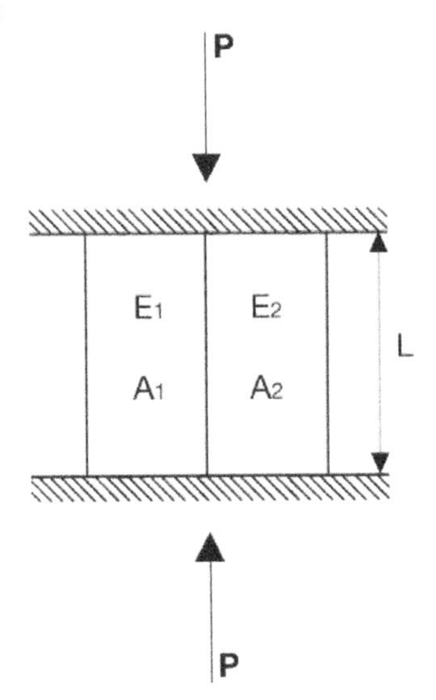

Based on the information above, the overall modulus of elasticity of the joined sections (E_{joined}) equals to:

(A) $(E_1 + E_2)/2$

(B) $2/3\ E_1$

(C) $4/3\ E_1$

(D) $(E_1 + 2E_2)/2$

PROBLEM 26 ***Material Properties***

The below figure shows two different sections placed under pressure with force P on top of each other. The two sections share a similar cross-sectional area (i.e., $A_1 = A_2$), and a similar length (L). Their moduli of elasticity differ as follows: $E_1 = 2E_2$

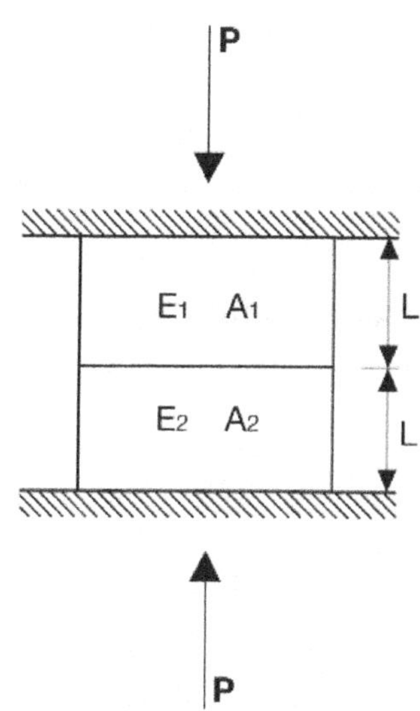

Based on the information given above, the overall modulus of elasticity of the joined sections (E_{joined}) equals to:

(A) $(E_1 + E_2)/2$

(B) $2/3\ E_1$

(C) $4/3\ E_1$

(D) $(E_1 + 2E_2)/2$

PROBLEM 27 ***Retention Pond Sizing***

The size of a retention pond in ft^3 situated in a sloped zone that has 50 $Acres$ of parks and cemeteries and the following data:

- Outflow rate of 35 cfs.
- Water flows in and out the pond in 15 $minutes$.
- Average rainfall intensity is 3.5 in/hr.

is most nearly:

(A) 7,875

(B) 131

(C) 39,375

(D) 31,500

PROBLEM 28 ***Environmental Chemical Reactions***

The equation for chemical phosphorus removal using Ferric Chloride from ponds is:

(A) $2FeCl_3 + 3Ca(HCO_3)_2 \leftrightarrow 2Fe(OH)_3 + 3CaCl_2 + 6CO_2$

(B) $Fe_2(SO_4)_3 + 3Ca(HCO_3)_2 \leftrightarrow 2Fe(OH)_3 + 3CaSO_4 + 6CO_2$

(C) $FeCl_3 + PO_4^{3-} \rightarrow FePO_4 (\downarrow) + 3\ Cl^-$

(D) $3FeCl_2 + 2PO_4^{3-} \rightarrow Fe_3(PO_4)_2 (\downarrow) + 6\ Cl^-$

PROBLEM 29 ***Concrete Design/Wall Thickness Determination***

The below is the plan and cross-section of a deep concrete shaft. Assume ground water depth is zero and concrete density is 150 pcf.

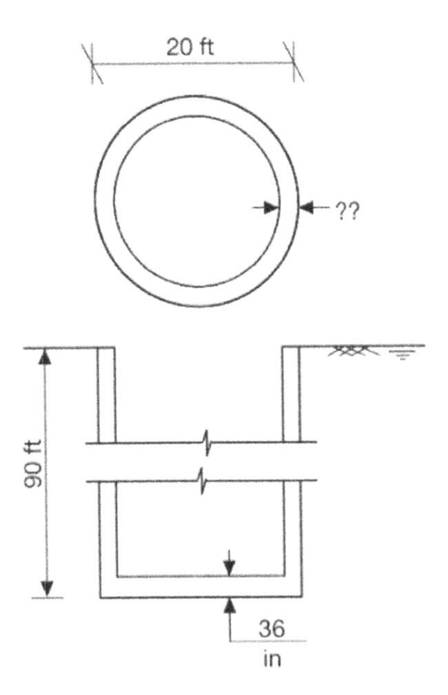

To achieve a floatation Safety Factor of '1.1', the minimum wall thickness for the shaft should be:

(A) 30 *in*

(B) 24 *in*

(C) 52 *in*

(D) 27 *in*

PROBLEM 30 ***Soil Properties***

The over consolidation ratio for a soil with a '0.33' normally consolidated at rest Rankine coefficient and a '0.85' over consolidated at rest coefficient is most nearly:

(A) 4.1

(B) 1.9

(C) 0.3

(D) 2.6

PROBLEM 31 ***Soil Classification System***

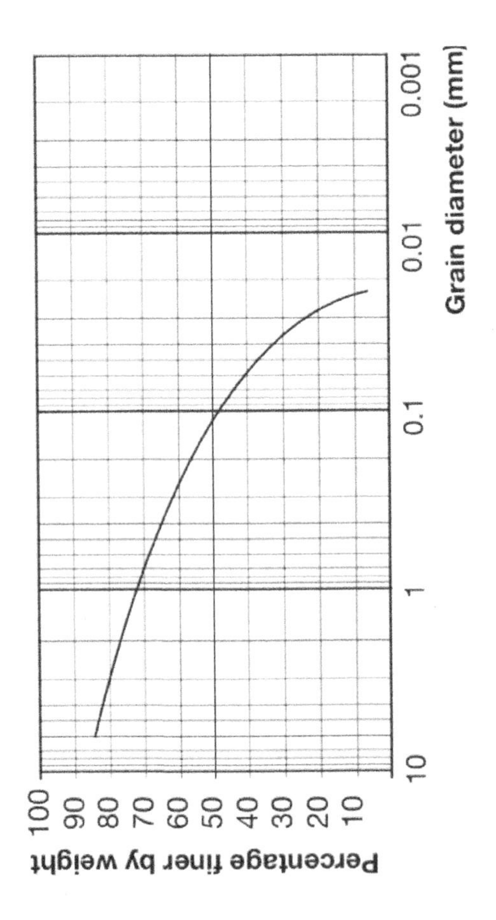

Using the Unified Soil Classification System, the above gradation sample is:

(A) GW

(B) GP

(C) SW

(D) SP

PROBLEM 32 ***Geotechnical***

A soil sample that has a total volume of 1 ft^3 and a total mass of 100 *lb* is removed from the ground. The water content of this sample is 20% and Specific Gravity '2.7'.

Based on the above information the following attributes are as follows:

The dry density of the sample is ______

The Degree of Saturation is _______

Porosity is _______

PROBLEM 33 ***Foundation Settlement***

The below is a $6\,ft \times 6\,ft$ square concrete footing laid in fine medium dense sand with a maximum load applied on it of $100\,kip$.

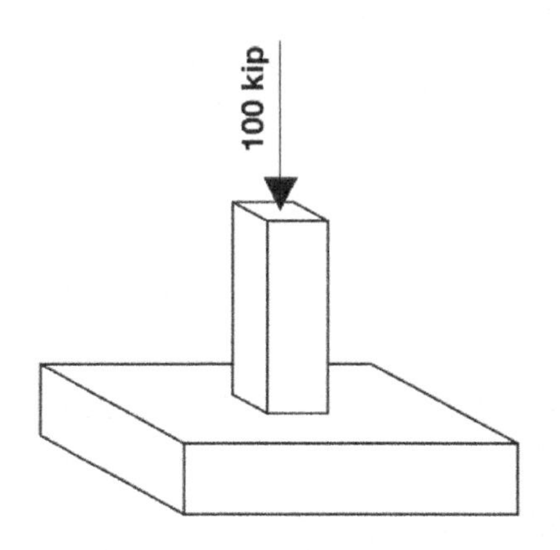

The maximum initial elastic vertical settlement this footing will experience is most nearly:

(A) $0.77\,in$

(B) $0.064\,in$

(C) $1.55\,in$

(D) $0.46\,in$

PROBLEM 34 ***Soil Classification System***

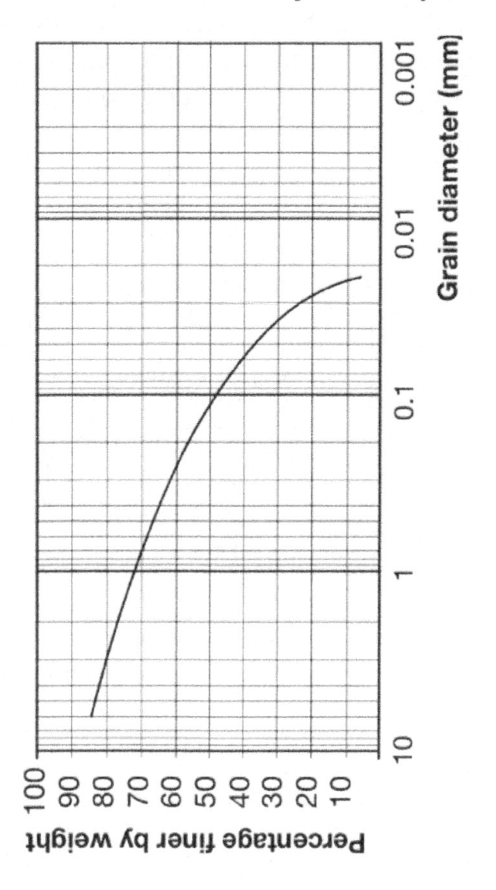

Based on the above soil sample gradation, and the following fines attributes:

- Liquid Limit (LL) = 50
- Plastic Limit (PL) = 35

Using the AASHTO classification system, the following represents the best group classification for this sample:

(A) A-7-5 (4)

(B) A-7-6 (4)

(C) A-7-4

(D) A-7-5

PROBLEM 35 ***Effective Stress Over Time***

The below matt foundation has a uniform weight of $1\,ksf$ and was built and loaded linearly over a period of 6 months on top of a layer of sand and clay as shown. The ground water level is $5\,ft$ below the footing. The density of sand and clay layers are both $120\,pcf$.

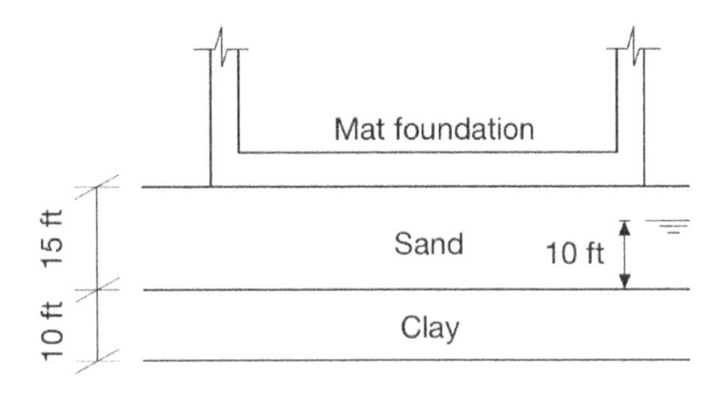

The profile that represents the change in effective stress over time at the bottom of each layer is:

(A) Profile A

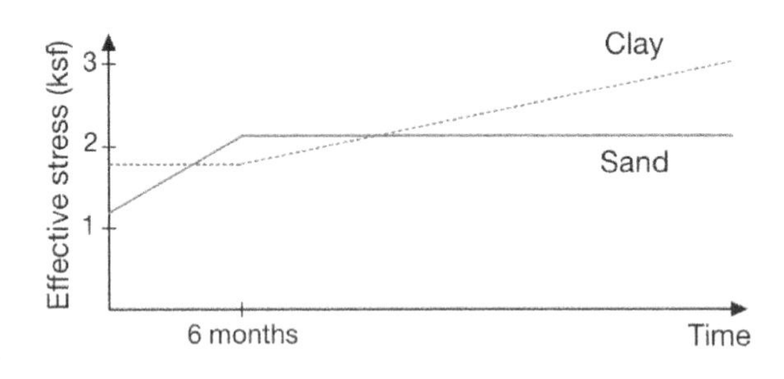

(B) Profile B

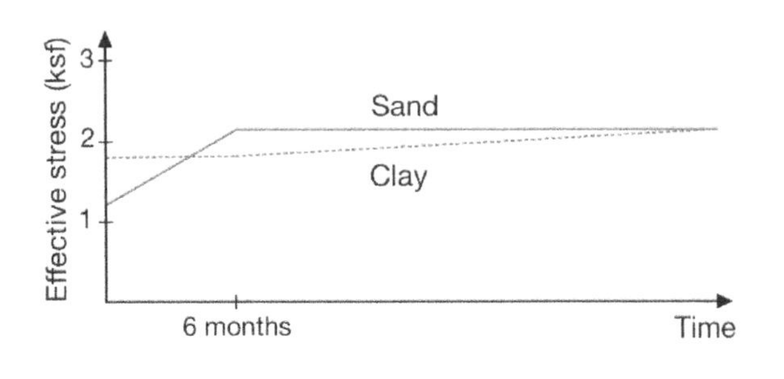

(C) Profile C

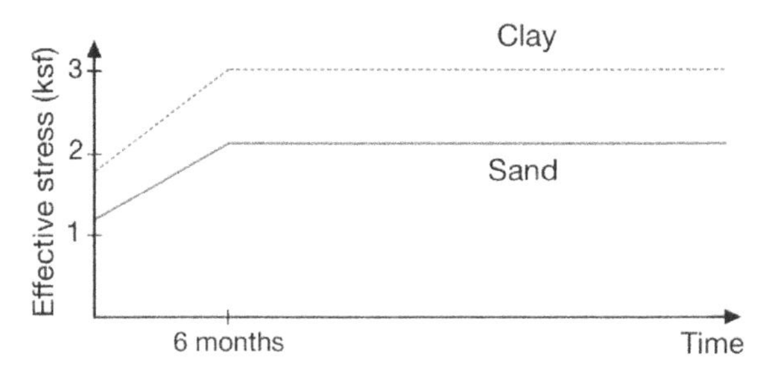

(D) Profile D

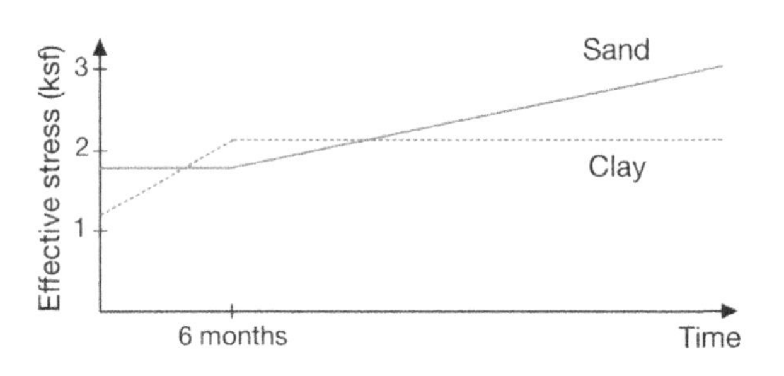

PROBLEM 36 ***Retaining Wall Safety Factor***

The below reinforced concrete cantilever retaining wall has the following properties:

- Concrete density $\gamma_{concrete} = 150\,pcf$
- Soil density $\gamma_{soil} = 130\,pcf$
- Water density $\gamma_{water} = 62.4\,pcf$
- Soil's friction angle $\emptyset'_{soil} = 37^{o}$
- Ground water level $6\,ft$ below surface

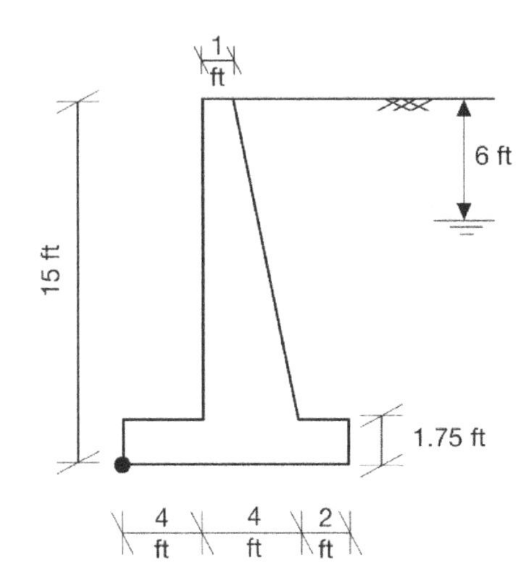

Based on the above information, wall overturning Safety Factor (*) is most nearly:

(A) 2.1

(B) 2.7

(C) 3.7

(D) 15.9

* Consider overturning will occur around left most bottom concrete edge.

PROBLEM 37 ***Budgetary Cost Acceptable Range***

The city is conceptualizing a large infrastructure project with an expected cost of nearly $25 million. The city appointed a consultant to develop the feasibility and determine the budget for authorization from city council.

Assuming the project would in fact cost the city $25 million upon completion, what is an acceptable range of a budgetary estimate the consultant can provide the city with?

(A) $17.5 million to $37.5 million

(B) $20 million to $32.5 million

(C) $21.25 million to $30 million

(D) $20 million to $30 million

PROBLEM 38 ***Head Losses***

The head loss in ft for a 20-year-old cast iron pipe, 200 ft long, 20 in dia, with a slope of 2%, is most nearly:

(A) 9

(B) 4

(C) 20

(D) 2

PROBLEM 39 ***Concrete Mix Design***

A concrete mix with a specified w/c ratio of 0.45, a mix design of 1:1.5:3, assuming aggregate (*) will consume 5% of its weight to achieve an SD condition, requires ____ liters of water to produce 1 ft^3 of yield.

(A) 5.5

(B) 5.0

(C) 10.5

(D) 19.3

* Assume a density of 165 lb/ft^3 and 195 lb/ft^3 for aggregate and cement respectively.

PROBLEM 40 ***2D Truss Reactions***

In the below tower crane, the cable at 'B' is connected just external to the support, the diameters for pulleys at 'C' and 'D' is 1 ft.

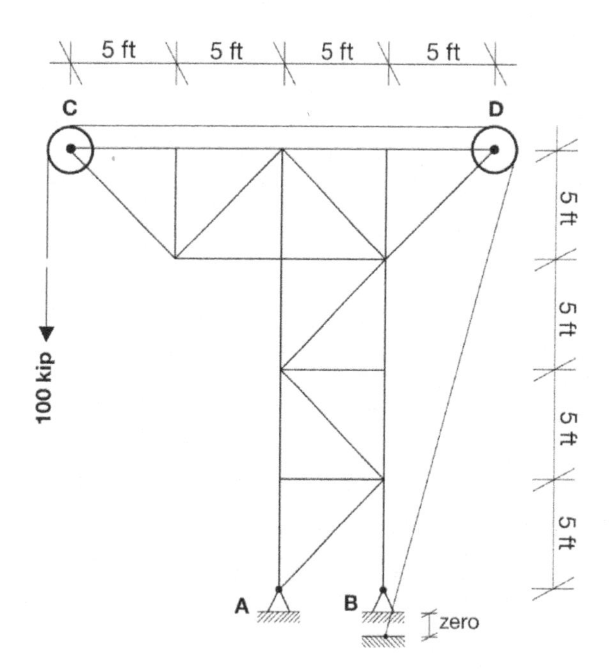

Reactions at hinged supports at 'A' and 'B' in kip are as follows (*):

(A) R_A: 310 ↑ & 27 ←
R_B: 114 ↓

(B) R_A: 300 ↑
R_B: 210 ↓

(C) R_A: 300 ↑ & 27 →
R_B: 210 ↓

(D) R_A: 310 ↑ & 27 →
R_B: 114 ↓

* Connection of cable close to support 'B' is not to be considered as part of the reaction at 'B' in this question).

PROBLEM 41 ***Roadway Density***

The Density of a highway measured in $veh/mile$ that has a flow rate of 1,500 veh/hr and an average travel speed of 65 mph is most nearly:

(A) 105

(B) 34

(C) 40

(D) 23

PROBLEM 42 ***Points on Vertical Curve***

The following points fall on a vertical curve profile:

Point	Station	Elevation
PVC	0 + 025	72.5
PVI	0 + 275	65
A point on the curve	0 + 150	70

Using the above data, the initial and final grade for this section are as follows:

(A) −0.03 , 0.05

(B) −0.03 , 0.06

(C) −0.06 , 0.10

(D) −0.06 , 0.08

PROBLEM 43 ***Frame Moment Diagram***

Which of the following diagrams most nearly represents the moment generated from the lateral load shown below:

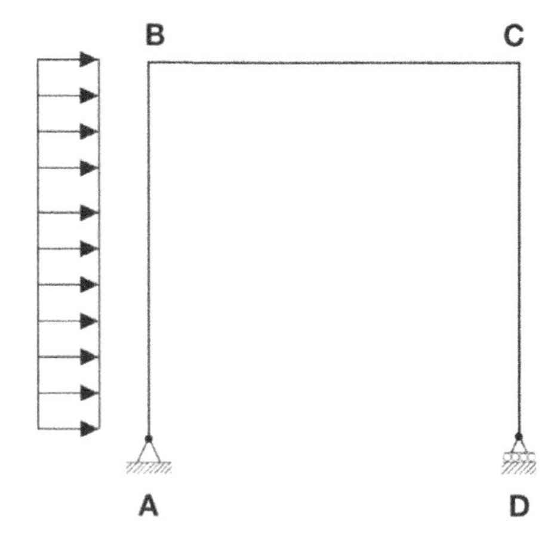

(A) Diagram A

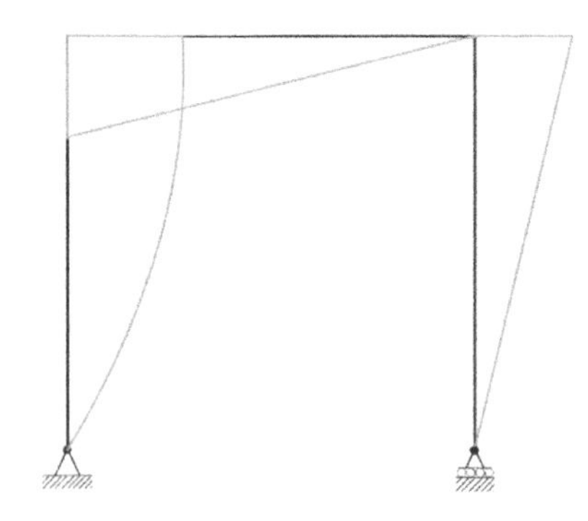

(B) Diagram B

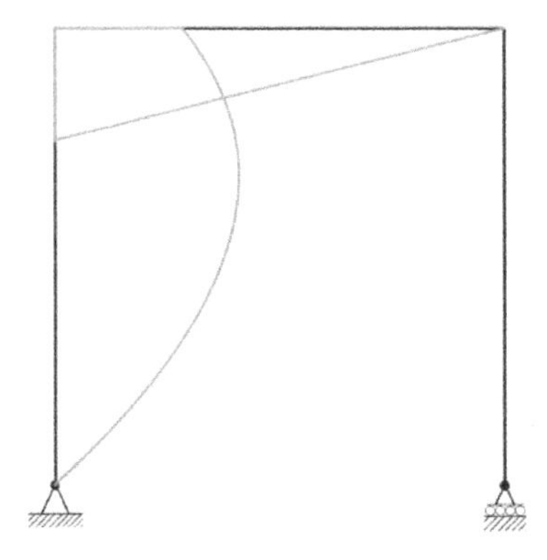

(C) Diagram C

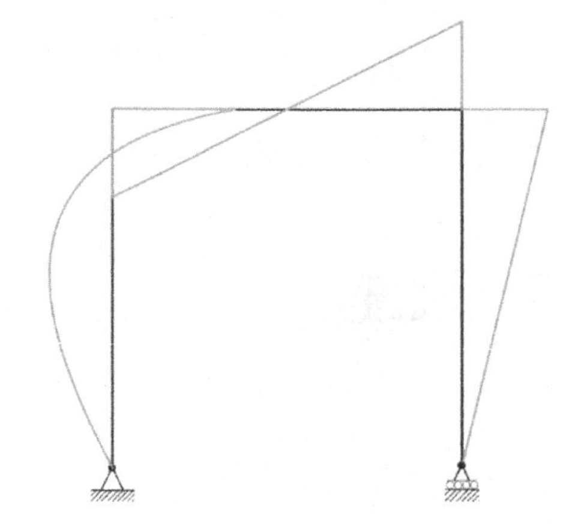

(D) Diagram D

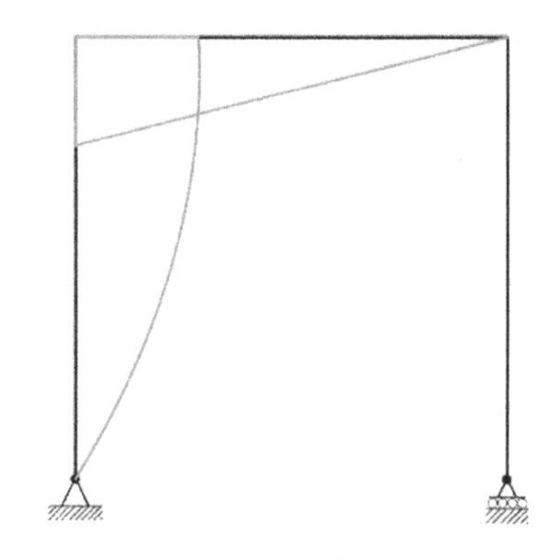

PROBLEM 44 *Benefit Cost Analysis*

The following roadway segment is 5 *miles* long and experiences an Average Annual Daily Traffic (*AADT*) of 80,000 vehicles per day.

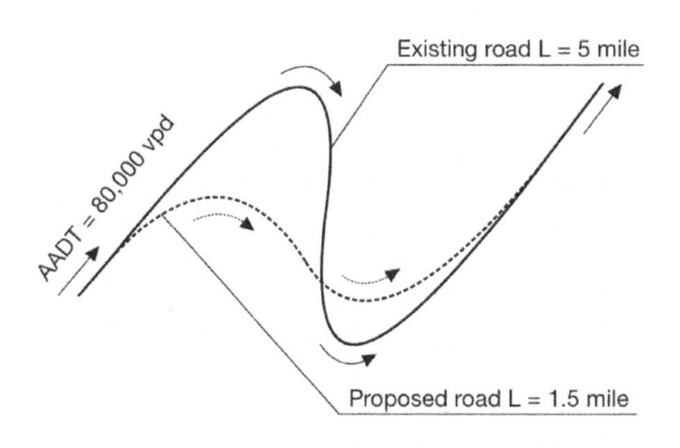

The dashed line is a proposed improvement/shortening of this segment which will cost \$120.0 *million* to erect and is 3.5 *miles* shorter.

Considering the following factors:

- Average fuel cost along with operation & maintenance cost is \$ 0.14 per vehicle per mile.
- Road maintenance cost is \$22,000 per mile per year.
- Inflation rate is estimated at 2%

The Benefit/Cost ratio for this project over a period of 25 years is most nearly:

(A) 0.91

(B) 1.17

(C) 2.33

(D) 1.31

PROBLEM 45 *Shallow Flow*

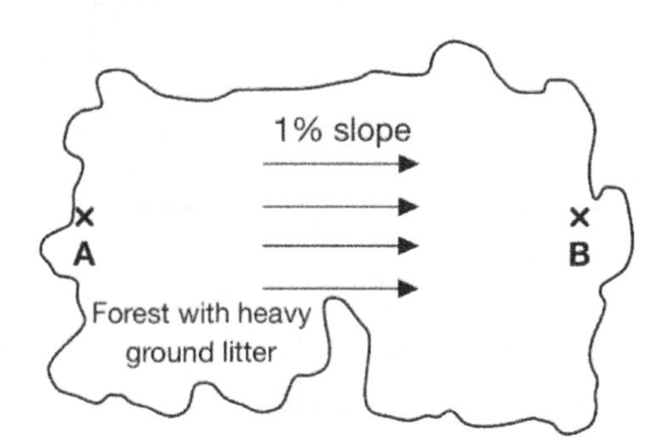

The above land is a forest characterized by heavy ground litter. This forest will be converted into short grass pasture. Land is sloped at a grade of 0.01 ft/ft. The expected improvement in velocity of water flowing from point A to point B is most nearly:

(A) 0.74 ft/sec

(B) 0.25 ft/sec

(C) 0.50 ft/sec

(D) 1.00 ft/sec

PROBLEM 46 *Resources Histogram*

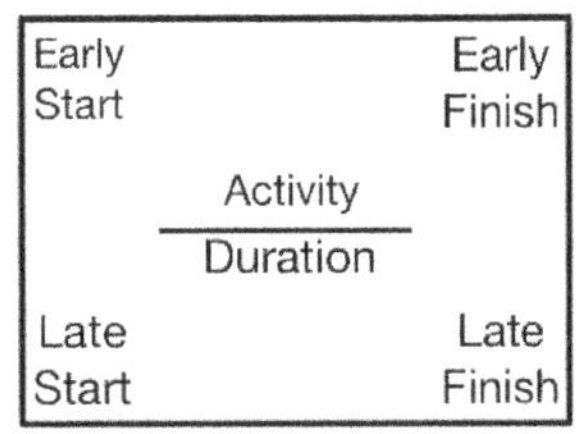

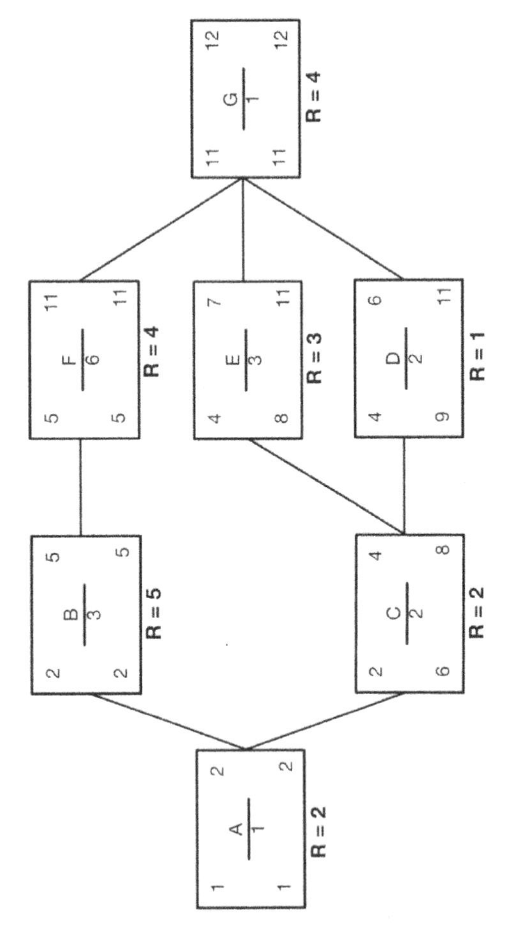

Assuming the activities in the above network diagram cannot be split, resources are of the same discipline, the below resources histogram is the representation of the best possible leveled histogram (*):

* Activities start at the beginning of a day and finish at an end of a day – e.g., 'G' starts at the beginning of day 11 and ends at the end of day 11.

(A) Leveled Histogram A

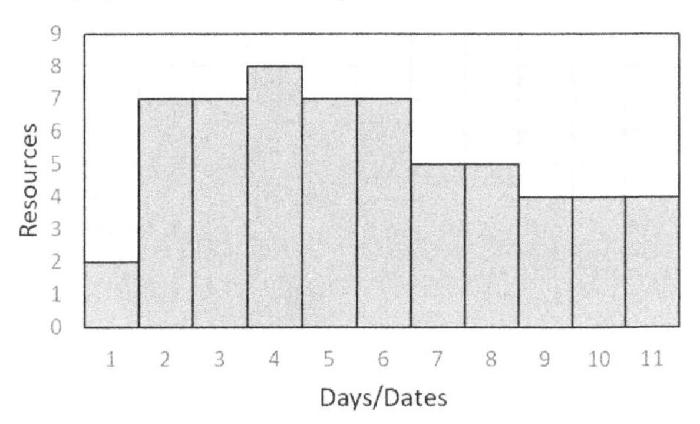

(B) Leveled Histogram B

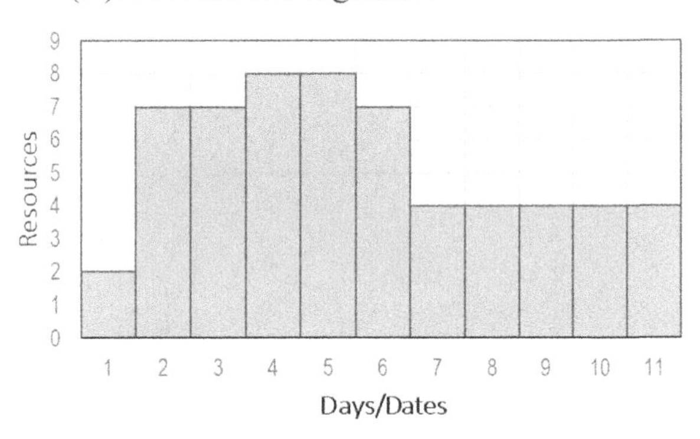

(C) Leveled Histogram C

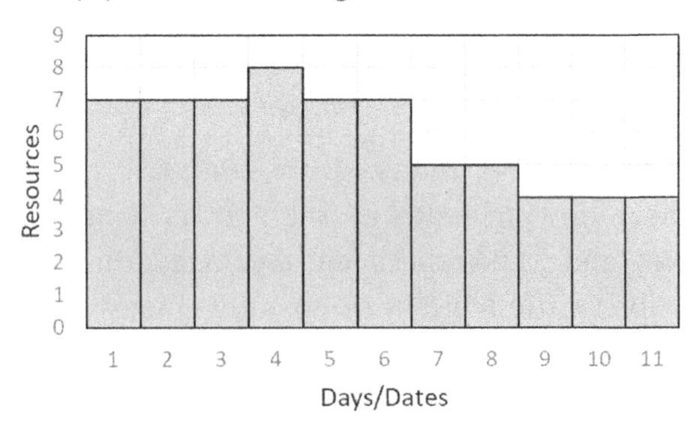

(D) Leveled Histogram D

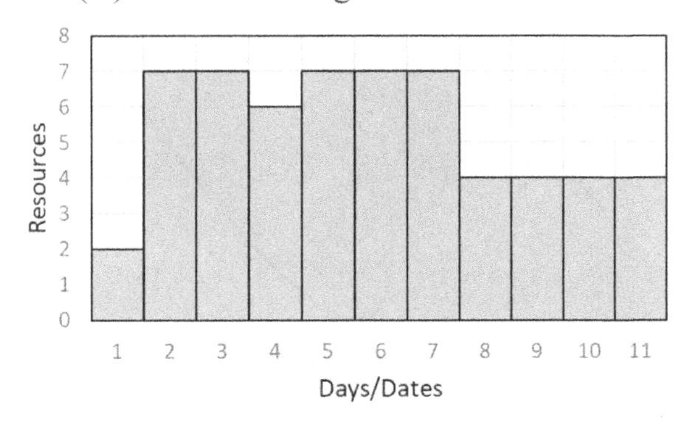

PROBLEM 47 ***Double Horizontal Curve***
The below figure depicts two identical horizontal curves connecting two parallel roads as shown with station $St.1 = 1 + 00$

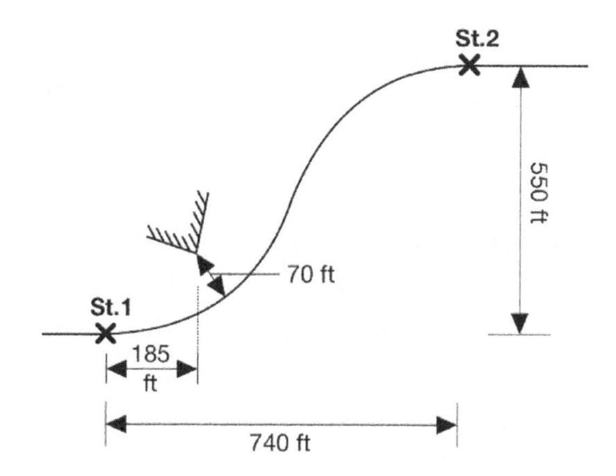

An obstruction is located at the middle ordinate of the bottom curve with a distance of $M = 70\ ft$.

Provided the above, station $St.2$ most nearly sits at:

(A) $8 + 40$

(B) $10 + 38$

(C) $10 + 78$

(D) $11 + 18$

PROBLEM 48 ***Mass Haul Diagram***
The below Mass Haul Diagram belongs to a highway with soil shrinkage factor of 14.5%

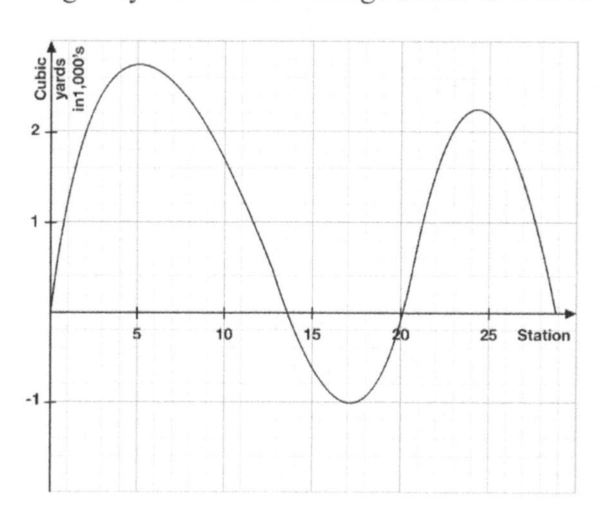

Considering a Free Haul Distance of $500\ ft$ and a Free Haul Volume of $1{,}600\ yard^3$, wastage in $yard^3$ for this project is most nearly:

(A) 800

(B) 400

(C) 1,400

(D) 685

PROBLEM 49 ***Peak Hour Factor***
In the below traffic flow count, the Peak Hour Factor PHF for the SW-NE direction is estimated to be 5% more than that of the NE-SW direction.

Time	SW-NE direction $Veh/15\ m$	NE-SW direction $Veh/15\ m$
4:00 PM	205	195
4:15 PM	195	180
4:30 PM	196	165
4:45 PM	165	X
5:00 PM	207	197
5:15 PM	210	134
5:30 PM	195	125

Given the information above, the estimated traffic 15 min flow in the NE-SW direction at 4:45 PM identified as 'X' is most nearly:

(A) 180 Veh

(B) 216 Veh

(C) 197 Veh

(D) 236 Veh

PROBLEM 50 ***Volume of Excavation***
The below cut and fill diagram belongs to a highway project. The project's existing ground consists of soil with 20% shrinkage factor and 15% swell factor.

The diagram's negative y-axis represents cross section areas to be cut and the positive y-axis represents cross areas sections to be filled.

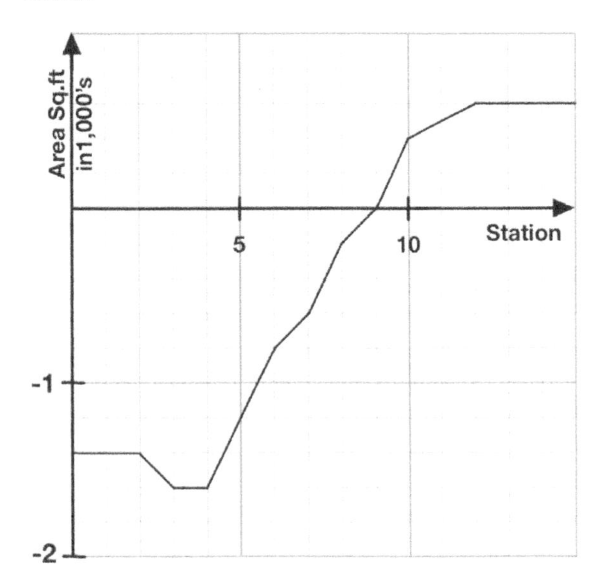

The project shall balance its cut and fill material and the rest goes to waste.

Considering dump trucks can haul up to 11 $yard^3$ per trip, the number of trips expected to haul waste outside the project is most nearly:

(A) 2,470

(B) 2,190

(C) 1,240

(D) 2,680

PROBLEM 51 ***Skid Marks Distance***
Two cars travelling in opposite directions to each other, the first car is travelling at a speed of 90 mph and the second car at a speed of 60 mph.

Due to reduced vision they both came into this realization when they were only 1,340 ft apart. With a perception time of 2 $seconds$, they both applied breaks and came into a stop leaving a gap of 35 ft between them.

The distance travelled by the first car and the second car respectively after applying breaks is most nearly:

(A) $907\ ft, 468\ ft$

(B) $599\ ft, 266\ ft$

(C) $519\ ft, 349\ ft$

(D) $540\ ft, 325\ ft$

PROBLEM 52 ***Unconfined Aquifer***
A 100 ft thick unconfined aquifer has a 12 in diameter well that pumps ground water from it at a rate of 65 gpm.

Assuming the radius of influence is 450 ft and permeability is $4 \times 10^{-4} ft/sec$, the drawdown at the well is most nearly:

(A) 96 ft

(B) 97.5 ft

(C) 4 ft

(D) 2.5 ft

PROBLEM 53 ***Transit Average Speed***
The average speed for a rapid transit motorized vehicle which accelerates from a station at a rate of 5 ft/sec^2 to reach a maximum speed of 90 mph, stays on this speed for 1.5 $minutes$ before it decelerates at a rate of 4.5 ft/sec^2 into its next stop, is most nearly:

(A) 60 mph

(B) 56 mph

(C) 73 mph

(D) 51 mph

PROBLEM 54 ***Retention Basin Design***

A slightly sloped 7 *acres* piece of land that used to be an old cemetery is to be fully redeveloped as follows:

- 45% downtown areas
- 35% playgrounds
- 15% asphalt roads
- 5% concrete walkways

Given the rainfall intensity in this area is $2\ in/hr$, the depth of a $250\ ft \times 150\ ft$ retention basin to be constructed to store the excess runoff from this redevelopment for one day is most nearly:

(A) $14\ ft$

(B) $12\ ft$

(C) $16.5\ ft$

(D) $18\ ft$

PROBLEM 55 ***Elevation of Water Surface in Reservoir***

A cast iron (*) pipeline with two 45° bends connects two large reservoirs as shown below. The diameter of the pipe is 12 *in* and is discharging $80^o\ F$ water at a rate of $20\ cfs$.

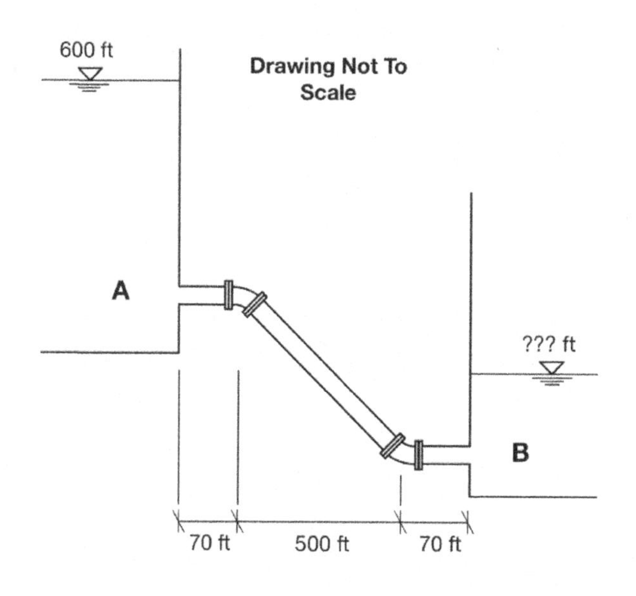

Given the elevations and distances in the above figure, elevation of the water surface at reservoir B is most nearly:

(A) $414\ ft$

(B) $190\ ft$

(C) $553\ ft$

(D) $50\ ft$

* Use lower range for roughness (ε) for cast iron pipes.

PROBLEM 56 ***Concrete Beam Deflection***

The below beam is a concrete section with reinforcements, properties and dimensions as shown.

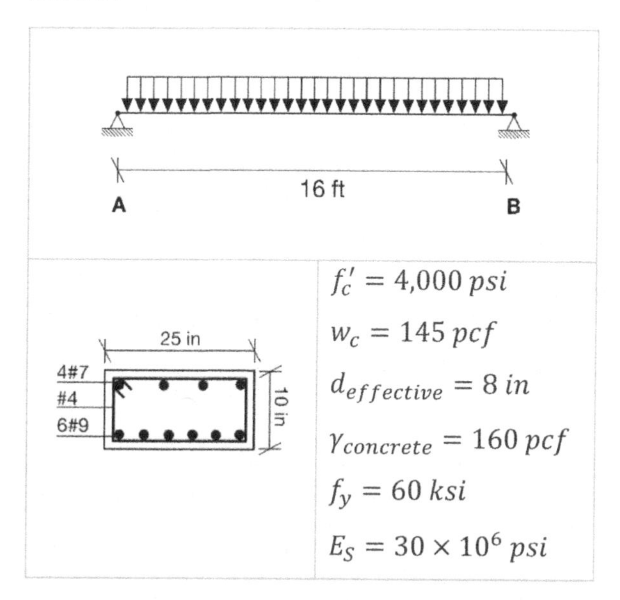

$f'_c = 4{,}000\ psi$

$w_c = 145\ pcf$

$d_{effective} = 8\ in$

$\gamma_{concrete} = 160\ pcf$

$f_y = 60\ ksi$

$E_S = 30 \times 10^6\ psi$

The maximum deflection that can occur due to self-weight in *inches* – assuming that the section will remain uncracked from its own weight – is most nearly:

(A) 0.0026

(B) 0.11

(C) 0.011

(D) 0.055

PROBLEM 57 ***Salvage Value***

A company procured a fleet of high end, specially customized, old trucks for *five* of its employees to perform supervision services for a road construction contact $100 million in value at a cost of $55,000 per truck.

The company, which is a third-party contractor, priced its supervision services for this contract at 2.5% of the road construction fees.

The below is some of the key assumptions made by the company to support this agreement:

Item	Contract price
Average salary per employee	$\$80{,}000\ /\ year$
Fringe benefits per employee	30%
Overhead cost	12.5% of company revenue
Allocations	7.5% of company revenue
Fuel & maintenance per truck	$\$0.25\ /\ mile$
Expected miles driven for supervision purposes	$95\ mile/day\ /employee$
Working days per year (excluding vacation time)	$240\ days$
Contract duration	$2.5\ years$

In order to maintain an Operating Income OI of 15%, and considering those trucks are pretty much worn out after being driven over rough land for $2.5\ years$ in a row, they should be salvaged at an average minimum sales price of:

(A) $0 per truck

(B) $21,250 per truck

(C) $4,250 per truck

(D) $9,250 per truck

PROBLEM 58 ***Soil Erodibility***

Which of the following contributes to the amount of soil loss caused by erosion:

I. Porosity: porosity affects soil structure. The more porous the soil, the weaker its structure becomes, leading to increased erodibility.
II. Reduction in vegetation cover contributes to erosion.
III. An increase in kinetic energy from both wind and rainfall leads to higher erosion rates.
IV. Runoff distance plays a role. Longer runoff routes (i.e., longer lands) result in reduced erosion.
V. Runoff slope. The steeper the land slope, the greater the expected erosion.

(A) I + II + III + IV + V

(B) II + III + V

(C) I + II + III + V

(D) II + III + IV + V

PROBLEM 59 ***Soil Loss Prevention***

The following bare land properties are provided for soil loss calculation and prevention:

- $300\ ft$ long sloped at 2%
- Type of soil is Ontario Loam
- Rainfall intensity index is $200\ ton.in/acre.yr$

Given a permissible soil loss limit of $11\ tons/hectare.yr$, the conservation factor should be:

(A) 0.3

(B) 0.74

(C) 0.014

(D) 0.033

PROBLEM 60 ***Optimum Moisture Content***
A proctor test was performed on *four* soil samples from the same batch using a proctor standard mold.

The weight of the moist samples after applying the test blows were: 4.3 lb, 3.92 lb, 3.92 lb and 4.36 lb, and their moisture content 11.6%, 17%, 8.8% and 14.8% respectively.

The optimum moisture content for this batch is most nearly:

(A) 14.0%

(B) 11.3%

(C) 12.5%

(D) 14.8%

PROBLEM 61 ***Slope Stability/ Slope Safety Factor***
The below slope belongs to an excavation in a soil with cohesion $c = 58\,psf$ and density $\gamma = 95\,pcf$ along with a friction angle of $\emptyset = 20^o$.

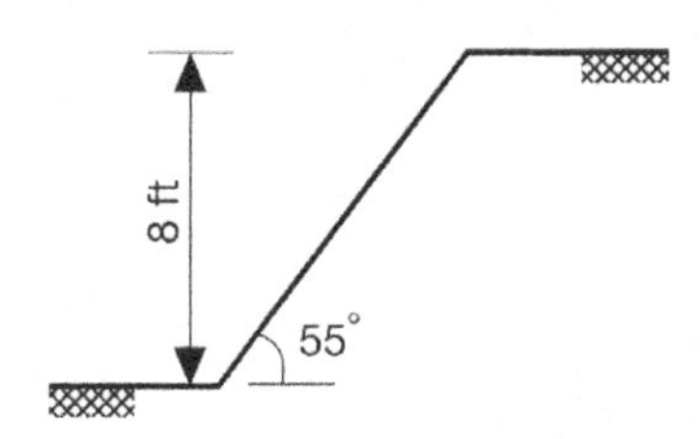

Using Taylor soil stability charts, the slope safety factor for this excavation is most nearly:

(A) 0.9

(B) 1.5

(C) 3.0

(D) 0.3

PROBLEM 62 ***Permissible Noise Exposure***
Workers are exposed to the following noise during their *eight* hours typical working day in a construction site:

Noise Level dBA	Exposure Time hr
80	2.0
85	2.0
90	3.0
95	1.0

The amount of noise dose those workers are exposed to during their typical working day is most nearly:

(A) 350 dBA

(B) 86.88 dBA

(C) 87.50 dBA

(D) 81.25 dBA

PROBLEM 63 ***Consolidation Settlement***
The below graph plots the results of an odometer test for a clay sample where x-axis represents the logarithm of pressure, and y-axis is the void ratio (e).

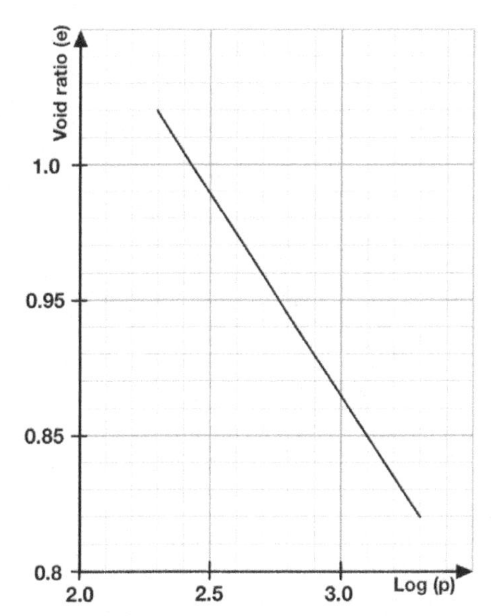

The expected settlement for a 4 ft thick layer of this clay when pressure increases from an initial pressure of 500 psf to a final pressure of 1,000 psi is most nearly:

(A) 2.4 in

(B) 0.2 in

(C) 3.8 in

(D) 0.3 in

PROBLEM 64 ***Excavation Construction Safety***

A mobile crane is required to lower pipes into the trench shown below. The distance from the center of rotation of the boom to the edge of excavation is 7.5 ft. Also, the center of rotation is 7 ft above ground.

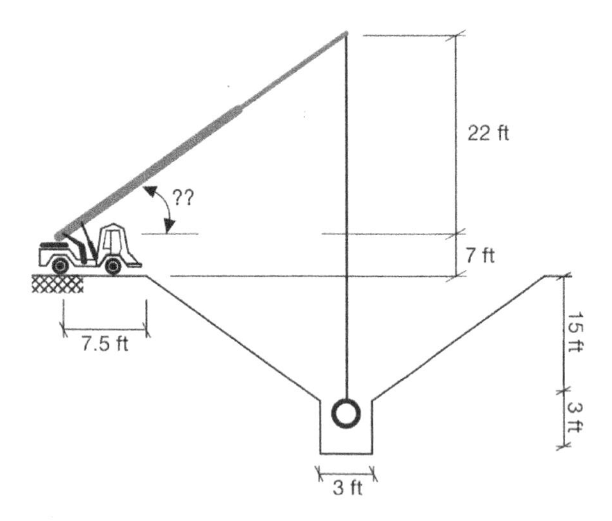

Knowing that the soil type is C per OSHA's definition of soil types, the boom angle is most nearly:

(A) 42.6°

(B) 34.9°

(C) 32.3°

(D) 44.0°

PROBLEM 65 ***Distresses in Flexible Pavements***

The cause of a series of closely spaced ridges and valleys (ripples) in flexible pavements that are perpendicular to the traffic direction is usually caused by:

I. Insufficient base stiffness and strength.
II. Insufficient subgrade stiffness and strength.
III. Moisture and drainage problems.
IV. Freezing and thawing.

(A) I

(B) I + II

(C) I + II + III

(D) I + II + III + IV

PROBLEM 66 ***Formwork Design***

The following figure represents a cross section in a slab formwork system where the thickness of plywood (the sheathing) is ¾ in. The plywood has a modulus of elasticity of $E = 1{,}500\ ksi$ and a self-weight of 8 psf.

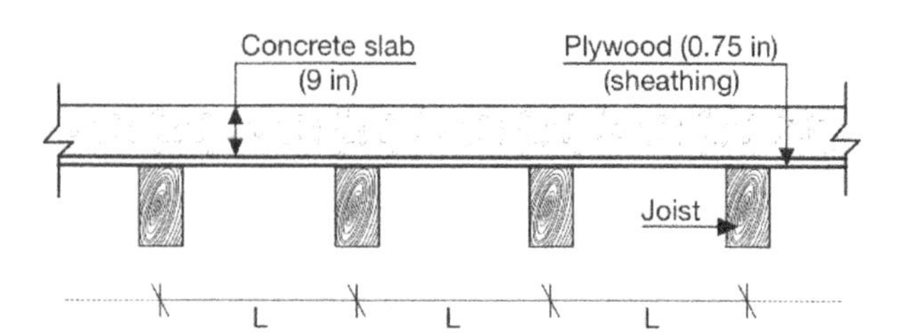

Assuming that deflection controls, with a deflection limit of $L/360$ (*), and construction loads of:

- Fresh reinforced concrete = 145 pcf
- Live load = 50 psf

Spacing $'L'$ between middle joists should nearly be:

(A) 12.0 in

(B) 18.2 in

(C) 62.4 in

(D) 26.5 in

* Use the following equations to calculate deflection in continuous beams/slabs:

$$\Delta_{end\ span} = \frac{wL^4}{148\ EI}$$

$$\Delta_{mid\ span} = \frac{wL^4}{1{,}923\ EI}$$

PROBLEM 67 ***Dam Site Rock Quality***

Drilling was carried out for a dam site investigation project for a depth of $100\ ft$. Total length of recovered core pieces for samples $> 4\ in$ add up to $70\ ft$.

The below best describes the quality of rock:

(A) Very Poor

(B) Poor

(C) Fair

(D) Good

PROBLEM 68 ***Concrete Section Design***

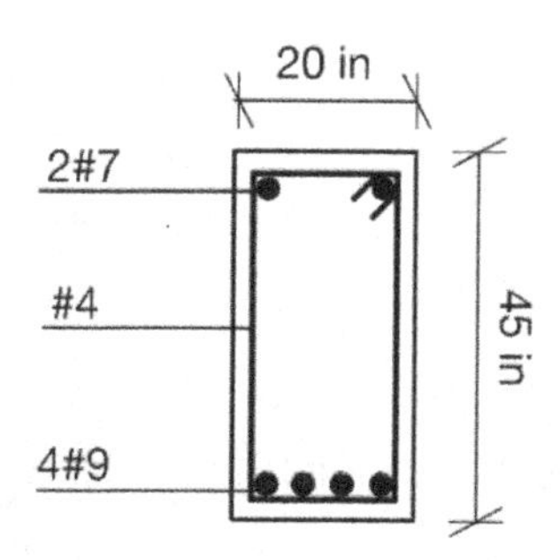

Using $f_y = 60\ ksi$ for the main positive steel/ reinforcements, and using a concrete strength of $f_c' = 4\ ksi$, along with a cover to the center of the bottom reinforcements of 1 ½ in. The nominal moment this section can withstand in $kip.ft$ is most nearly :

(A) 9,936.5

(B) 860

(C) 830

(D) 250

PROBLEM 69 ***Combined Stresses***

The below point load is applied to the square column section on its Y-Z plane at an angle of 30° as shown, and at a distance of $4\ in$ from edge AD and $9\ in$ from edge AB.

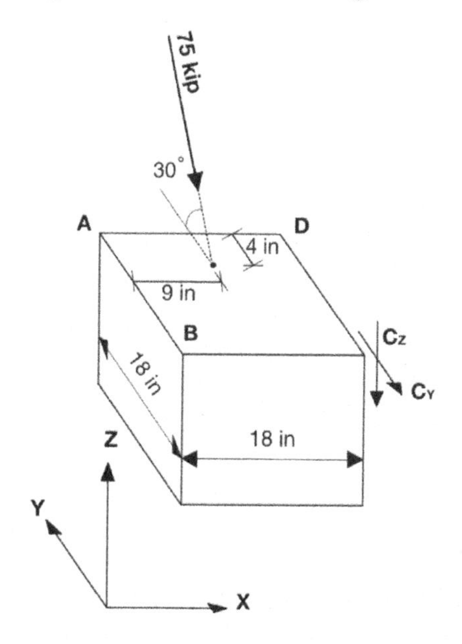

The Z and Y stresses generated at corner C (i.e., C_z & C_y) are most nearly:

(A) $+\ 0.08\ ksi$ & $Zero$

(B) $-0.19\ ksi$ & $Zero$

(C) $+0.08\ ksi$ & $-\ 0.20\ ksi$

(D) -0.31 & $-\ 0.2\ ksi$

PROBLEM 70 ***Fastener Group in Shear***

The below is a steel corbel connected to an H column using shear bolts as shown.

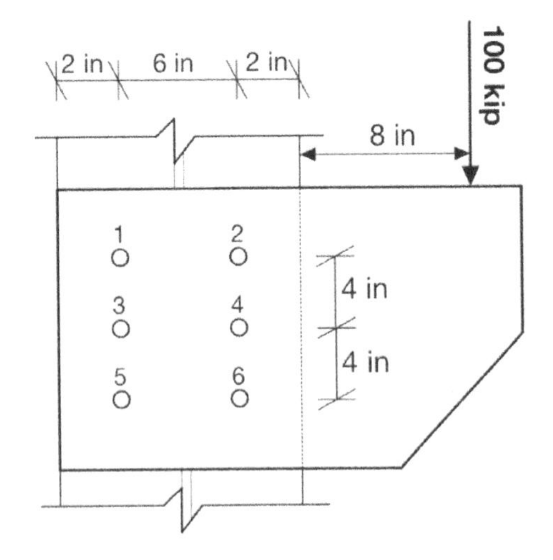

The total shear force which bolt No.1 experiences due to the external loading of 100 kip in kip is most nearly:

(A) 16

(B) 63

(C) 47

(D) 70

PROBLEM 71 ***Steel Props Spacing***

The below is a concrete 9 in floor/slab plan view with locations of future 12 in × 12 in columns showing.

The floor/slab formwork is to be designed and the number of steel props that will support the slab's formwork/plywood needs to be determined (*).

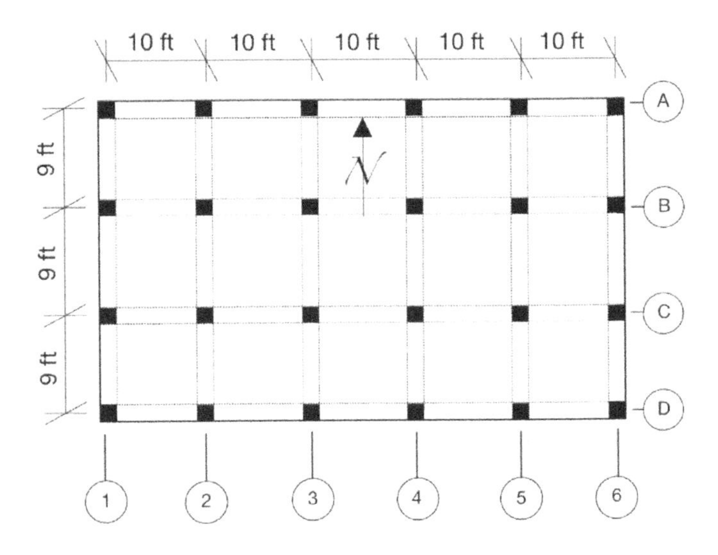

With the following material properties and specifications, the number of props needed to support this slab is most nearly:

- Fresh reinforced concrete = 145 pcf
- Live load = 50 psf
- Plywood self-weight of 8 psf
- Steel prop service capacity = 7 kip
- Maximum spacing (**) between props to keep deflection under control is 80 in.
- Dimensions shown are center-to-center for all columns.

(A) 24

(B) 34

(C) 40

(D) 56

(*) Columns locations are not considered in the calculations.

(**) This represents joists spacing center-to-center and stringers spacing center-to-center.

PROBLEM 72 ***Rate of Concrete Placement***
The below figure is a plan view for a 14 *in* thick 11 *ft* high shear wall combination prepared for concrete placement.

The building under consideration has eight of those wall combinations and concrete should be placed in sequence starting from wall combination 1 to wall combination 8.

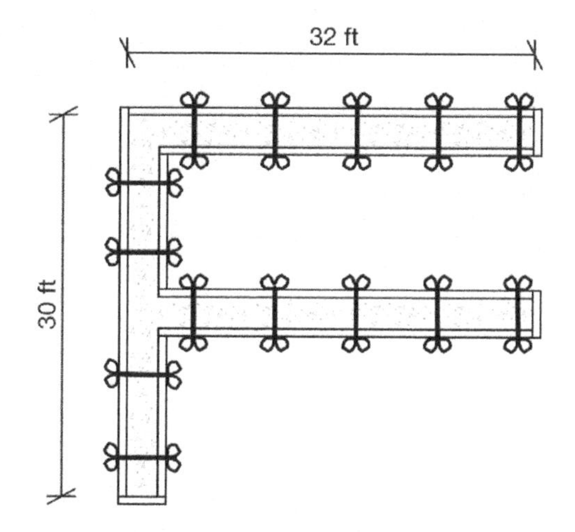

Concrete placement starts at 8:00 AM and must stop at 12:00 AM.

Assuming concrete is pumped continuously at an average rate of 70 $yard^3/hr$ inclusive of lost time, a horizontal construction joint, if needed, will be placed at:

(A) Wall combination 6 at a height of 0.5 *ft*

(B) Wall combination 6 at a height of 5 *ft*

(C) Wall combination 7 at a height of 5 *ft*

(D) Wall combination 7 at a height of 2.5 *ft*

PROBLEM 73 ***85th Percentile Speed***
The below table has 21 speed readings for vehicles moving passed a monitoring point:

Speed *mph*
55
45
72
59
32
25
46
12
56
45
21
45
69
59
73

The 85th percentile speed for the above data set is:

(A) 69 *mph*

(B) 48 *mph*

(C) 25 *mph*

(D) 64 *mph*

PROBLEM 74 ***Safety Incidence Rate***
The number of safety incidents on a 2 *year* construction project with 150 workers spending 260 *hrs* per month along with an absence rate of 15% is 250.

Half the workers moved into a new 1.5 *year* project with the same working conditions.

Assuming 25% improvement on safety, the expected number of incidents in the second project is most nearly:

(A) 117

(B) 70

(C) 150

(D) 120

PROBLEM 75 ***Form Prop Loading***

The below is a formwork system that supports a freshly placed 9 in thick square 30 $ft \times 30\ ft$ concrete slab.

Joists are placed below the plywood @ 60 in center to center and they are continuous. Stringers are placed below joists @120 in center to center and they are simply supported as shown (*).

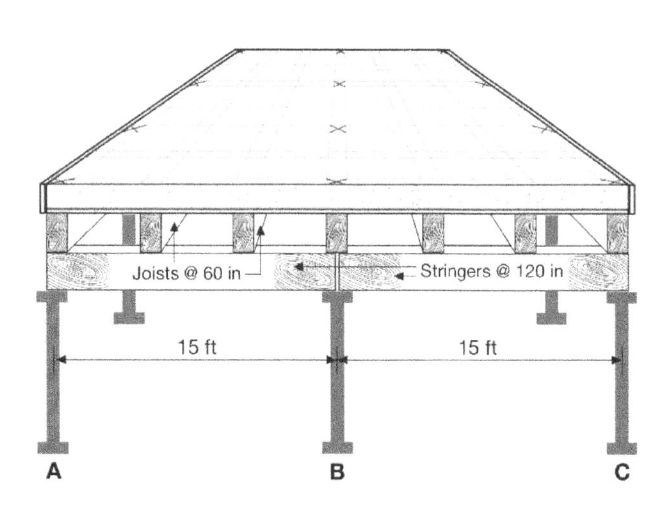

Assuming fresh concrete density is 145 pcf, a live load of 50 psf, and ignoring the weight of the formwork system, the service load that is transferred to prop A is most nearly:

(A) 4.8 kip

(B) 6.0 kip

(C) 1.6 kip

(D) 6.4 kip

(*) Dashed lines on the concrete slab represent locations of joists and stringers. 'X' marks the location of steel props.

PROBLEM 76 ***Bearing Capacity for a Square Foundation***

The below is a concrete square foundation placed 5 ft below surface in soil that has the following properties:

- Cohesion = 450 psf
- Friction angle = 35°
- Density = 130 pcf
- Ground water 5 ft deep

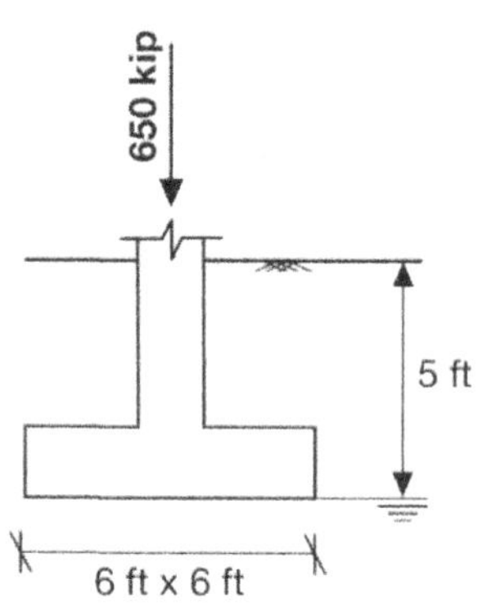

Considering the single footing carries a load of 650 kip, safety factor for this footing against shear failure is most nearly:

(A) 4.3

(B) 2.9

(C) 3.6

(D) 4.6

PROBLEM 77 ***Construction Operation Over a Single Footing***

During an operation of removing soil from over an embedded foundation for an existing building, the following can happen:

(A) Stress relief as the load which was exerted by the soil on top is removed.

(B) Bearing capacity reduction with possible shear failure.

(C) Reduction in stresses at the bottom reinforcements of the footing.

(D) None of the above.

PROBLEM 78 ***Estimated Cost and Profitability***

You are a formwork/wood supplier who has been asked to provide material for one of the contractors to erect a building project. The material will cost you $100,000 to supply. The contractor asked you for a monthly payment plan over a year for cash flow reasons. The monthly charge for the contractor that gets you to maintain an overall profit margin of 20% on all your expenses, knowing that bank charges a fixed interest of 6% per year, is most nearly:

(A) $10,600

(B) $10,952

(C) $10,332

(D) $10,720

PROBLEM 79 ***Ground Anchor Capacity***

The below reinforced concrete cantilever retaining wall has the following properties:

- Concrete density $\gamma_{concrete} = 150\ pcf$
- Soil density $\gamma_{soil} = 130\ pcf$
- Soil's friction angle $\emptyset'_{soil} = 35^o$
- Anchor(s) shown below placed at $3\ ft$ intervals along the length of the wall.

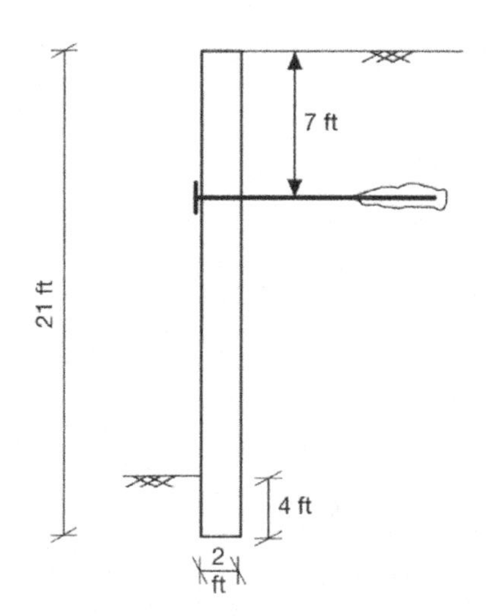

The expected Anchor loading using the tributary area method is most nearly (*):

(A) $8\ kip$

(B) $2\ kip$

(C) $24\ kip$

(D) $16\ kip$

(*) Use $p = k_a\, H\, \gamma_{soil}$ as the maximum ordinate for the tributary area method.

PROBLEM 80 ***Pile Depth Calculation***

The below $12\ in$ diameter concrete pile has been placed into a soil with skin friction of $150\ psf$ and a potential bearing capacity of $200\ psf$ at the expected depth.

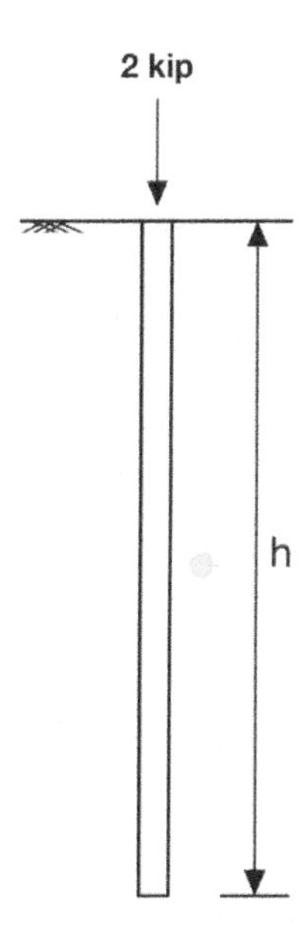

Pile depth h that should resist a pile loading of $2\ kip$ with a safety factor of 2.5 is most nearly:

(A) $5\ ft$

(B) $10\ ft$

(C) $15\ ft$

(D) $20\ ft$

GENERAL BREADTH

SOLUTIONS

SOLUTION 1

The *Thiessen method* is a weighing method with a weight assigned to each of the gauging stations. Straight lines are drawn to connect all the stations as shown in the figure below. Perpendicular bisectors of those connecting lines are then extended to form polygons around each station.

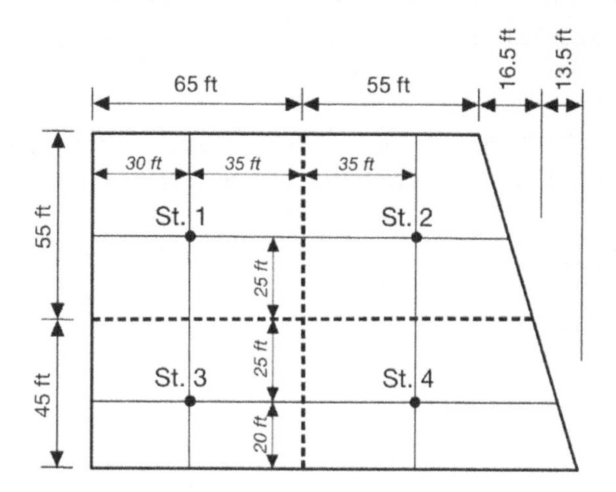

The area of each polygon is considered as the effective area per station. A weighted average is then used to compute the required average rain depth. This method is referenced and explained in further detail in the *NCEES Handbook*.

St.	Effective area (A) ft^2	Gauge depth in	A × depth
1	3,575.00	7.5	26,812.50
2	3,478.75	5.5	19,133.13
3	2,925.00	11.5	33,637.50
4	3,521.25	7.5	26,409.38
	13,500.00		**105,992.50**

Average depth $= 105{,}995.5/13{,}500 = 7.85\ in$

Correct Answer is (B)

SOLUTION 2

The most accurate method for averaging precipitation over an area is the *Isohyetal method.* In this method, the amount of precipitation measured at each station is placed on the watershed map at gauges locations. Contours of equal precipitation, which are called Isohyets, are then drawn. The area between Isohyets is determined accurately using planimetry. Based on those measures, the average precipitation for each area between those Isohyets is estimated by taking the average of two Isohyets sharing a boundary. These values are then multiplied by each area's percentage and added together to obtain the weighted average.

The *NCEES Handbook version 1.2,* Section 6.5.6.2 has a detailed elaboration for this method, the *Thiessen method,* and the *arithmetic averaging method* as well.

Correct Answer is (B)

SOLUTION 3

The *shrinkage limit* represents the water content that corresponds to transitioning between a brittle state and a semi-solid state.

The *plastic limit* represents the water content that corresponds to transitioning between a semi-solid state and a plastic state.

The *liquid limit* represents the water content that corresponds to transitioning from a plastic state to a liquid state.

The *plasticity index* represents the range within which soil remains in its plastic state bounded with its *plasticity limit* as the lower limit, and the *liquid limit* as the upper limit to this range.

This renders the *liquid limit* the maximum water content any soil can get to.

Correct Answer is (A)

SOLUTION 4

Manning Equation is used to solve this question.

$$Q = \frac{1.486}{n} A R_H^{2/3} S^{1/2}$$

Q is the discharge or flow rate (cfs).

A is the cross-sectional area of the flow (ft^2).

RR_H is the hydraulic radius, can be calculated by dividing the area of the flow (A) by the wetted Perimeter (P), or by using the tables provided in the *NCEES Handbook* Section 6.4.5.4.

S slope (ft/ft).

n is the manning roughness coefficient found in the *NCEES Handbook version 1.2* for concrete lined channel $'0.015'$.

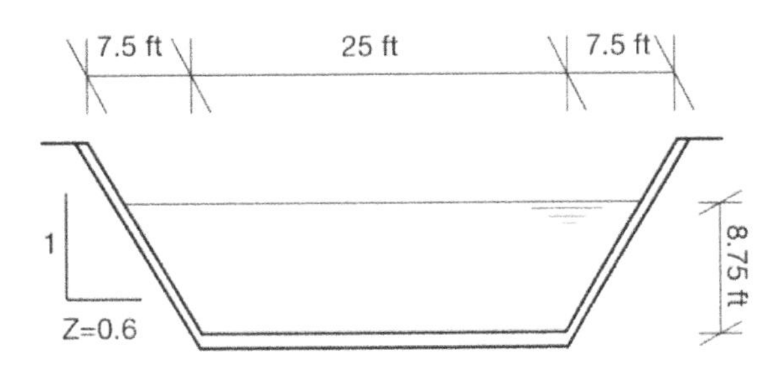

Using equations provided by *NCEES Handbook* and sourced by Chow (1959):

$$R_H = \frac{(b+zy)y}{b+2y\sqrt{1+z^2}}$$

$$= \frac{(25+0.6\times8.75)\times8.75}{25+2\times8.75\times\sqrt{1+0.6^2}}$$

$$= 5.83\ ft$$

$$A = (b + zy)y$$

$$= (25 + 0.6 \times 8.75) \times 8.75 = 264.7\ ft^2$$

$$Q = \frac{1.486}{0.015} \times 264.7 \times 5.83^{2/3} \times 0.02^{1/2}$$

$$= 12{,}012.75\ ft^3/sec = 7{,}764\ MGD$$

Correct Answer is (B)

SOLUTION 5

Reynold number R_e is calculated as follows:

$$R_e = \frac{\nu R_H}{\upsilon}$$

v is the mean velocity of the flow (given as 0.3 ft/sec).

υ is the kinematic viscosity for water which equals to $1.059 \times 10^{-5}\ ft^2/sec$ for water at $70^o\ F$.

R_H is the hydraulic radius and equals to A/P which is calculated as follows for such a channel cross section (refer to the *NCEES Handbook*).

$$R_H = \frac{zy}{2\sqrt{1+z^2}}$$

$$= \frac{0.75\times3}{2\sqrt{1+0.75^2}}$$

$$= 0.9\ ft$$

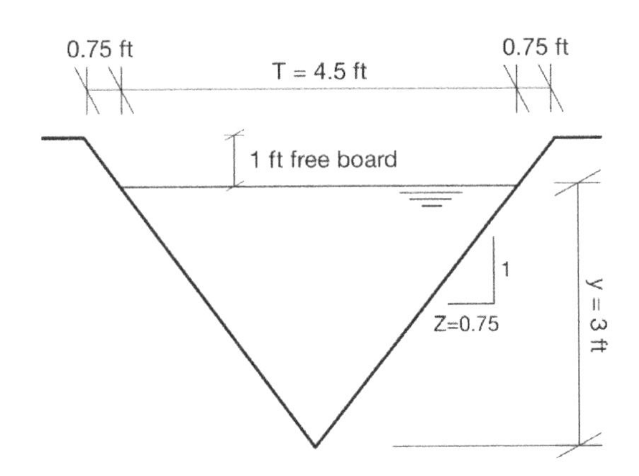

$$R_e = \frac{0.3\times0.9}{1.059\times10^{-5}}$$

$$= 25{,}496$$

Correct Answer is (D)

SOLUTION 6

Reference is made to the earned value management equations in *NCEES Handbook* based upon which the below tables are constructed:

		I	II	III
Activity	**Duration** (Weeks)	**Budget** (USD)	**Actual Cost** (USD)	**Progress**
A	2.0	10,000	4,850	85%
B	1.0	2,000	-	0%
C	6.0	18,000	8,500	60%
D	1.0	10,000	-	0%
	8.0	**40,000**	**13,350**	

IV	I × IV	II	I × III
Planned completion wk3	**BCWS** Planned Budget (USD)	**ACWP** Actual Cost (USD)	**BCWP** Earned Value (USD)
100%	10,000	4,850	8,500
100%	2,000	-	-
33.3%	6,000	8,500	10,800
0	-	-	-
	18,000	**13,350**	**19,300**

$$CPI = BCWP/ACWP$$
$$= 19{,}300/13{,}350$$
$$= 1.445$$

$$SPI = BCWP/BCWS$$
$$= 19{,}300/18{,}000$$
$$= 1.07$$

$$ETC = (BAC - BCWP)/CPI$$
$$= (40{,}000 - 19{,}300)/1.445$$
$$= \$\,14{,}318$$

$$EAC = ACWP + ETC$$
$$= \$13{,}350 + \$14{,}318$$
$$= \$\,27{,}668$$

The above indicates that the project is ahead of schedule with SPI >1.0 and cost performance is positive with CPI >1.0 and the estimate to completed is $27,668.

Correct Answer is (A)

SOLUTION 7

Clay is an undrained layer, which means that when loaded, water will not drain immediately. Rather, water, due to excess pressure, will drain/get expelled slowly and over a long period of time. The slow expulsion of water from the voids between clay particles causes the layer to lose its structure which leads into its ultimate and slow settlement over time.

When clay is loaded, and due its undrained property, pore water pressure increases. Upon water expulsion, pore water pressure decreases over time resulting in a gradual long-term increase in effective stress.

Correct Answer is (B)

SOLUTION 8

Using the critical path method, start with building an activity network. *Activity on nodes* was the preference in this solution with the following nomenclature and further steps:

Step 1: A *forward pass* to determine early dates (Early Start ES and Early finish EF). When two activities feed forward into the same successor, the latest (longest) EF prevails.

Step 2: A *backward pass* to determine late dates (Late Start LS and Late finish LF). When two activities feed backward into the same predecessor, the earliest (shortest) LF prevails.

Early Start		Early Finish
	Activity	
	Duration	
Late Start		Late Finish

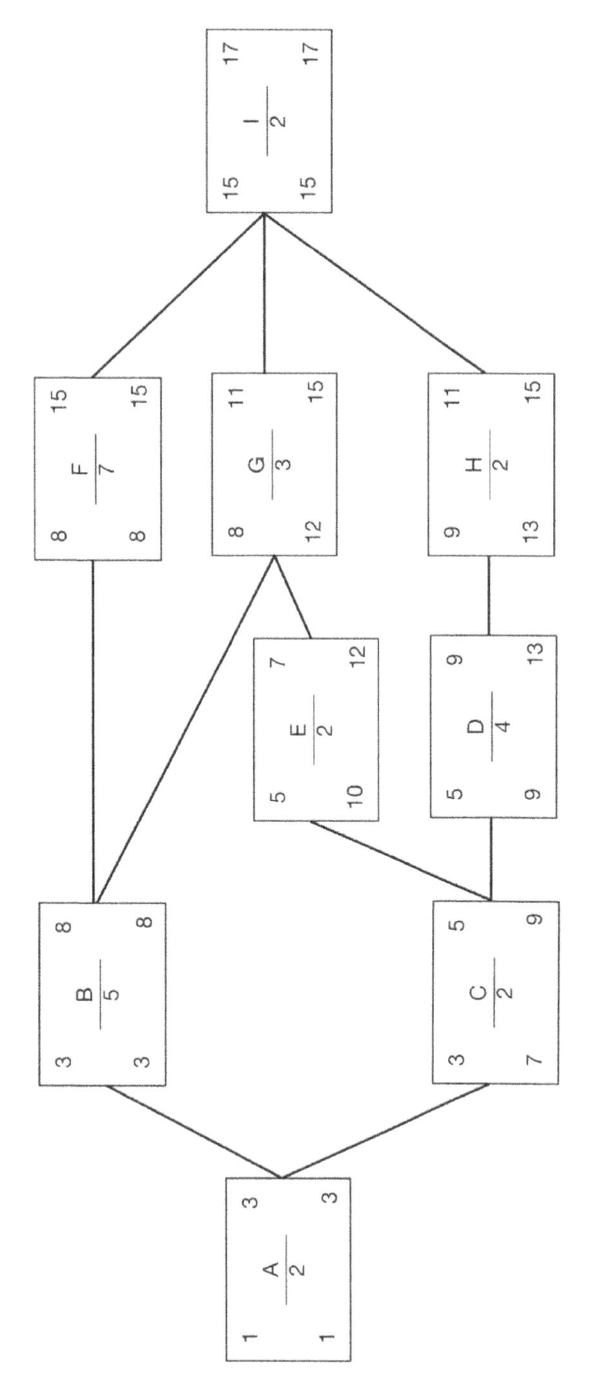

The following equations were applied to activity 'G':

$$Total\ Float\ =\ LS - ES\ =\ 12 - 8\ =\ 4\ days$$

$$Free\ Float\ =\ Earliest\ ES_{successor} - EF$$

$$=\ 15 - 11\ =\ 4\ days$$

Correct Answer is (B)

SOLUTION 9

NCEES Handbook version 1.2 Section 3.7 provides a guideline for the minimum number of exploration points and their depth, as published by the *Federal Highway Administration* FHWA 2002.

The guideline stipulates that the minimum number of boreholes required for retaining walls $> 100\ ft$ in length should be spaced $100\ ft$ to $200\ ft$ with locations alternating from the front of the wall to behind it. The depth of these exploration points should be 1 to 2 times the wall height or a minimum of $10\ ft$ below the bedrock.

Correct Answer is (B)

SOLUTION 10

Generally, well graded Sands have higher friction angles compared to gap graded sands. From the presented chart, Sand 2 seems to have gaps in its gradation around particle sizes 0.1 mm to 1.0 mm, and this will have a detrimental effect on its friction angle.

Shear strength is proportional to cohesion and to the friction angle, see below:

$$\tau = c + \sigma_n tan\emptyset$$

τ Shear strength

c Total cohesion

σ_n Normal stress

$\emptyset$ Friction angle

There is a higher chance of having a higher friction angle for Sand 1 compared to Sand 2, which renders Sand 1 stronger in shear during normal loading.

Correct Answer is (B)

SOLUTION 11

This question can be solved in two methods:

(1) *Pythagorean theorem*: with the obvious triangular symmetry, the vertical deflection can be calculated by considering the shortening of hypotenuses AC and BC.

(2) *Virtual work* or *unit load* method.

<u>Using the Pythagorean Theorem:</u>

Due to triangular symmetry, internal forces

$AC = BC$

Vertical loads' sum at joint 'C' is assessed as follows:

$$300\ kip = \left(\frac{4}{5}\right) \times AC + \left(\frac{4}{5}\right) \times BC$$

$$AC = BC = 187.5\ kip\ in\ compression$$

$$\delta_{AC} = \frac{PL}{AE} = \frac{187.5\ kip \times 5\ ft \times \left(12\ \frac{in}{ft}\right)}{6.45\ in^2 \times 29{,}000\ ksi} = 0.06\ in$$

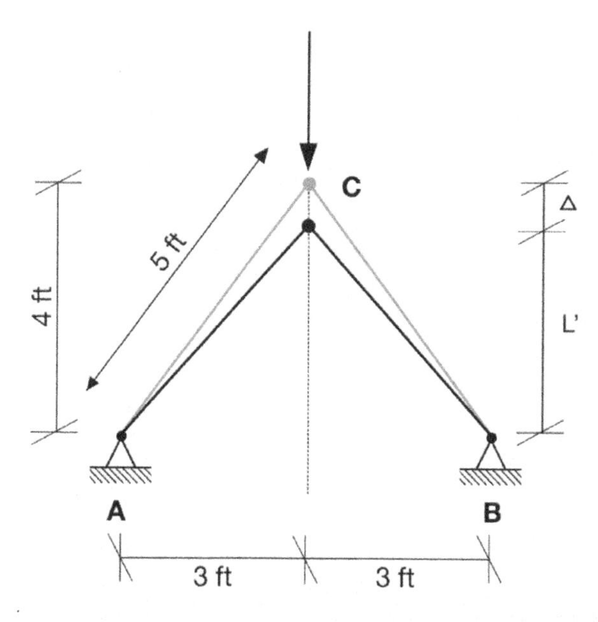

Referring to the Pythagorean theorem, the hypotenuse is shortened by 0.06 *in* as shown below:

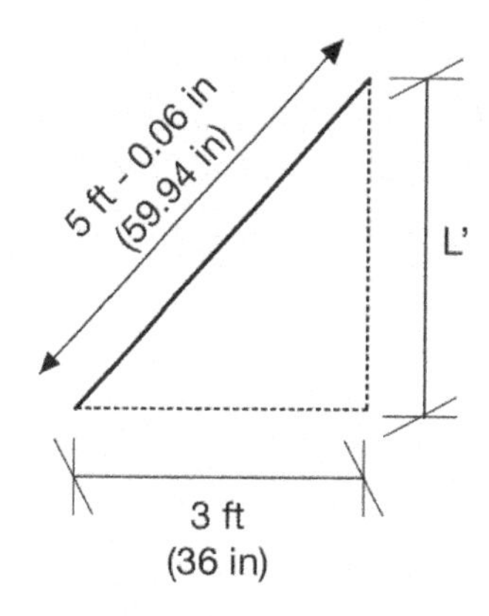

The new vertical leg measurement L' is:

$$L' = \sqrt{59.94^2 - 36^2} = 47.92\ in$$

$$\Delta_c = 47.92\ in - 4\ ft \times 12\ \frac{in}{ft}$$

$$= -0.08\ in \downarrow$$

<u>Using the Unit Load Method:</u>

Displacement using the unit load method is determined as follows:

$$\Delta_{joint} = \sum f_i(\delta)_i$$

f_i is the force in each member due to unit load applied to the direction of the wanted displacement.

$$f_{AC} = f_{BC} = -\frac{187.5}{300} = -0.625\ kip$$

δ is the deflection per member from the applied external load which has been determined before as:

$$\delta_{AC} = \delta_{BC} = \frac{PL}{AE} = \frac{187.5\ kip \times 5\ ft \times \left(12\ \frac{in}{ft}\right)}{6.45\ in^2 \times 29{,}000\ ksi}$$

$$= 0.06\ in$$

$$\Delta_C = -0.625 \times 0.06 + -0.625 \times 0.06$$

$$= -\ 0.08\ in \downarrow$$

Correct Answer is (A)

SOLUTION 12

The requested setback is identified in the *NCEES Handbook version 1.2* as distance M and is measured to the center of the road.

$$M = R - R\cos\left(\frac{\Delta}{2}\right)$$

$$(12 + 0.5 \times 11)\, ft = R \times \left(1 - \cos\left(\frac{48}{2}\right)\right)$$

$$\rightarrow R = 202.4\, ft$$

$$L = \frac{R\,\Delta\,\pi}{180} = \frac{202.4 \times 48 \times \pi}{180} = 169.5\, ft$$

Correct Answer is (D)

SOLUTION 13

The pressure right below the footing is calculated as follows:

$$q_o = \frac{P}{A} = \frac{65}{6{\times}6} = 1.8ksf(1{,}800psf)$$

<ins>Boussinesq's method:</ins>
Using the square footing part of Boussinesq's Isobars chart – copied below for ease of reference, the horizontal and vertical coordinates of the chart are determined as portions of ***B*** (i.e., the footing width) as follows:

$$Horizontal\ axis = \frac{hor.\ location\ from\ edge\ of\ footing}{Width\ of\ the\ square\ footing}$$

$$= \frac{7.5\, ft}{6\, ft} = 1.25\, B$$

$$Vertical\ axis = \frac{Ver.location\ from\ bottom\ of\ footing}{Width\ of\ the\ square\ footing}$$

$$= \frac{12\, ft}{6\, ft} = 2\, B$$

Interpolating those coordinates using the Isobar chart:

$$\Delta P = 0.06q_o = 0.06 \times 1{,}800 = 108\, psf$$

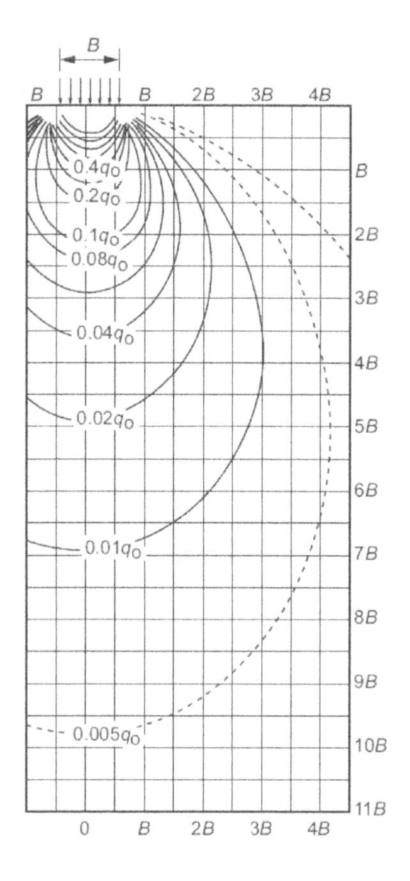

<ins>The 2:1 method:</ins>
The 2:1 method assumes a 2:1 trapezoidal distribution of the load as shown in the following figure:

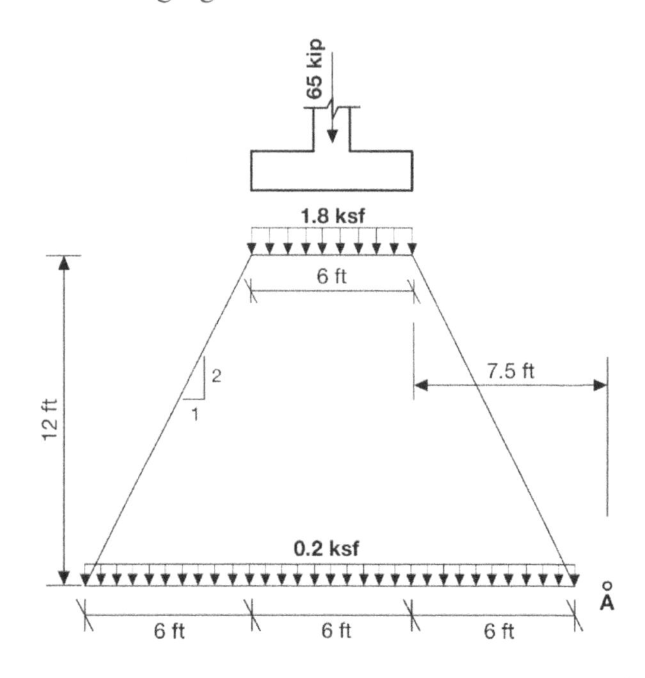

Based on the above distribution, a vertical depth of $Z = 12\,ft$ corresponds to a horizontal distance from the face of the footing of $6\,ft$. Point 'A' however sits at $7.5\,ft$ from the edge of the footing, i.e., $1.5\,ft$ away.

The pressure at point 'A' using this method is therefore Zero.

Correct Answer is (A)

SOLUTION 14

Type K cement contains aluminate that expands during reaction and fills out air voids and offsets any shrinkage that may take place with a net volume change of zero. Due to its expansive nature, type K cement reduces permeability and provides increased abrasion resistance. Long-term performance is observed in traditional cast-in-place and post-tension designs as well.

Correct Answer is (C)

SOLUTION 15

This question requires more of a hands-on field experience and is not a typical question that can be found in a handbook.

Construction machinery used for the purpose of asphalting are as follows:

✓ *(Double) Steel wheel roller*
Steel wheel rollers are self-propelled compaction devices that use steel drums to compress hot asphalt mixes. Usually one, two or three drums. Two drums are most used.

✓ *Paver screed*
Paver screed receives the hot mix from dumping trucks, distributes it and paves it to the required thickness. Pavers have enough power to push the truck as it empties its content into the paver's receiving shaft. Pavers control the speed by which the truck moves, and it can achieve 70-80% density of the layer.

✓ *Pneumatic tire roller*
Those are self-propelled compaction devices that have pneumatic tires which provide proper and smooth compaction to the layer beneath as those tires have no threads in them.

The following machinery are not commonly used in a normal asphalting operation:

✗ *Milling machine*
Milling machines are used to cut existing asphalt or remove a top layer of the existing pavement to create a bed surface. In this case it cannot be claimed to be useful during an asphalting operation unless specifically required on the field.

✗ *Single smooth drum roller*
Smooth drum rollers use static and vibratory pressure to compact rough material such as gravel, rocks, and sand. They are more effective in granular material, and they would not normally produce a smooth surface. A single drum roller has tires at the back and those tires come with deep threads which are not suitable for finishing asphalt layers.

✗ *Motor grader*
Motor graders have an adjustable blade that can be used to complete various construction activities such as surface leveling, fine grading, creating slopes, creating ditches and earth moving. Graders have tires with deep threads which makes it challenging to work on top of a hot mix asphalt. Graders can however be used for asphalting during

abnormal circumstances, such as during a breakdown of the asphalt paver, or paving steep slopes, or locations with narrow access that screeds cannot enter.

SOLUTION 16

Shear stress is calculated as follows:

$$\tau = \frac{VQ}{IB}$$

Q is the first moment of area above or below the point where shear stress is to be determined. It equals to the distance from the area's centroid to the centroid of the shape multiplied by the area itself.

The bottom portion below plane 'B' will be taken for further analysis.

Location of the neutral axis for the L section:

$$\bar{y} = \frac{12 \times 30 \times 15 + 12 \times 18 \times 24}{12 \times 30 + 12 \times 18} = 18.4\ in$$

Moment of inertia calculation for the L section:

$$I = \frac{12 \times 30^3}{12} + 30 \times 12 \times (18.4 - 15)^2$$
$$+ \frac{18 \times 12^3}{12} + 18 \times 12 \times (24 - 18.4)^2$$
$$= 40{,}527\ in^4$$

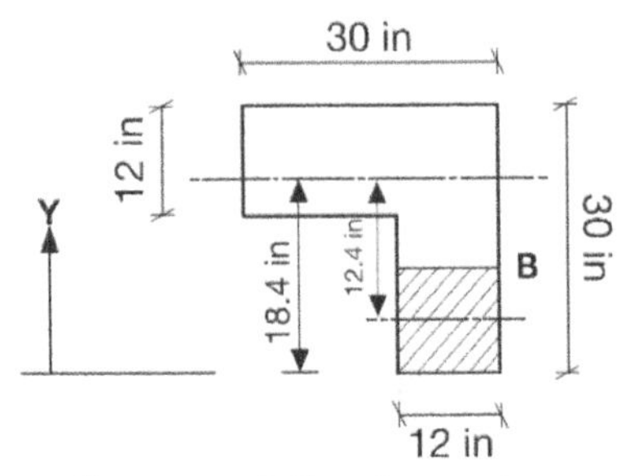

Moment of area and the shear stress:

$$Q = 12 \times 12 \times 12.4 = 1{,}785.6\ in^3$$

$$\tau = \frac{100\ kip \times 1{,}785.6\ in^3}{40{,}527\ in^4 \times 12\ in} = 0.37\ ksi$$

Correct Answer is (B)

SOLUTION 17

The flexible wall permeameter test is used when the tested materials' permeability is lower than $1 \times 10^{-3} cm/sec$. The specimen in this case is encased in a membrane, and with the proper amount of pressure, flow through the specimen is recorded with time.

Correct Answer is (D)

SOLUTION 18

The impulse momentum principle is used to determine forces F_A and F_B which corresponds with the reactions at the support.

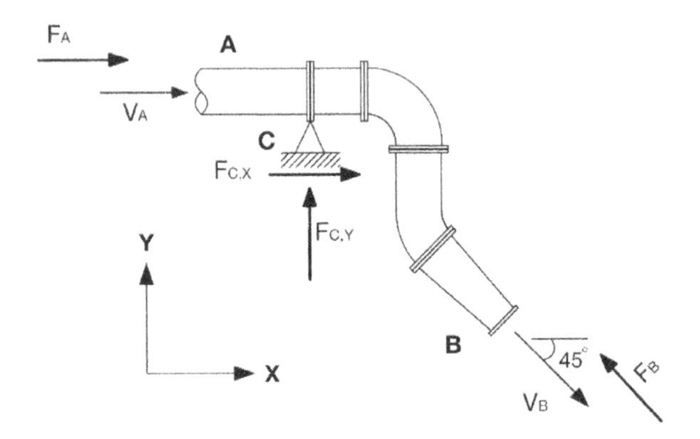

$$V_A = \frac{Q}{A_A} = \frac{1\ cfs}{\pi \times (1.75/12)^2\ ft^2} = 15.0\ ft/sec$$

$$V_B = \frac{Q}{A_B} = \frac{1\ cfs}{\pi \times (0.75/12)^2\ ft^2} = 81.5\ ft/sec$$

$$\sum F_x = (\rho \times Q \times \Delta V_x)/g_c$$
$$= \left(\rho \times Q \times \left(V_{B,x} - V_{A,x}\right)\right)/g_c$$
$$= 62.4\ lb/ft^3 \times 1 ft^3/sec \times$$
$$(81.5\ cos(45)\ ft/sec - 15.0\ ft/sec)/$$
$$32.174\ lbm.ft/lbf.sec^2$$
$$= 82.7\ lb$$

$$F_A - F_B \times cos(45) + F_{C,x} = 82.7\ lb$$

$$F_{C,x} = F_B \times cos(45) - F_A + 82.7\ lb$$
$$= P(A_B \times cos(45) - A_A) + 82.7\ lb$$

$= 100 \frac{lb}{in^2} (\pi\ (0.75\ in)^2\ cos(45) -$

$\pi\ (1.75\ in)^2) +\ 82.7\ lb$

$= -754.5\ lb \leftarrow$

$\sum F_y = (\rho \times Q \times \Delta V_y)/g_c$

$= (\rho\ \times\ Q\ \times\ (V_{B,y}\ - V_{A,y}))\ /g_c$

$= 62.4\ lb/ft^3 \times 1ft^3/sec$

$\times (-81.5\, sin(45)\ ft/sec\ - zero)/$

$32.174 lbm.ft/lbf.sec^2$

$=\ -111.8\ lb$

$F_B \times sin(45) + F_{C,y} = -111.8\ lb$

$F_{C,y} = -111.8lb - F_B \times si\, n(45)$

$= -111.8lb - P \times A_B \times si\, n(45)$

$= -111.8lb - 100 \frac{lb}{in^2} \times \pi(0.75\ in)^2 \times$
$si\, n(45)$

$= -236.75\ lb \downarrow$

Correct Answer is (A)

SOLUTION 19

This is a sheet flow, which is a flow over plane surfaces at very shallow depths (about 0.1 ft). Per *NCEES Handbook*, sheet flow travel time in minutes over a flow length L (ft), and a slope S measured in (ft/ft)is calculated as follows:

$$T_{ti} = \frac{K_u}{I^{0.4}} \left(\frac{nL}{\sqrt{S}}\right)^{0.6}$$

$$= \frac{0.933}{2.0^{0.4}} \left(\frac{0.011 \times 240\ ft}{\sqrt{0.005}}\right)^{0.6} = 6.2\ minutes$$

Correct Answer is (C)

SOLUTION 20

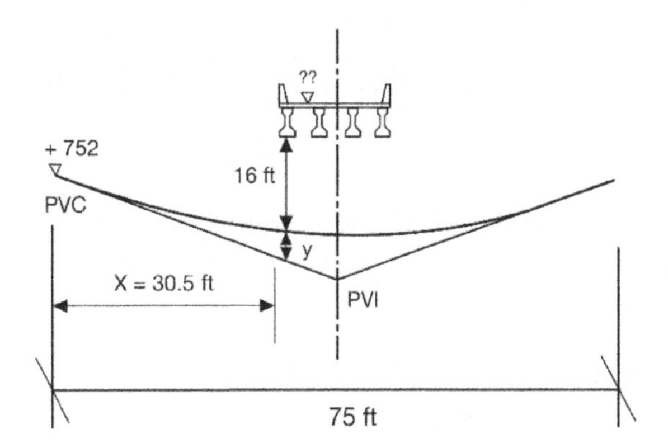

$Y_{PVC} = +752$

$Curve\ elevation = Y_{PVC} + g_1 x + ax^2$

$$a = \frac{g_2 - g_1}{2L} = \frac{0.05 - (-0.06)}{2 \times 75} = 7.33 \times 10^{-4}$$

'x' is calculated to the edge of the bridge:

$$x = \frac{75}{2} - \frac{14}{2} = 30.5ft$$

Curve Elevation

$= 752 + (-0.06) \times 30.5\ +\ 7.33 \times 10^{-4} \times 30.5^2$

$=\ +750.85$

Bridge Deck Elevation

$= 750.85\ +\ 16\ + \frac{53}{12} =\ +771.3$

Correct Answer is (B)

SOLUTION 21

Cement stabilization is considered when *plasticity index* is less than 10 and is used to strengthen granular soils by mixing in Portland cement, typically 3 – 5% of the soil dry weight. Check *NCEES Handbook* for more information.

Lime modification is used to improve fine grained soils with the addition of 0.5 – 3% time to the soil dry weight.

However, with a plasticity index > 10 for subbase or base materials with clay gravel soils, lime stabilization is considered the best option. Typically, 3 – 5% should be enough to dry up the mud contained in the subbase layer.

Correct Answer is (D)

SOLUTION 22

Although this seems a slightly longer question than an exam question would be, it provides an explanation of the concept of NPV and options analysis for a real, and frequent, engineering problem.

One of the means of performing options analysis is the *Net Present Value* (NPV) method. With this, future costs shall be determined first using average inflation. After that, NPV is calculated per option using the given *discount rate* as *interest rate*.

Step 1: Calculate inflation and inflation factors as follows:

Option A

Event	Year	n	Cost (I)	Inflation Symbol	Inflation Factor (II)
New pumps	0		$25,000	-	1.0000
New pumps	10	10	$25,000	(F/P, 5%, 10)	1.6289
New pumps	20	20	$25,000	(F/P, 5%, 20)	2.6533
Maintenance	1→9 & 10→19 & 20→30	28	$3,000	(F/A, 5%, 30) *minus* 2 x (F/P, 5%, 30)	66.4388 *minus* 2 x 4.3219

Option B

Event	Year	n	Cost (I)	Inflation Symbol	Inflation Factor (II)
New pumps	0		$27,500	-	1.0000
New pumps	10	10	$27,500	(F/P, 5%, 10)	1.6289
New pumps	20	20	$27,500	(F/P, 5%, 20)	2.6533
Maintenance	1→9 & 10→19 & 20→30	28	$2,000	(F/A, 5%, 30) *minus* 2 x (F/P, 5%, 30)	66.4388 *minus* 2 x 4.3219

Option C

Event	Year	n	Cost (I)	Inflation Symbol	Inflation Factor (II)
New pumps	0		$40,000	-	1.0000
New pumps	15	15	$40,000	(F/P, 5%, 15)	2.0789
Maintenance	1→14 & 15→30	29	$1,500	(F/A, 5%, 30) *minus* 1 x (F/P, 5%, 30)	66.4388 *minus* 1 x 4.3219

Option D

Event	Year	n	Cost (I)	Inflation Symbol	Inflation Factor (II)
New pumps	0		$47,500	-	1.0000
New pumps	15	15	$47,500	(F/P, 5%, 15)	2.0789
Maintenance	1→14 & 15→30	29	$700	(F/A, 5%, 30) *minus* 1 x (F/P, 5%, 30)	66.4388 *minus* 1 x 4.3219

Step 2: Perform NPV analysis:

Option A

Event	Inflated Cost I x II	NPV Symbol	NPV Factor	NPV
New pumps	$25,000.00	NA	1.0000	$25,000.00
New pumps	$40,722.50	(P/F, 7%, 10)	0.5083	$20,699.25
New pumps	$66,332.50	(P/F, 7%, 20)	0.2584	$17,140.32
Maintenance	$173,385.00	(P/F, 7%, 30)	0.1314	$22,782.79
				$85,622.35

Option B

Event	Inflated Cost I x II	NPV Symbol	NPV Factor	NPV
New pumps	$27,500.00	NA	1.0000	$27,500.00
New pumps	$44,794.75	(P/F, 7%, 10)	0.5083	$22,769.17
New pumps	$72,965.75	(P/F, 7%, 20)	0.2584	$18,854.35
Maintenance	$115,950.00	(P/F, 7%, 30)	0.1314	$15,235.83
				$84,359.35

Option C

Event	Inflated Cost I x II	NPV Symbol	NPV Factor	NPV
New pumps	$40,000.00	NA	1.0000	$40,000.00
New pumps	$83,156.00	(P/F, 7%, 15)	0.3624	$30,135.73
Maintenance	$93,175.35	(P/F, 7%, 30)	0.1314	$12,243.24
				$82,378.98

Option D

Event	Inflated Cost I x II	NPV Symbol	NPV Factor	NPV
New pumps	$47,500.00	NA	1.0000	$47,500.00
New pumps	$98,747.75	(P/F, 7%, 15)	0.3624	$35,786.18
Maintenance	$43,481.83	(P/F, 7%, 30)	0.1314	$5,713.51
				$88,999.70

Option (A) explained:

- Year zero: new pumps to be procured at a cost of $25k. The $25k is the present value of this initial engineering decision and hence no factor will be applied on it and this cost shall be taken to the finish line as is.

- Year 10: new pumps to be procured at a cost of $25k *now*. In 10 years however, the cost of those pumps is expected to be $(F/P, 5\%, 10)$ more – i.e.,
  $\$25{,}000 \times 1.6289 = \$40{,}722.5$

  The $40,722.5 is brought back to year zero using the city discounted rate (i.e., the best estimate for their return on investment on capital projects) using $(P/F, 7\%, 10)$ – i.e., $\$40{,}722.5 \times 0.5083 = \$20{,}699.25$

- Year 20: new pumps to be procured at a cost of $25k *now* as given in the question. In 20 years however, the cost of those pumps is expected to be $(F/P, 5\%, 20)$ more – i.e., $66,332.5. This amount is brought back to year zero using $(P/F, 7\%, 20)$ as:
  $(25{,}000 \times 2.6533) \times 0.2584 = 17{,}140.3$

- Years 1 to 30: yearly maintenance cost of $3,000. With inflation, this cost is brought forward to year 30 in one lump with the factor $(F/A, 5\%, 30)$ as:
  $\$3{,}000 \times 66.4388 = \$199{,}316.4$

  There are two years however during when maintenance costs are not required which is when pumps are installed at year 10 and at year 20. Those two years' payments are deducted in year 30 with $2 \times (F/P, 5\%, 30)$ as:
  $2 \times 4.3219 \times \$3{,}000 = \$25{,}931.4$

  ➔ $199,319.4 – $25,931.4 = $173,385.0

  The overall net maintenance resultant in year 30 is therefore expected to be $173,385.0. This amount is discounted (i.e., brought back to year zero) with the factor $(P/F, 7\%, 30)$ to:
  $\$173{,}385.0 \times 0.1314 = \$22{,}782.79$

The total NPV for option A in this case is:

$$\$25{,}000 + \$20{,}699.25 + \$17{,}140.32 + \$22{,}782.79 = \$85{,}622.36$$

A similar process is applied for all options and option C was found the cheapest of all.

Correct Answer is (C)

SOLUTION 23

The below equation can be found in the *NCEES Handbook* Section 2.3.3.1.

Compacted cubic yards per hour:

$$= \frac{1}{n}(16.3 \times W \times S \times L \times efficiency)$$

$$= \frac{1}{5}\left(16.3 \times \frac{66\ in \times 1\ ft}{12\ in} \times 3.5\ mph \times 12 \times \frac{50\ min}{60\ min}\right)$$

$$= 628\ yard^3/hour\ (6{,}280\ yard^3/day)$$

A 6 $mile$, 12 in thick, 11 ft wide stretch has the following volume:

$$\frac{\left(6\ mile \times 5{,}280\ \frac{ft}{mile}\right) \times \left(12\ in \times \frac{1}{12}\frac{ft}{in}\right) \times 11\ ft}{27\frac{ft^3}{yard^3}} = 12{,}903\ yd^3$$

$$No. of\ rollers\ req'd = \frac{12{,}903}{6{,}280} \cong 2$$

Correct Answer is (B)

SOLUTION 24

Shear flow is calculated using the following equation:

$$q = \frac{VQ}{I}$$

V is the shear load, and ***I*** is the moment of inertia for the overall shape, and both are constant and do not contribute to how the final shear flow profile will look like. This indicates that the first moment of area ***Q*** controls the requested profile.

Q equals to: (1) the distance from the overall shape's centroid to the area's centroid directly above or below the plane under consideration which is bounded by the plane where shear is to be calculated, multiplied by, (2) the area of that part.

(1) Distance from shape's centroid to the area's centroid that is bounded by the plane where shear is to be calculated:

The centroid of the overall shape is constant and does not change, whereas the centroid of each of the selected areas (above or below the plane under consideration) are ever changing as the plane under consideration moves up or down. Therefore, the distance from this centroid to the centroid of the entire shape – expressed in the below figure as '*d*' – is variable and not constant.

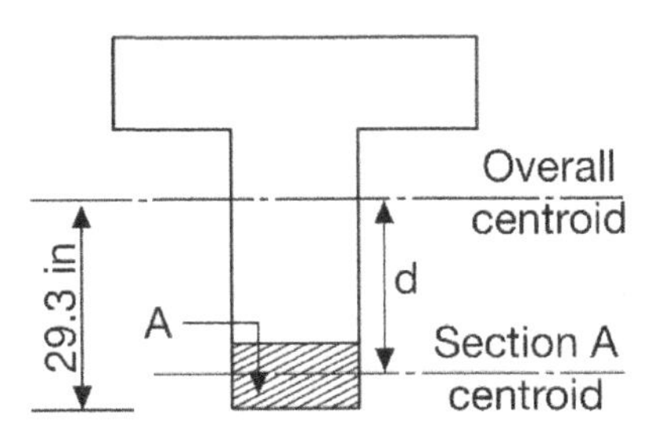

(2) The maximum area above or below the plane under consideration:

The maximum moment of area occurs at a section that strikes a balance – i.e., where the area above it and below it are both maximum, and this can only happen when they are both equal. See below:

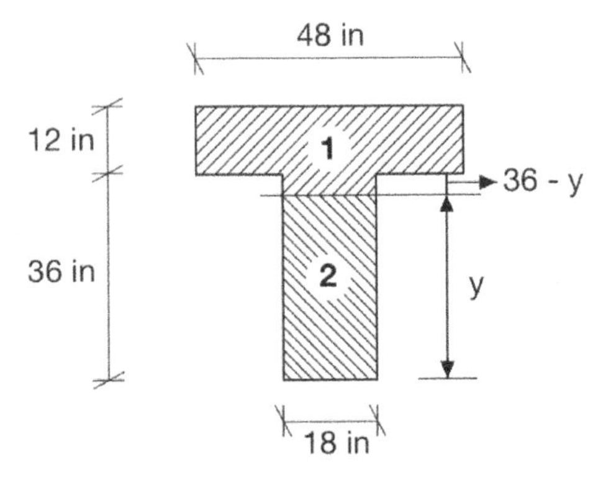

Solving for '*y*' – which DOES NOT represent the location of the centroid of the overall shape – by equating Area 1 to Area 2 gives us the location of the horizonal plane that the moment of area is maximum at Q_{max} (in which case the shear flow is maximum as well). See below:

$$A1 = A2$$

$$48 \times 12 + 18 \times (36 - y) = 18 \times y$$

$$\rightarrow y = 34\ in$$

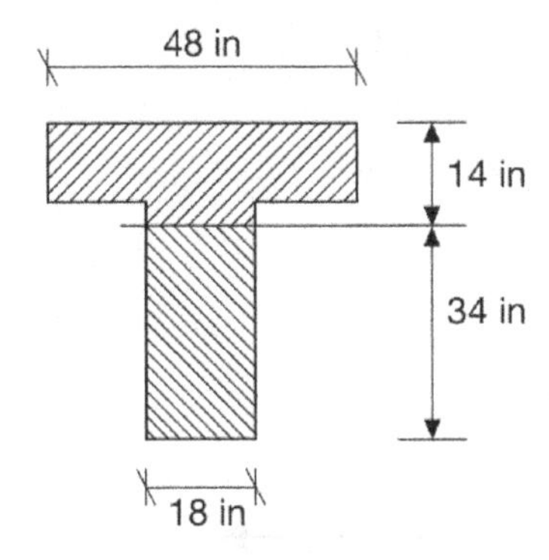

Starting from the top or the bottom of the overall shape, area 'A' starts as zero when 'd' is max, hence moment of area is zero, which means that the shear profile starts at zero and builds up to its maximum value located at 34 in as shown in the following figures.

Centroid of the Overall Shape taking datum from the bottom:

$$centroid = \frac{48\times12\times\left(36+\frac{12}{2}\right)+36\times18\times\left(\frac{36}{2}\right)}{48\times12+36\times18}$$

$$= 29.3\ in$$

Centroid of Area 2 taking datum from the bottom (Area 1 works as well):

$$centroid = 34/2 = 17\ in$$

Although not requested in the question, distance 'd' and the maximum moment of area 'Q_{max}' are calculated as follows:

$$d = 29.3 - 17 = 12.3\ in$$

$$Q_{max} = A2 \times d$$

$$= (34 \times 18) \times 12.3$$

$$= 7{,}527.6\ in$$

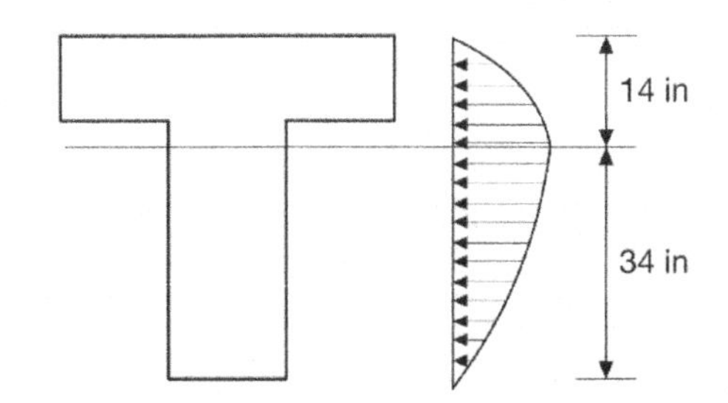

Correct Answer is (C)

SOLUTION 25

Given the two sections are loaded in parallel, they share the same strain $\varepsilon_1 = \varepsilon_2 = \varepsilon$. They do not share the same stress though.

$$A_1 = A_2 = 0.5A$$

$$\sigma_1 = \frac{P_1}{0.5A} = \varepsilon\, E_1 \qquad \text{Equation 1}$$

$$\sigma_2 = \frac{P_2}{0.5A} = \varepsilon\, E_2 = \varepsilon \times 0.5E_1 \qquad \text{Equation 2}$$

Dividing Equation 1/Equation 2 concludes that $P_1 = 2P_2$

Provided that:

$P = P_1 + P_2$ and $P = 3P_2$ or $P = (3/2)P_1$

$$\sigma_{joined} = \frac{P}{A} = \varepsilon \times E_{joined}$$

$$E_{joined} = \frac{P}{\varepsilon\times A} \quad \text{but} \quad \varepsilon = \frac{P_1}{0.5AE_1} = \frac{\frac{2}{3}P}{0.5AE_1}$$

$$E_{joined} = \frac{P}{\left(\frac{\frac{2}{3}P}{0.5AE_1}\right)\times A} = \frac{0.5E_1}{2/3} = \frac{3}{4}E_1$$

$$or\ = 1.5E_2$$

Which is equivalent to $\left(\frac{E_1+E_2}{2}\right)$

A graphical representation for the above relationship is as follows:

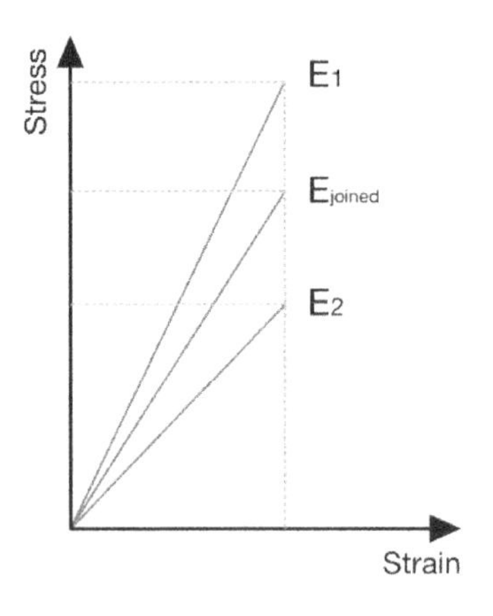

Correct Answer is (A)

SOLUTION 26

Given the two sections are placed on top of each other, they do not share the same strain, they share the same stress though, i.e.,

$$\sigma_1 = \sigma_2 = \sigma_{joined}$$

Also:

$$\varepsilon_1 = \Delta L_1/L \quad \text{and} \quad \varepsilon_2 = \Delta L_2/L$$

$$\varepsilon_{joined} = \frac{\Delta L_1 + \Delta L_2}{2L} = \frac{1}{2}(\varepsilon_1 + \varepsilon_2)$$

$$\sigma_1 = \varepsilon_1\, E_1$$

$$\sigma_2 = \varepsilon_2\, E_2 = \varepsilon_2 \times 0.5E_1$$

$$\varepsilon_1\, E_1 = \varepsilon_2 \times 0.5E_1 \rightarrow \varepsilon_1 = 0.5\varepsilon_2$$

$$\sigma_{joined} = \varepsilon_{joined}\, E_{joined}$$

$$= \frac{1}{2}(\varepsilon_1 + \varepsilon_2)\, E_{joined}$$

$$\frac{1}{2}(\varepsilon_1 + \varepsilon_2)\, E_{joined} = \varepsilon_1\, E_1$$

$$or = \varepsilon_2\, E_2$$

$$E_{joined} = \frac{2\,\varepsilon_1}{\varepsilon_1 + \varepsilon_2} E_1 \quad \text{or} \quad = \frac{2\,\varepsilon_2}{\varepsilon_1 + \varepsilon_2} E_2$$

$$= \frac{2\,\varepsilon_1}{\varepsilon_1 + 2\varepsilon_1} E_1 \quad \text{or} \quad = \frac{2\,\varepsilon_2}{0.5\varepsilon_2 + \varepsilon_2} E_2$$

$$= \frac{2}{3}\, E_1 \quad \text{or} \quad = \frac{4}{3}\, E_2$$

A graphical representation for the above relationship is as follows:

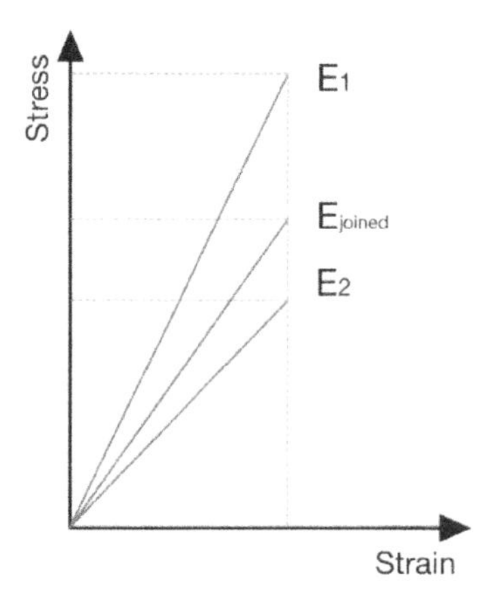

Correct Answer is (B)

SOLUTION 27

Referring to Section 6.5.9.1 of the *NCEES Handbook*, the size of a retention pond using the rational method is calculated as follows – considering that the runoff coefficient C for sloped parks and cemeteries is '0.25' per Section 6.5.2.1:

$$V_s = V_{in} - V_{out}$$

$$= (i \sum AC - Q_o) \times t$$

$$= \left(3.5 \frac{in}{hr} \times 50 Acres \times 0.25 - 35 cfs\right) \times \left(15\, min \times 60 \frac{sec}{min}\right)^*$$

$$= 7{,}875\, ft^3$$

Correct Answer is (A)

* Although the *NCEES Handbook versions 1.1 and 1.2* identify $'t'$ as time in minutes, given the units consistency in this question, $'t'$ is measured in seconds.

SOLUTION 28

Section 6.8.7.2 *Chemical Phosphorus Removal* of the *NCEES Handbook version 1.2* provides the chemical formula for phosphorus removal using Ferric Chloride '$FeCl_3$' as follows:

$$FeCl_3 + PO_4^{3-} \rightarrow FePO_4 (\downarrow) + 3\,Cl^-$$

Correct Answer is (C)

SOLUTION 29

Floatation/Buoyancy Safety Factor is calculated as follows:

$$S.F. = \frac{Concrete\ Weight + Friction}{Buoyancy\ (upward) force}$$

$$= 1.1$$

Weight of the structure/ downward force:

$$W_{Base} = \gamma_{concrete} \times \pi \times r_{out}^2 \times t_{base}$$

$$= 150\,pcf \times \pi \times (10\,ft)^2 \times \left(\frac{36\,in \times 1\,ft}{12\,in}\right)$$

$$= 141{,}300\,lb$$

$$W_{wall} = \gamma_{concrete} \times \pi \times (r_{out}^2 - r_{in}^2) \times height$$

$$= 150\,pcf \times \pi \times (10^2 - r_{in}^2) \times 87\,ft$$

$$= 40{,}977 \times (100 - r_{in}^2)$$

$$= 4{,}097{,}700 - 40{,}977 \times r_{in}^2$$

**Taking height in this case without the base*

$$W_{Total} = W_{Base} + W_{Wall}$$

$$= 141{,}300 + (4{,}097{,}700 - 40{,}977\,r_{in}^2)$$

$$= 4{,}239{,}000 - 40{,}977\,r_{in}^2$$

Buoyancy/ upward force:

$$Buoyancy = \gamma_{water} \times \pi \times r_{out}^2 \times full\ height$$

$$= 62.4\,pcf \times \pi \times (10\,ft)^2 \times 90ft$$

$$= 1{,}763{,}424\,lb$$

$$S.F. = \frac{4{,}239{,}000 - 40{,}977 \times r_{in}^2}{1{,}763{,}424} = 1.1$$

$$4{,}239{,}000 - 40{,}977 \times r_{in}^2 = 1{,}939{,}766.4$$

$$r_{in} = \sqrt{\frac{4{,}239{,}000 - 1{,}939{,}766.4}{40{,}977}}$$

$$\rightarrow r_{in} = 7.5\,ft$$

$$t_{wall} = r_{out} - r_{in}$$

$$= 10\,ft - 7.5\,ft$$

$$= 2.5\,ft\ (30\,in)$$

Correct Answer is (A)

SOLUTION 30

Reference is made to the *NCEES Handbook version 1.2.*

At rest Rankine Coefficient for normally consolidated soils:

$$K_{o,NC} = 1 - \sin\emptyset'$$

$$0.33 = 1 - \sin\emptyset' \qquad \rightarrow \sin\emptyset' = 0.67$$

For over consolidated soils:

$$K_{o,OC} = (1 - \sin\emptyset') \times OCR^{\sin\emptyset'}$$

$$= K_{o,NC} \times OCR^{\sin\emptyset'}$$

$$OCR = \left(\frac{K_{o,OC}}{K_{o,NC}}\right)^{\frac{1}{\sin\emptyset'}}$$

$$= \left(\frac{0.85}{0.33}\right)^{\frac{1}{0.67}}$$

$$= 4.1$$

Correct Answer is (A)

SOLUTION 31
The following information is gathered from the gradation chart:

Nearly 45% is finer than 0.075 mm (No. 200 sieve), which means 55% is retained on this Sieve classifying the sample as either sand or gravel.

Nearly 82% is finer than 4.75 mm (No. 4 sieve), which means that 18% is retained on this sieve classifying the sample as sand.

$D_{10} = 0.025\ mm$
$D_{30} = 0.037\ mm$
$D_{60} = 0.27\ mm$

Coefficient of uniformity:

$C_u = D_{60}/D_{10} = 0.27/0.025 \approx 10$

Coefficient of curvature:

$C_c = (D_{30})^2/(D_{60}\ D_{10})$

$= (0.037)^2/(0.025 \times 0.27) \approx 0.2$

Based on the USCS classification system found in the *NCEES Handbook version 1.2* Table 3.7.2, for soils with > 50% retained on sieve No. 200, and > 50% passes No. 4 sieve, $C_u < 6$ and/or $C_c < 1$, soil would be classified as Poorly Graded Sand (SP).

Correct Answer is (D)

SOLUTION 32
Referring to the *NCEES Handbook version 1.2,* Section 3.8.3:

Dry Density:

$$\gamma_d = \frac{\frac{W_t}{1+w}}{V}$$

$$= \frac{100lb/(1+0.2)}{1\ ft^3} = 83.33\ lb/ft^3$$

Degree of saturation:

$$S = \frac{w}{\left(\frac{\gamma_w}{\gamma_d} - \frac{1}{G}\right)} = \frac{0.2}{\left(\frac{62.4}{83.33} - \frac{1}{2.7}\right)} = 0.53$$

Porosity:

$$n = 1 - \frac{W_s}{GV\gamma_w}$$

$$= 1 - \frac{\gamma_d}{G\ \gamma_w} = 1 - \frac{83.33}{2.7 \times 62.4} = 0.51$$

SOLUTION 33
Referring to the *NCEES Handbook version 1.2,* Section 3.5.2, the vertical elastic settlement is calculated as follows:

$$\delta_v = \frac{C_d \Delta p B_f (1 - v^2)}{E_m}$$

$\boldsymbol{C_d}$ is the rigidity factor and is looked up from the table in the same section as '0.99' for rigid (i.e., concrete) square shaped foundations. $\boldsymbol{\Delta p}$ is the increase in pressure right below the foundation which is $(100kip/36ft^2)$. $\boldsymbol{B_f}$ is the footing dimension which is 6 ft for squared footings in this case.

For fine medium dense sand, Poisson's ratio $\boldsymbol{v}$ is '0.25' as collected from the same Section. The young modulus $\boldsymbol{E_m}$ shall be the lowest of the range provided in the guide – 120 tsf in this case – as the question is looking for maximum settlement.

$$\delta_v = \frac{0.99 \times \left(\frac{100}{36}\right) kip/ft^2 \times 6ft \times (1 - 0.25^2)}{120\frac{ton}{ft^2} \times 2\frac{kip}{ton}}$$

$$= 0.064\ ft(0.77in)$$

Correct Answer is (A)

SOLUTION 34
Using the AASHTO classification system found in the *NCEES Handbook version 1.2* Section 3.7.3, nearly 45% is finer than 0.075 mm (i.e., passing No. 200 sieve). This

indicates that the sample is not granular and falls within the Silt-Clay material categories of: A-4, 5, 6 or 7.

The characteristics of the fines – material finer than 0.425 mm (i.e., passing No. 40 sieve) were given as follows:

$$LiquidLimit(LL) = 50$$

$$Plasticlimit(PL) = 35$$

$$Plasticity\ Index\ (PI) = 50 - 35 = 15$$

Based on this, the material is classified as either A-7-5 or A-7-6.

Using the comment section of the classification table and provided that $LL - 30 = 20 > PI$, classification of the material would be that of A-7-5.

The group index GI for this category is calculated as follows:

GI = (F – 35) [0.2 + 0.005 (LL – 40)] + 0.01 × (F – 15)(PI – 10)

= (45 – 35) [0.2 + 0.005 (50-40)] + 0.01 × (45 – 15)(15 – 10)

= 4

Final classification is A-7-5 (4)

Correct Answer is (A)

SOLUTION 35

The effective stress $\boldsymbol{\sigma'}$ is the total stress $\boldsymbol{\sigma}$ removed from it the pore/water pressure $\boldsymbol{u}$. In which case, the following effective stress changes at the bottom of each layer shall occur over the indicated period of 6 months and the age of the project.

<u>Month zero prior to constructing the foundation:</u>

$$\sigma_{sand} = 120\ pcf \times 15\ ft$$

$$= 1{,}800 psf (1.8 ksf)$$

$$u_{sand} = 62.4\ pcf\ \times 10\ ft$$

$$= 624 psf (0.62 ksf)$$

$$\sigma'_{sand} = \sigma_{sand} - u_{sand}$$

$$= 1.8 - 0.62 \cong 1.2\ ksf$$

$$\sigma_{clay} = 120\ pcf \times (10 + 15) ft$$

$$= 3{,}000 psf (3.0 ksf)$$

$$u_{clay} = 62.4\ pcf\ \times 20\ ft$$

$$= 1{,}248 psf (1.25 ksf)$$

$$\sigma'_{clay} = \sigma_{clay} - u_{clay}$$

$$= 3.0 - 1.25 \cong 1.8\ ksf$$

<u>Month 6 after constructing the foundation:</u>

When the foundation is constructed, the sand layer will drain the excess (now pressurized) water immediately and hence no increase shall occur in the sand pore pressure. This will reflect in an increase in the effective pressure of the sand, and given the loading of the matt foundation took place linearly over a period of 6 months, the increase in effective pressure for sand will be linear as well.

$$\sigma_{sand} = 0.12\ kcf \times 15\ ft\ + 1\ ksf$$

$$= 2.8\ ksf$$

$$u_{sand} = 62.4\ pcf\ \times 10\ ft$$

$$= 624\ psf (0.62 ksf)$$

$$\sigma'_{sand} = \sigma_{sand} - u_{sand}$$

$$= 2.8 - 0.62 = 2.2\ ksf$$

When it comes to the pore pressure of the clay layer, clay will not drain the excess (now pressurized) water right away (unlike sand). Drainage in this case shall occur over a long period of time instead. The pore pressure of the clay layer will increase due to this by the amount of the added load, and this keeps the effective stress unchanged.

$\sigma_{clay} = 0.12\ kcf \times 25\ ft + 1\ ksf$

$= 4.0\ ksf$

$u_{clay} = 0.0624\ kcf\ \times 20\ ft + 1\ ksf$

$= 2.25\ ksf$

$\sigma'_{clay} = \sigma_{clay} - u_{clay} = 4.0 - 2.25$

$= 1.8\ ksf$

Over a long period of time after constructing the foundation:

The sand effective stress will not change as the water has already drained from it long ago.

$\sigma'_{sand} = \sigma_{sand} - u_{sand}$

$= 2.8 - 0.62$

$= 2.2\ ksf$

As for the clay layer, and over a long period of time, the excess (pressurized) water would have been drained then and this should bring the pore pressure down to:

$u_{clay} = 0.0624\ kcf\ \times 20\ ft$

$= 1.25\ ksf$

$\sigma'_{clay} = \sigma_{clay} - u_{clay}$

$= 4.0 - 1.25$

$= 2.8\ ksf$

This makes the Profile A the most representative profile.

Correct Answer is (A)

SOLUTION 36

Based on the information given in the question, Rankine's active coefficient is calculated as follows:

$k_a = tan^2\left(45 - \frac{\phi'}{2}\right)$

$= tan^2\left(45 - \frac{37^o}{2}\right) = 0.25$

The resultant soil pressure equation:

$p_a = \frac{k_a\, h^2 \gamma_{soil}}{2}$ triangular shaped pressure

$p_a = k_a\ h_{above\ water}\ h_{below\ water} \gamma_{soil}$ rectangular shaped pressure

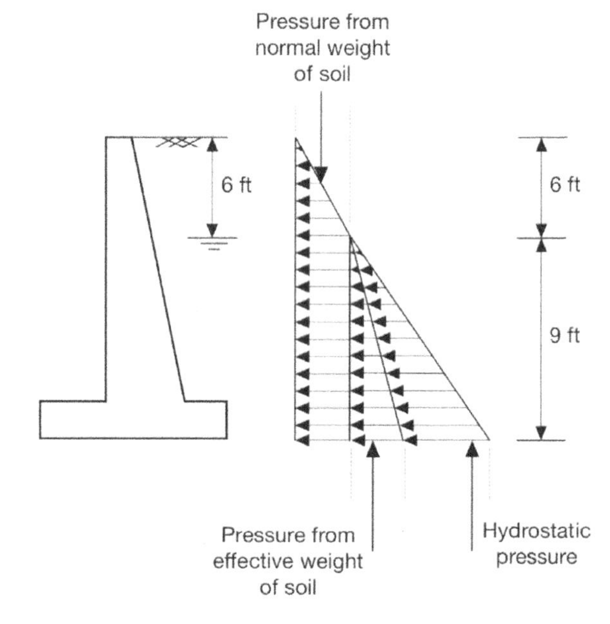

Based on this, lateral/overturning pressures are calculated per linear ft as follows:

Pressure resultant force from normal weight of soil:

$p_{a,(0-6ft)} = \frac{1}{2} \times 0.25\ \times (6ft)^2 \times 130 \frac{lb}{ft^3}$

$= 585\ lb/ft$

$p_{a,(6-15ft)} = 0.25\ \times (9ft)^2\ \times 130 \frac{lb}{ft^3}$

$= 2{,}632.5\ lb/ft$

Pressure resultant force from effective weight of soil:

$p_{a,(6-15ft)} = \frac{1}{2} \times 0.25 \times (9\ ft)^2 \times (130 - 62.4) \frac{lb}{ft^3}$

$= 684.5\ lb/ft$

Pressure resultant force from hydrostatic pressures:

$$p_{water,(6-15ft)} = \frac{1}{2} \times (9ft)^2 \times 62.4\frac{lb}{ft^3}$$

$$= 2{,}527.2\ lb/ft$$

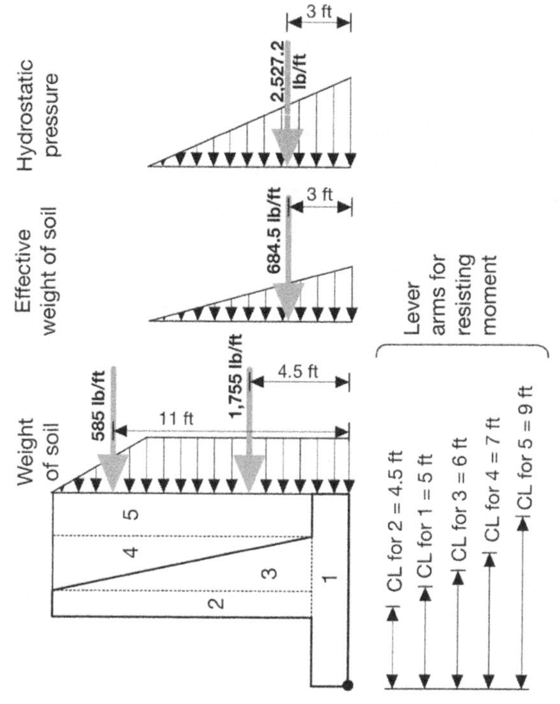

Overturning moments (O.M.) around the marked left most point:

Description	Force lb	Lever arm ft	Overturning Moment $lb.ft$
Soil 0-6 ft	585.0	11.0	6,435.0
Soil 6-15 ft	1,755	4.5	7,897.5
Eff. soil 6-15ft	684.5	3.0	2,053.5
Hydrostatic 6-15 ft	2,527.2	3.0	7,581.6
			23,967.5

The sum of overturning moments:
$\sum O.M. = 23{,}967.5\ lb.ft/ft$

Resisting moments (R.M.) with items 4 and 5 belong to the soil , $\gamma_{soil} = 130\ pcf$:

Sec	Area ft^2	Weight lb	Lever arm ft	Resisting Moment $lb.ft$
1	17.5	2,625.0	5.0	13,125.0
2	13.25	1,987.5	4.5	8,943.8
3	19.9	2,985.0	6.0	17,910.0
4	19.9	2,587.0	7.0	18,109.0
5	26.5	3,445.0	9.0	31,005.0
				89,092.8

The sum of resisting moments

$\sum R.M. = 89{,}092.8\ lb.ft/ft$

Safety Factor calculation:

$$S.F._{OT} = \frac{\sum R.M.}{\sum O.M.} = \frac{89{,}092.8}{23{,}967.5} = 3.7$$

Correct Answer is (C)

SOLUTION 37

This is a common request in real life projects and is usually backed up with principles from the *Association for the Advancement of Cost Engineering* AACE.

The *Cost Estimate Matrix* adopted by the AACE is found in the *NCEES Handbook version 1.2* Section 2.2.2. This matrix classifies estimates for *Design Development, Budget Authorization and Feasibility* as Class 3 Estimates.

The expected accuracy range for an AACE Class 3 estimate is:

L: $-$ 5% to $-$ 15%
(i.e., $\$\mathbf{21.25}\ million$ to $\$23.75\ million$)

H: + 10% to + 20%
(i.e., $\$27.5\ million$ to $\$\mathbf{30.0}\ \boldsymbol{million}$)

The range would therefore be constituted from the lowest which is $\$21.25\ million$ to the highest being $\$30.0\ million$.

Correct Answer is (C)

SOLUTION 38

Using the Hazen-Williams equation and coefficients provided in Section 6.3 of the *NCEES Handbook version 1.2*, see below:

$$h_f = \frac{4.73\ L}{C^{1.852}\ D^{4.87}}\ Q^{1.852}$$

C is the Hazen-Williams Coefficient which equals to *100* for 20-year-old pipes. ***L*** and ***D*** are length and diameter in ft.

$Q = 1.318\, C\, A\, R_H^{0.63}\, S^{0.54}$

R_H is the hydraulic radius, and it equals to the *area of flow* (A) / *wetted perimeter* (P). For pipelines running in full capacity, $R_H = r/2$

$$Q = 1.318 \times 100 \times \pi \times \left(\frac{10}{12} ft\right)^2 \times \left(\frac{5}{12} ft\right)^{0.63} \times 0.02^{0.54}$$

$$= 20 ft^3/sec$$

$$h_f = \frac{4.73 \times 200}{100^{1.852} \times \left(\frac{20}{12}\right)^{4.87}} \times 20^{1.852}$$

$$= 4.0\ ft$$

A quicker way for determining the answer is by using the slope ***S***.

$S = h_f / L$

$\rightarrow h_f = S \times L = 0.02 \times 200 = 4\ ft$

Correct Answer is (B)

SOLUTION 39

The weight of cement consumed to produce $1\, ft^3$ is calculated. Based upon which, the quantity of water can be determined based on the *w/c* ratio provided.

One bag of cement is 94 lb and is sufficient to produce the following yield:

Component	Ratio	Weight lb	Volume ft^3
Cement	1	94	0.48
Fine agg.	1.5	141	0.85
Coarse agg.	3	282	1.70
Water		42.3	0.68
TOTAL			**3.71**

Volume of water required for $1\, ft^3$
$= 0.68/3.71 = 0.183\, ft^3$ (5.2 *litres*)

Add an extra 5% to account for aggregates:

$= 5.2$ *litres*

Correct Answer is (A)

SOLUTION 40

The cable is connected external to support 'B' and not to the member, because of this, forces exhibited on pulley 'D' resemble external forces and can push the crane to the left.

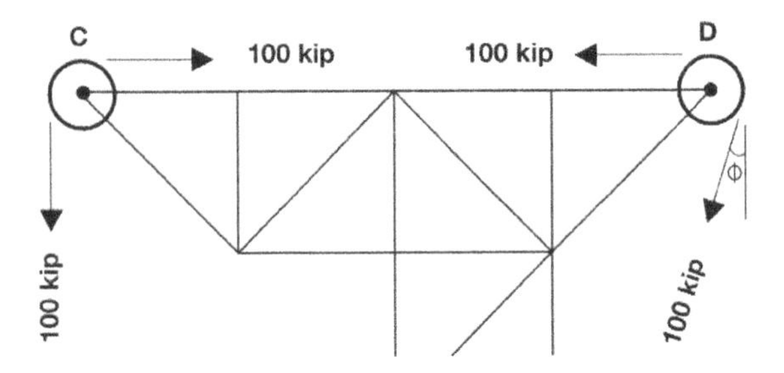

Forces at pulley 'D' are analyzed using trigonometry as follows:

$\emptyset = tan^{-}\left(\frac{5.5}{20}\right) = 15.4^o$

$D_{horizontal} = 100\, sin(15.4) = 26.6\ kip \leftarrow$

$D_{vertical} = 100\, cos(15.4) = 96.4\ kip \downarrow$

Those two forces act external to the body of the crane in addition to the external load at 'C', while the two horizontal forces that equal to 100 kip shall cancel each other and therefore have no contribution to the reactions at 'A' and 'B'.

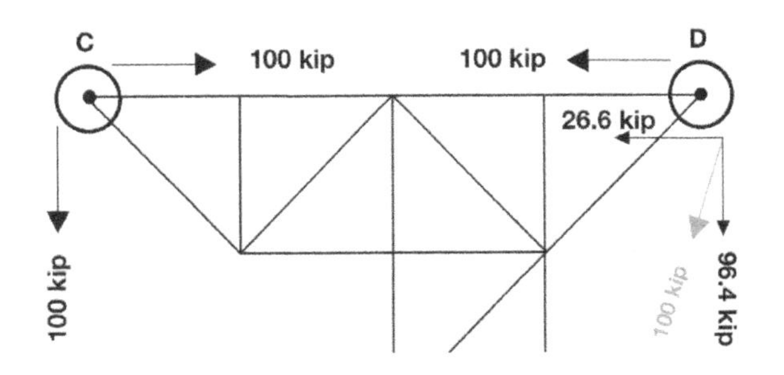

Taking moment around 'B' equals to zero:

$\Sigma M_B = 0$

$100 \times 15.5 - 96.4 \times 5.5 + 26.6 \times 20.5 - R_{A,vert} \times 5 = 0$

$\rightarrow R_{A,vertical} = 313\ kip \uparrow$

Taking moment around 'A' equals to zero:

$\Sigma M_A = 0$

$100 \times 10.5 - 96.4 \times 10.5 + 26.6 \times 20 - R_{B,vert} \times 5 = 0$

$\rightarrow R_{B,vertical} = 114\ kip \downarrow$

Taking the sum of horizontal forces equals to zero:

$\Sigma X = 0$

$\rightarrow R_{A,horizontal} = 26.6\ kip \rightarrow$

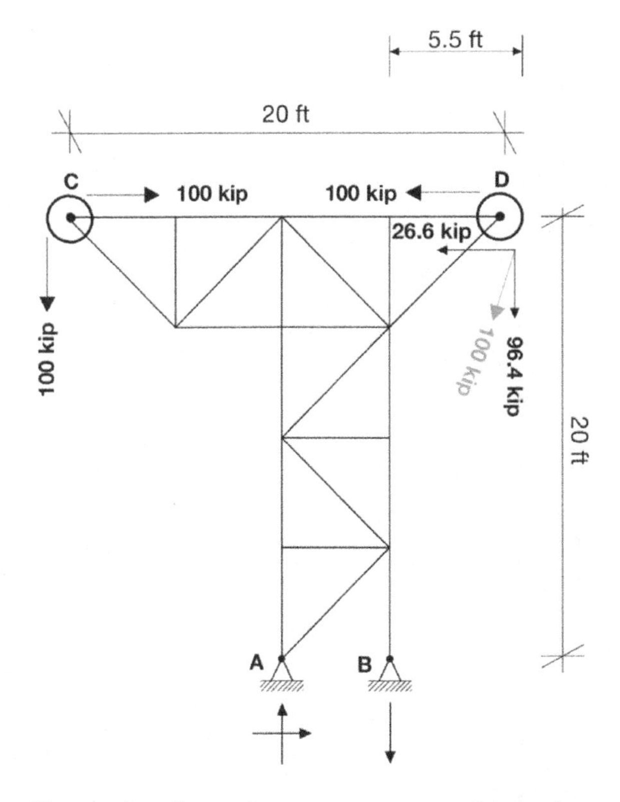

From the first glance, one may think that there should be no horizontal reactions. Given the cable is connected external to the body of the truss/crane, the entire crane would move to the left if it was not for the horizontal restraint provided at support 'A'.

Correct Answer is (D)

SOLUTION 41

A quick method can be applied by taking advantage of the units provided as follows:

$$Density(veh/mile) = \frac{flow\ rate\ (veh/hr)}{av.travelspeed(mph)}$$

$$= \frac{1{,}500\ veh/hr}{65\ mile/hr}$$

$$= 23\ veh/mile$$

Another method of solving this problem can be found in Chapter 4 of the *HCM Manual* provided you have access to it during the exam:

$$flow\ rate\ (veh/hr) = \frac{3{,}600\ sec/hr}{av.\ headway\ (sec/veh)}$$

$$\rightarrow Av.headway\ (sec/veh) = \frac{3{,}600\ sec/hr}{Flow\ rate\ (veh/hr)}$$

$$= \frac{3{,}600\ sec/hr}{1{,}500\ veh/hr}$$

$$= 2.4\ sec/veh$$

$$Av.headway\ (sec/veh) = \frac{av.\ spacing\ (ft/veh)}{av.\ travel\ speed\ (ft/sec)}$$

$$2.4\ s/veh = \frac{av.\ spacing\ (ft/veh)}{65\ mph \times \frac{5{,}280\ ft/mile}{3{,}600\ sec/hr}}$$

$$\rightarrow Av.spacing\ (ft/veh)$$

$$= 2.4\ sec/veh \times 65\ mph \times \frac{5{,}280\ ft/mile}{3{,}600\ sec/hr}$$

$$= 228.8\ ft/veh$$

$$Density\ (veh/mile) = \frac{5{,}280\ ft/mile}{av.\ spacing\ (ft/veh)}$$

$$= \frac{5{,}280\ ft/mile}{228.8\ ft/veh}$$

$$= 23\ veh/mile$$

Correct Answer is (D)

SOLUTION 42

In reference to the *NCEES Handbook version 1.2,* Section 5.3.1 Symmetrical Vertical Curve Formula, and the table given in the question, a vertical curve can be constructed as follows:

Point	Station	Elevation
PVC	$0 + 025$	72.5
PVI	$0 + 275$	65
$A\ point\ on\ the\ curve$	$0 + 150$	70

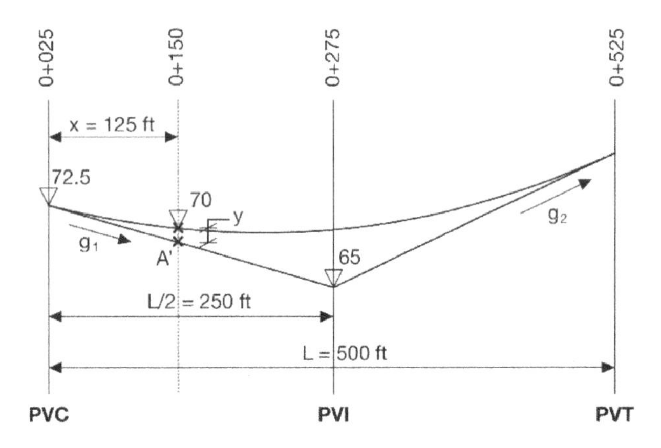

$$g_1 = \frac{65-72.5}{275-25} = -0.03$$

Determine Elevation of the point on the Back Tangent at station $0 + 150$ located at $x = 125\ ft$.

$$A'_{Elevation} = 72.5 - 0.03 \times 125 = 68.75$$

$$y = 70 - 68.75 = 1.25$$

$$y = ax^2$$

$$\rightarrow a = \frac{y}{x^2} = \frac{1.25}{125^2} = 8 \times 10^{-5}$$

$$a = \frac{g_2 - g_1}{2L}$$

$$\rightarrow g_2 = 2aL + g_1$$

$$= 2 \times 8 \times 10^{-5} \times 500 + (-0.03)$$

$$= 0.05$$

Correct Answer is (A)

SOLUTION 43

The following suggested steps can be followed to visually construct a moment diagram:

Step 1: Determine external reactions

This frame is a determinate structure as it has three unknown reactions along with three equilibrium equations to compute those reactions. The directions of those reactions are as follows:

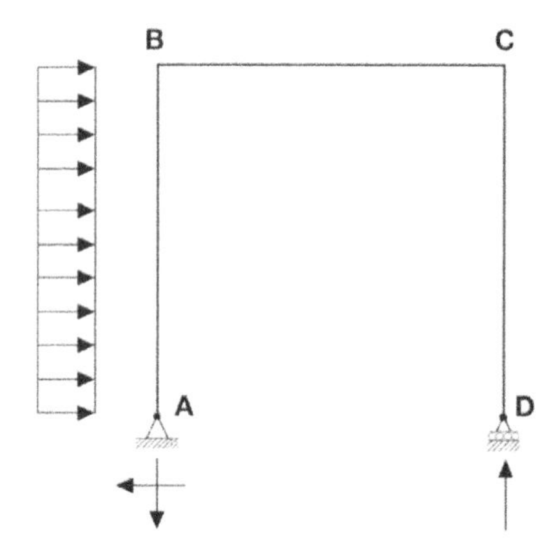

Step 2: Determine internal forces

Dismantle all members and understand the direction of internal forces acting at the ends of members as follows:

Step 3: Establish the shear diagram
At this stage the requested moment diagram can be computed, however, it is always a good strategy to start with the shear diagram first to:

- Determine the location of maximum moments as they occur at zero shear locations.
- Determine whether there is a point of inflection in the moment diagram as it occurs when shear changes signs.
- Determine the shape function of the moment diagram – i.e., the moment diagram represents the area under the shear diagram and its shape function is the integral of that of the shear diagram.

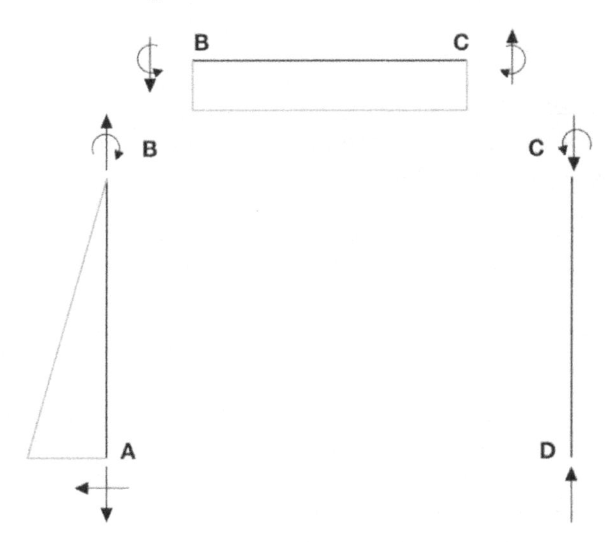

It is important to observe that the shear diagram for member AB is a linear function. The shear diagram represents the area under the load diagram, this indicates that its shape function is the integral of that of the load diagram. The load diagram in this case is represented by a straight/constant equation, consequently, the shear diagram is represented by a linear equation.

Step 4: Establish the moment diagram:
The moment diagram is the integral of the shear function. See below:

- Member AB: the shear diagram has a linear shape; the moment diagram in this case would be best represented by a parabolic function.
- The maximum moment occurs at the location where shear equals to zero (i.e., joint B).
- Member BC: the shear diagram is represented by a constant function; hence, moment diagram would be represented by a linear equation.
- Member CD has no shear hence no moment is acting on it.

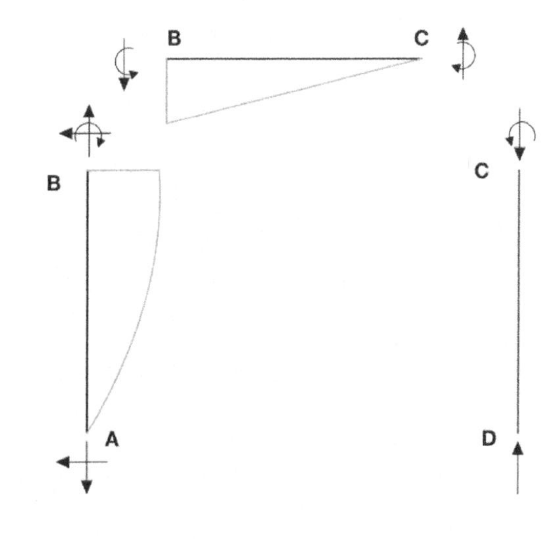

Finally:

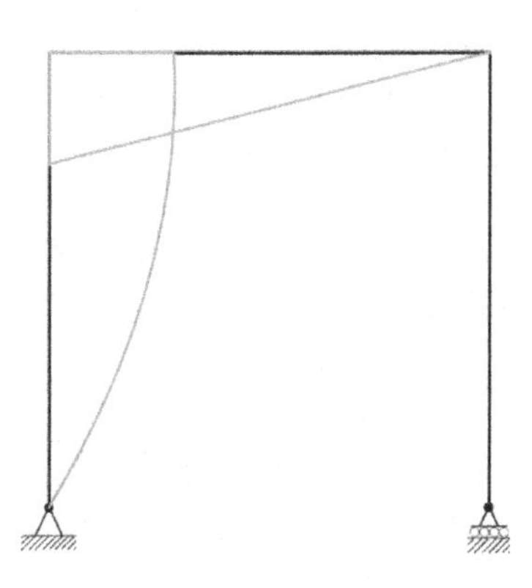

Correct Answer is (D)

SOLUTION 44

To calculate the Benefit/Cost ratio for the road improvements project, we need to determine the total benefits and total costs over the 25 years period.

Costs:

Annual operation & maintenance and fuel costs are brought to their **Present Values (*PV*)** using Table 1.7.10 of the *NCEES Handbook version 1.2* with an interest rate of $i = 2\%$ and $n = 25\ years$ represented as $(P/A, 2\%, 25) = 19.5$, then calculate costs as follows:

(A) Annual cost before improvements is calculated as follow:

Road maintenance cost:
$= 5\ miles \times \$\ 22{,}000 \times 19.5$
$= \$\ 2{,}145{,}000\ (\$\ 2.145\ million)$

Estimate the annual Vehicle-Miles Traveled (VMT) using the *NCEES Handbook* Section 5.1.3.2:

$$\begin{aligned} VMT_{365} &= AADT \times L \times 365 \\ &= 80{,}000 \times 5 \times 365 \\ &= 146 \times 10^6\ miles\ per\ year \end{aligned}$$

Vehicle fuel and O&M cost:
$= 146 \times 10^6\ miles \times \$\ 0.14 \times 19.5$
$= \$\ 398.58\ million$

Total fuel and vehicle O&M cost ($TC_{O\&M\ before}$):
$= \$\ 398.58\ million + \$\ 2.145\ million$
$= \$\ 400.73\ million$

(B) Annual cost after improvements is calculated as follow:

Road maintenance cost:
$= 1.5\ miles \times \$\ 22{,}000 \times 19.5$
$= \$\ 643{,}500\ \ (\$\ 0.64\ million)$

Estimating the annual Vehicle-Miles Traveled (VMT) using the *NCEES Handbook* Section 5.1.3.2:

$$\begin{aligned} VMT_{365} &= AADT \times L \times 365 \\ &= 80{,}000 \times 1.5 \times 365 \\ &= 43.8 \times 10^6\ miles\ per\ year \end{aligned}$$

Vehicle fuel and O&M cost:
$= 43.8 \times 10^6\ miles \times \$\ 0.14 \times 19.5$
$= \$\ 119.574\ million$

Total fuel and vehicle O&M cost ($TC_{O\&M\ after}$):
$= \$\ 119.57\ million + \$\ 0.64\ million$
$= \$\ 120.21\ million$

Total Cost after improvement (TC_{after}):
$= \$\ 120.21\ million + \$\ 120.00\ million$
$= \$\ 240.21\ million$

Benefits:

The benefits arise from the cost savings due to reduced fuel and O&M for vehicles, along with maintenance cost for the improved road segment over the specified duration of 25 years.

$$= TC_{O\&M\ before} - TC_{O\&M\ after}$$
$$= \$\ 400.73\ million - \$\ 120.21\ million$$
$$= \$\ 280.52\ million$$

Benefit/Cost ratio:

B/C is calculated by dividing the total benefits (savings) by total cost:

$$\begin{aligned} B/C &= \frac{Savings}{Total\ Cost(TC_{after})} \\ &= \frac{\$\ 280.52\ million}{\$\ 240.21\ million} \\ &= 1.17 \end{aligned}$$

A B/C of 1.17 suggests that for every dollar invested in the road improvement, you would receive approximately \$ 1.17 in benefits over the 25 years period. Typically, a B/C greater than $'1.0'$ is considered economically viable, as it indicates that the benefits outweigh the costs.

Correct Answer is (B)

It is essential to consider other factors in the Benefit Cost analysis such as potential future savings, environmental impacts, improved safety, and any additional intangible benefits that may not be captured in this simple analysis. A more comprehensive evaluation in real life examples may be necessary to make well-informed decisions.

SOLUTION 45

Water *Velocity Versus Slope for Shallow Concentrated Flow* diagram of the *NCEES Handbook version 1.2,* Section 6.5.5 Hydrograph Development and Applications, page 405, also adopted from the *National Engineering Handbook: Part 630 Hydrology*, can be used to solve this problem.

The graph is pasted below for ease of reference.

A horizontal line is constructed from y-axis at slope 0.01 ft/ft which intersects with the desired two velocity graphs as shown:

$v_{forest} = 0.25\ ft/sec$

$v_{short\ grass} = 0.74\ ft/sec$

$\Delta v = 0.74 - 0.25 = 0.49\ ft/sec$

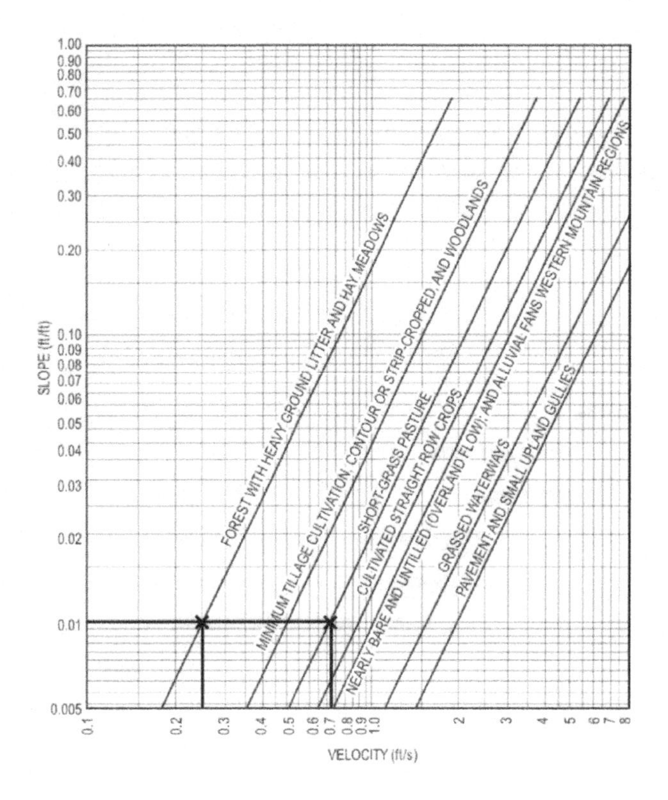

Correct Answer is (C)

SOLUTION 46

It is important to understand which activities can change their start and finish dates so that they can be used to level resources. Those activities are the ones with positive free float available. Free float can be determined using the following equation:

$Free\ Float = Earliest\ ES_{successor} - EF$

$FF_D = 11 - 6 = 5\ days$

$FF_E = 11 - 7 = 4\ days$

The following table summarizes the network diagram along the available free floats:

Activity	Duration	Resources	ES	EF	FF
A	1	2	1	2	
B	3	5	2	5	
C	2	2	2	4	
D	2	1	4	6	5
E	3	3	4	7	4
F	6	4	5	11	
G	1	4	11	12	

This table is then converted into the base Gantt chart shown below, out of which a resources histogram is established by assigning resources to the bar chart and summing them altogether in the following histogram.

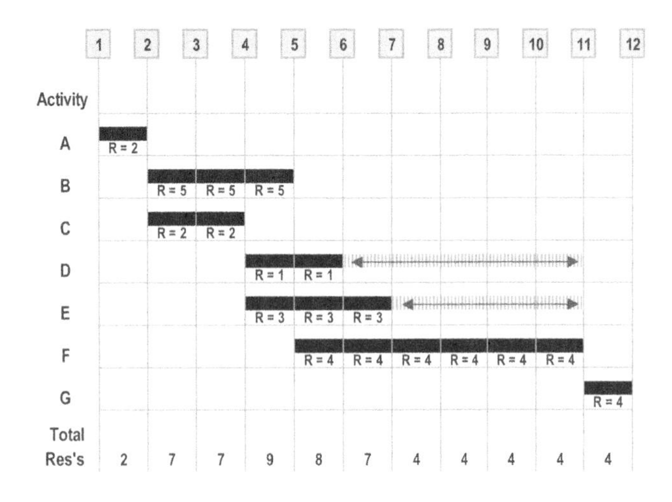

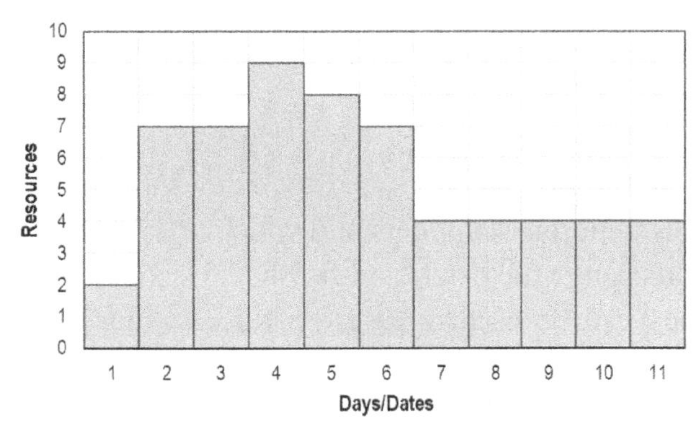

As pointed out in the above Gantt chart, activities D and E can move freely as indicated by the two-sided arrows for *five* and *four* days respectively without affecting the completion date of the project.

In this case, the solution becomes a matter of trial and error. Activities D and E can move horizontally within their float to match the best available option from the four options provided in this question.

Based on this, the following Gantt chart was established by sliding activity D *three* days to the right which could provide the most leveled usage for those resources, or at least match one of the diagrams provided in the question.

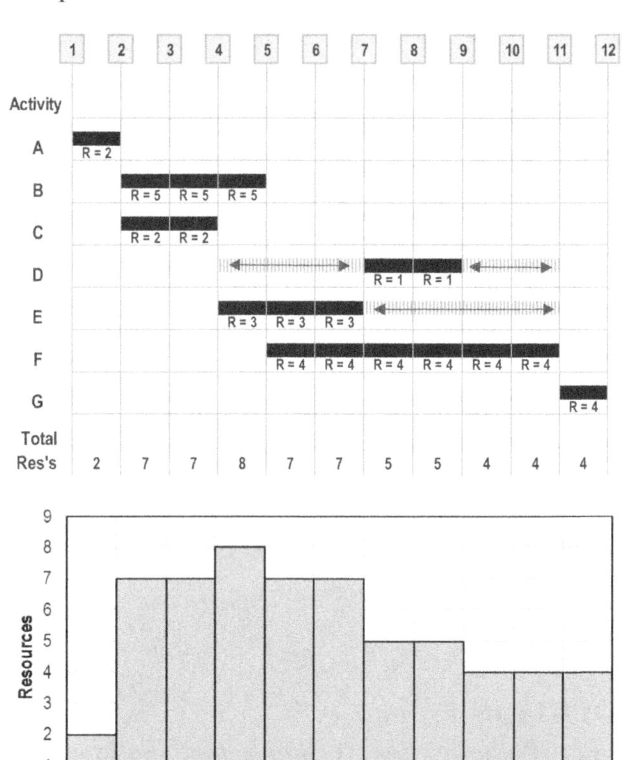

The rest of the diagrams are simply incorrect.

Correct Answer is (A)

SOLUTION 47

This solution is provided with reference to the Transportation Section / Horizontal Design of the *NCEES Handbook*.

Given the two curves are identical, the chord C for each curve can be found using trigonometry as follows:

$$C = \sqrt{370^2 + 275^2} = 461\ ft$$

Also, the obstruction is located at the middle ordinate $\rightarrow M = 70\ ft$.

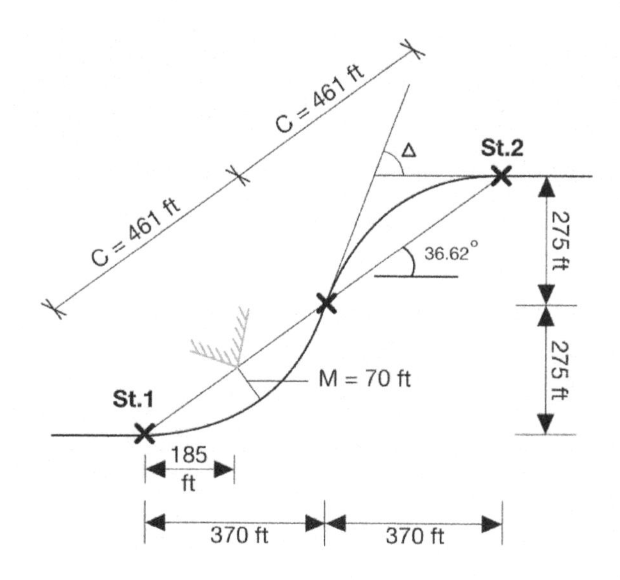

The above two distances C & M can be used to calculate the length of curve L which shall be used to determine Station $St.2$ location as follows:

$$M = R\left(1 - \cos\frac{\Delta}{2}\right)$$

$$C = 2\,R\,sin\frac{\Delta}{2}$$

Divide the above two equation to cancel common coefficients:

$$\frac{M}{C} = \frac{R\left(1 - cos\frac{\Delta}{2}\right)}{2\,R\,sin\frac{\Delta}{2}}$$

$$\rightarrow \frac{70}{461} = \frac{1 - cos\frac{\Delta}{2}}{2\,sin\frac{\Delta}{2}}$$

$$0.3034\sin\frac{\Delta}{2} = 1 - \cos\frac{\Delta}{2}$$

This equation can be solved with the use of the following identity that can be found in the *NCEES Handbook* Section 1.3 Mathematics:

$$\left(sin\frac{\Delta}{2}\right)^2 + \left(cos\frac{\Delta}{2}\right)^2 = 1$$

$$sin\frac{\Delta}{2} = \sqrt{1 - \left(cos\frac{\Delta}{2}\right)^2}$$

Substituting this equation into the previous one provides the following:

$$0.3034 \times \sqrt{1 - \left(cos\frac{\Delta}{2}\right)^2} = 1 - cos\frac{\Delta}{2}$$

Raise the two sides of the above resultant equation to the power of *two* to get rid of the square root and then rearrange into a quadratic equation:

$$\left(0.3034 \times \sqrt{1 - \left(cos\frac{\Delta}{2}\right)^2}\right)^2 = \left(1 - cos\frac{\Delta}{2}\right)^2$$

$$\left(cos\frac{\Delta}{2}\right)^2 - 1.831\left(cos\frac{\Delta}{2}\right) + 0.831 = 0$$

The above is a quadratic equation with $a = 1$, $b = -1.831$ and $c = 0.831$ and can be solved as follows:

$$root = \frac{-b \mp \sqrt{b^2 - 4ac}}{2a}$$

$$\cos\frac{\Delta}{2} = \frac{+1.831 \mp \sqrt{1.831^2 - 4 \times 1 \times 0.831}}{2 \times 1}$$

$$= (0.831, 1)$$

The root of $'0.831'$ is used in this case as a root of $'1'$ generates a *zero*-deflection angle.

$$\frac{\Delta}{2} = cos^{-}(0.831) = 33.79^o$$

$$\Delta = 67.58^o$$

The radius of curvature R can be calculated using any of the previously used equations for either M or C:

$$R = \frac{M}{\left(1-\cos\frac{\Delta}{2}\right)} = \frac{70}{1-\cos 33.79} \cong 414.5\ ft$$

Or

$$R = \frac{C}{2\ sin\frac{\Delta}{2}} = \frac{461}{2\ sin\ 33.79} \cong 414.5\ ft$$

The length of the two curves, given they are identical, is calculated as follows:

$$2L = 2 \times \frac{R\Delta\pi}{180} = 2 \times \frac{414.5 \times 67.58 \times \pi}{180}$$

$$\cong 978\ ft$$

Station $St.2$ is therefore calculated as follows:

$$St.2 = '1 + 00' + 978\ ft$$

$$= 10 + 78$$

Correct Answer is (C)

SOLUTION 48

The Free Haul Distance (FHD) has been given in the problem as 500 ft. The FHD is the distance below which earthmoving is considered part of the contract and contractor cannot claim for extras for overhauling.

To identify stations which fall within the FHD, a 500 ft *to-scale* horizonal line is drawn and fit in position to intersect close to the peaks and troughs of the Mass Haul Diagram (MHD) curves as shown in the figure below. The y-axis generated values from the FHD intersection with the MHD curves represent the quantity which will be hauled as part of the contract price with no extras.

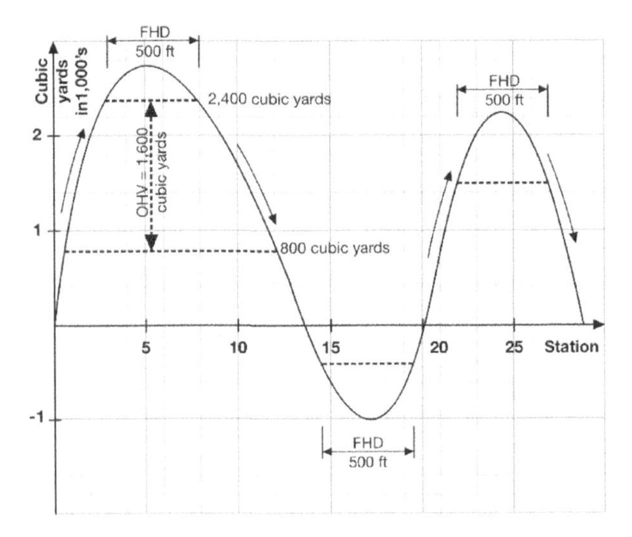

The Over Haul Volume (OVH) is the volume beyond which earthmoving can be claimed as extras by the contractor. OVH is the vertical distance from the FHD towards the x-axis (whether upwards or downwards) to an imaginary horizontal line that intersects with the two sides of the semi parabolic curves.

In this case, and as given in the question, the $OHV = 1{,}600\ yard^3$ measured as a vertical distance from the FHD down to an intersection of 800 $yard^3$ for the first MHD curve. This does not apply to the second or the third curves because there is no enough vertical distance to establish such ordinates.

The vertical ordinates generated form this intersection represent either: (1) the waste material when the curve is moving upwards (i.e., cut sections which is more economical to dump outside the project), or (2) the borrow material when the curve is moving downwards (i.e., fill sections which is more economical to supply its material from outside the project). The horizonal intersection is hence called the Limit of Economic Hall (LEH).

Shrinkage does not apply in the cut section therefore, wastage in this case will only be 800 $yard^3$.

Correct Answer is (A)

To put thing into perspective for this question, and as concluded from the MHD, the material that should go to waste is located between stations $'0+00'$ and $'0+70'$. If we assume this is a 60 ft wide highway construction project, the height of this cut section that should go to waste is nearly:

$$\frac{800\ yard^3 \times 27\dfrac{ft^3}{yard^3}}{60\ ft\ \times 70\ ft} \approx 5\ feet$$

SOLUTION 49

In the *NCEES Handbook version 1.2* is used to solve this problem, Section 5.1.3.3 Peak-Hour Factor as follows:

$$PHF_{SW-NE} = \frac{V}{V_{15} \times 4}$$

Where $\boldsymbol{V}$ represents the four consecutive 15 minutes period that give us the highest number – see below table. $\boldsymbol{V_{15}}$ represents the highest flow within that hour.

In the following table, X represents V_{15} at 4:45 PM :

Time	SW-NE direction *Veh/hr*	NE-SW direction *Veh/hr*
4:00 PM – 4:45 PM	761	540 + X
4:15 PM – 5:00 PM	763	**542 + X**
4:30 PM – 5:15 PM	**778**	496 + X
4:45 PM – 5:30 PM	777	456 + X

$$PHF_{SW-NE} = \frac{778}{210 \times 4} = 0.92$$

$$PHF_{NE-SW} = \frac{0.92}{1.05} = 0.876$$

At this stage, and in order to calculate PHF, $V_{15,max}$ could either be 197 veh or $'X'$ (only if $X > 197$). Both values will be used to determine $'X'$ as follows:

Taking $V_{15} = 197$:

$$PHF_{NE-SW} = \frac{542 + X}{197 \times 4} = 0.876$$

$$\rightarrow X = 148$$

Taking $V_{15} = X$:

$$PHF_{NE-SW} = \frac{542 + X}{X \times 4} = 0.876$$

$$\rightarrow X = 216$$

Although mathematically the two $'\boldsymbol{X}'$ values can work, applying logic and observing the data around the given hour, a value of $\boldsymbol{216\ veh}$ within that hour seems more logical compared to a value of $\boldsymbol{148\ veh}$.

Correct Answer is (B)

SOLUTION 50

The average area method is used to measure the volume for each two consecutive cross-sectional areas where the distance between consecutive areas is a station = 100 ft:

$$V = L\left(\frac{A_1 + A_2}{2}\right)$$

The following two tables represent cross-sectional areas measured at each station along with the volume between every two consecutive stations.

Bank (undisturbed) cut volume $V_{B,cut} = -950{,}000\ ft^3$ is converted to lose volume $V_{L,cut}$. Fill requirements $V_{B,fill}$ is also converted into loose volume $V_{L,fill}$. The balance of both goes to waste.

Station	Cut Area ft^2	Fill Area ft^2
0+00	− 1,400	
1+00	− 1,400	
2+00	− 1,400	
3+00	− 1,600	
4+00	− 1,600	
5+00	−1,200	
6+00	− 800	
7+00	− 600	
8+00	− 200	
9+00	0	0
10+00		400
11+00		500
12+00		600
13+00		600
14+00		600
15+00		600

Station	Cut Volume ft^3	Fill Volume ft^3
0+00 to 1+00	− 140,000	
1+00 to 2+00	− 140,000	
2+00 to 3+00	− 150,000	
3+00 to 4+00	− 160,000	
4+00 to 5+00	− 140,000	
5+00 to 6+00	−100,000	
6+00 to 7+00	− 70,000	
7+00 to 8+00	− 40,000	
8+00 to 9+00	− 10,000	0
9+00 to 10+00	0	20,000
10+00 to 11+00		45,000
11+00 to 12+00		55.000
12+00 to 13+00		60,000
13+00 to 14+00		60,000
14+00 to 15+00		60,000
Total	**− 950,000**	**300,000**

$$V_{L,cut} = \left(1 + \frac{S_w}{100}\right) V_{B,cut}$$

$$= \left(1 + \frac{15}{100}\right)(-950{,}000)$$

$$= -1{,}092{,}500\ ft^3$$

$$V_{L,fill} = \left(1 + \frac{S_h}{100}\right) V_{B,fill}$$

$$= \left(1 + \frac{20}{100}\right)(300{,}000)$$

$$= 360{,}000\ ft^3$$

$$V_{L,waste} = -1{,}092{,}500 + 360{,}000$$

$$= -732{,}500\ ft^3$$

$$No.of\ trips = \frac{732{,}500\ ft^3}{11\ yd^3 \times 27 \frac{ft^3}{yd^3}}$$

$$= 2{,}466\ trip$$

Correct Answer is (A)

SOLUTION 51

The *NCEES Handbook*, Chapter 5 Transportation, can be referred to for the skid marks equation stopping distance as follows:

$$d_b = \frac{v_1^2 - v_2^2}{30(f \mp G)}$$

For the first car travelling at a speed of 90 *mph*, the stopping distance is calculated as follows:

$$d_{b,1} = \frac{v_1^2 - 0}{30(f \mp 0)} = \frac{270}{f}$$

Distance travelled by the first car during the perception/reaction time of the driver:

$$d_{p,1} = v \times t$$

$$= 90\ \text{mph} \times \frac{5{,}280\ ft/mile}{3{,}600\ sec/hr} \times 2\ \text{sec}$$

$$= 264\ ft$$

For the second car travelling at a speed of 60 *mph*, the stopping distance is calculated as follows:

$$d_{b,2} = \frac{v_1^2 - 0}{30(f \mp 0)} = \frac{120}{f}$$

Distance travelled by the second car during the perception/reaction time of the driver:

$$d_{p,1} = v \times t$$

$$= 60 \text{ mph} \times \frac{5{,}280\ ft/mile}{3{,}600\ sec/hr} \times 2 \text{ sec}$$

$$= 176\ ft$$

For those two cars not to hit each other, and keep a distance of $35\ ft$ between them after stopping, the following equation should apply:

$$d_{b,1} + d_{b,2} + d_{p,1} + d_{p,2} + 35\ ft = 1{,}340\ ft$$

$$\frac{270}{f} + \frac{120}{f} + 264 + 176 + 35 = 1{,}340$$

$$\rightarrow f = 0.451$$

Reapply the friction factor to the initial equations to determine both $d_{b,1}$ and $d_{b,2}$:

$$d_{b,1} = \frac{270}{f} \approx 599\ ft$$

$$d_{b,2} = \frac{120}{f} \approx 266\ ft$$

Correct Answer is (B)

SOLUTION 52

The *NCEES Handbook*, Chapter 6 Water Resources and Environmental, Dupuit's Formula can be used to solve this question.

$$Q = \frac{\pi K(h_2^2 - h_1^2)}{ln\left(\frac{r_2}{r_1}\right)}$$

$\boldsymbol{Q}$ is the flow rate in ft^3/sec, $\boldsymbol{h_1}$ and $\boldsymbol{h_2}$ are heights of the aquifer measured from its bottom at the perimeter of the well (i.e., $r_1 = \frac{12}{2}\ in$) and at the influence radius of $r_2 = 450\ ft$ respectively.

Radius of influence defines the outer radius of the cone of depression, hence $h_2 = 100\ ft$. Check the below figure for more clarity.

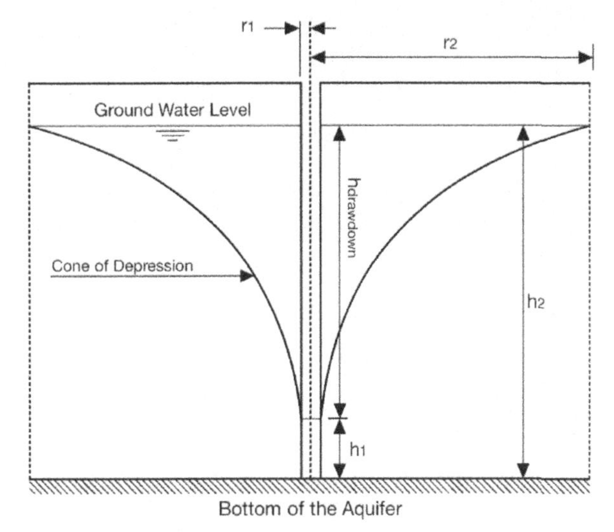

$$h_1 = \sqrt{\frac{\pi K h_2^2 - Q \times ln\left(\frac{r_2}{r_1}\right)}{\pi K}}$$

$$= \sqrt{\frac{\pi \times 4 \times 10^{-4} \frac{ft}{sec} \times (100\ ft)^2 - 65 \frac{gal}{min}\left(\frac{0.134\ ft^3}{gal}\right)\left(\frac{1\ min}{60\ sec}\right) \times ln\left(\frac{450\ ft}{0.5\ ft}\right)}{\pi \times 4 \times 10^{-4} \frac{ft}{Sec}}}$$

$$= 96\ ft$$

$$h_{drawdown} = 100\ ft - 96\ ft = 4\ ft$$

Correct Answer is (C)

SOLUTION 53

The *NCEES Handbook version 1.2,* Section 5.1.4.1 Acceleration, of the Transportation chapter can be referred to for the solution of this question.

In this question we shall determine the total distance travelled x_{total} and divide it by the total time it took the transit to travel this distance t_{total}, as follows:

$$x_{total} = x_a + x_c + x_d$$

$$t_{total} = t_a + t_c + t_d$$

$$S_{average} = \frac{x_{total}}{t_{total}}$$

1. *Acceleration distance 'x_a'*

$$x_a = ½\, at_a^2 + S_o t_a + x_o$$

Where x_a is the distance being sought, t_a is the time needed to accelerate to speed $S = 90\ mph$ calculated as follows where S_o is initial speed which is zero in this case:

$$S = a\, t_a + S_o$$

$$90\ mph \times \frac{\frac{5{,}280\ ft}{mile}}{\frac{3{,}600\ sec}{hr}} = 5.0\ \frac{ft}{sec^2} \times t_a + 0$$

$$t_a = 26.4\ sec$$

$$x_a = ½\, at_a^2 + 0 + 0$$
$$= ½\ (5.0\ \frac{ft}{sec^2})(26.4\ sec)^2$$
$$= 1{,}742.4\ ft$$

2. *Deceleration distance 'x_d'*

$$x_d = ½\, at_d^2 + S_o t_o + x_o$$

Will assume that datum starts at the point of deceleration, and hence S_o and x_o will be taken as *zeros* and those shall be brought up in the final equation.

$$S_{final} = a\, t_d + S = 0$$

$$0 = (-4.5\ \frac{ft}{sec^2})t_d + 90\ mph \times \frac{\frac{5{,}280\ ft}{mile}}{\frac{3{,}600\ sec}{hr}}$$

$$t_d = 29.33\ sec$$

$$x_d = ½\, at_a^2 + 0 + 0$$
$$= ½\ (4.5\ \frac{ft}{sec^2})(29.33\ sec)^2$$
$$= 1{,}936\ ft$$

3. *In-between (constant) Distance 'x_c'*

$$x_c = S\, t_c$$
$$= \left(90\ mph \times \frac{\frac{5{,}280\ ft}{mile}}{\frac{3{,}600\ sec}{hr}}\right)\left(1.5\ min \times 60\ \frac{sec}{min}\right)$$
$$= 11{,}880\ ft$$

$$x_{total} = x_a + x_c + x_d$$
$$= 1{,}742.4 + 11{,}880 + 1{,}936$$
$$= 15{,}558.4\ ft\ (2.95\ mile)$$

$$t_{total} = t_a + t_c + t_d$$
$$= 26.4 + 1.5 \times 60 + 29.33$$
$$= 145.73\ sec\ (0.0405\ hr)$$

$$S_{average} = \frac{2.95\ mile}{0.0405\ hr} = 72.8\ mph$$

Correct Answer is (C)

SOLUTION 54

The *NCEES Handbook version 1.2,* Section 6.5.2 Runoff Analysis can be used, and given the input provided in the question, the best method to be implemented in this question is the Rational Method.

$$Q = CIA$$

Q is the discharge in ft^3/sec, this value is calculated pre and post development, the difference of both shall be used to design the retention basin.

C is the runoff coefficient which can be obtained from the *NCEES Handbook*. Given the area is only slightly sloped, the lower range of C in the runoff table can be used.

I is rainfall intensity given in the question as $2\ in/hr$, and A is area which is $7\ acres$.

Pre-development:

For slightly sloped parks and cemeteries $C = 0.10$

$$Q_{pre} = CIA$$
$$= 0.1 \times 2\ in/hr \times 7\ acres$$
$$= 1.4\ ft^3/sec$$

$$V_{pre} = Q_{pre} \times 1\ day$$
$$= 1.4\ ft^3/sec \times 86{,}400\ sec\ per\ day$$
$$= 120{,}960\ ft^3/day$$

Post-development:

Calculate weighted runoff as follows:

- Downtown areas → $C = 0.70$
- Playgrounds → $C = 0.20$
- Asphalt roads → $C = 0.70$
- Concrete walkways → $C = 0.80$

$$C_w = 45\% \times 0.7 + 35\% \times 0.2 + 15\% \times 0.7 + 5\% \times 0.8$$
$$= 0.53$$

$$Q_{post} = CIA$$
$$= 0.53 \times 2\ in/hr \times 7\ acres$$
$$= 7.42\ ft^3/sec$$

$$V_{post} = Q_{post} \times 1\ day$$
$$= 7.42\ ft^3/sec \times 86{,}400\ sec\ per\ day$$
$$= 641{,}088\ ft^3/day$$

$$\Delta V = V_{post} - V_{pre}$$
$$= 641{,}088 - 120{,}960$$
$$= 520{,}128\ ft^3$$

Depth of the rectangular pond is therefore calculated as follows:

$$d = \frac{520{,}128\ ft^3}{250\ ft \times 150\ ft} = 13.87\ ft$$

Correct Answer is (A)

SOLUTION 55

The *NCEES Handbook,* Section 6.3 Closed Conduit Flow and Pumps can be used to solve this problem.

Given the inputs provided in this question, the following equations/sections will be referred to:

- The Energy Equation – Section 6.2.1.2
- Reynolds Number for circular pipes – Section 6.2.2.2
- Darcy-Weisbach Equation for head losses due to flow – Section 6.2.3.1
- Minor Losses in Pipe Fittings, Contractions and Expansions – Section 6.3.3

The Energy Equation:

$$\frac{P_A}{\gamma} + z_A + \frac{v_A^2}{2g} = \frac{P_B}{\gamma} + z_B + \frac{v_B^2}{2g} + h_f$$

Pressure at surface A and surface B is zero (atmospheric only). Also, velocity of the surface dropping can be neglected in large reservoirs. With this, the energy equation can be reduced as follows:

$$z_A = z_B + h_f$$

Calculating Head Losses h_f:

Head loss due to flow:

Using Darcy-Weisbach Equation for head losses:

$$h_{f,flow} = f\frac{L}{D}\frac{v^2}{2g}$$

f is a function of both Reynolds Number.

R_e and the relative roughness $\left(\frac{\varepsilon}{D}\right)$ taken from the Moody, Darcy, or Stanton Friction Factor Diagram page 319 of the *NCEES Handbook version 1.2.*

v in the denominator of Reynolds Number equation (shown below) is the kinematic viscosity taken from the physical properties of water table Section 6.2.1.6 as $0.93 \times 10^{-5}\ ft^2/sec$.

v in the numerator is velocity in the pipe $= \frac{Q}{A} = \frac{20\ ft^3/sec}{\pi\ (0.5\ ft)^2} = 25.5\ ft/sec$

$$R_e = \frac{vD}{v}$$

$$= \frac{25.5\ ft/sec \times 12\ in \times \frac{1\ ft}{12\ in}}{0.93 \times 10^{-5}\ ft^2/sec}$$

$$= 2.7 \times 10^6$$

$$\frac{\varepsilon}{D} = \frac{0.0006}{1\ ft} = 0.0006$$

Substitute R_e and $\frac{\varepsilon}{D}$ in Moody, Darcy, or Stanton Friction Factor Diagram page 319 of the *NCEES Handbook version 1.2* generates a friction value of $f = 0.019$

$$h_{f,flow} = f\frac{L}{D}\frac{v^2}{2g}$$

$$= 0.019 \times \frac{\left(70 + \frac{500}{sin(45)} + 70\ ft\right)}{1\ ft}\frac{\left(25.5\frac{ft}{sec}\right)^2}{2 \times 32.174\frac{ft}{sec^2}}$$

$$= 162.6\ ft$$

Head loss due to pipe fittings:

$$h_{f,fittings} = C\frac{v^2}{2g}$$

Where C equals to '1.0', '0.5' and '0.4' for sharp reservoir exit, sharp reservoir entrance and 45° elbow respectively taken from Sections 6.3.3.1 and 6.3.3.2.

$$h_{f,fittings} = C_{total}\frac{v^2}{2g}$$

$$= (1 + 2 \times 0.4 + 0.5) \times \frac{\left(25.5\frac{ft}{sec}\right)^2}{2 \times 32.174\frac{ft}{sec^2}}$$

$$= 23.2\ ft$$

Energy Equation and Total Head Losses:

$$z_A = z_B + h_f$$

$$z_B = z_A - h_f$$

$$= z_A - (h_{f,flow} + h_{f,fittings})$$

$$= 600 - (162.6 + 23.2)$$

$$= 414.2\ ft$$

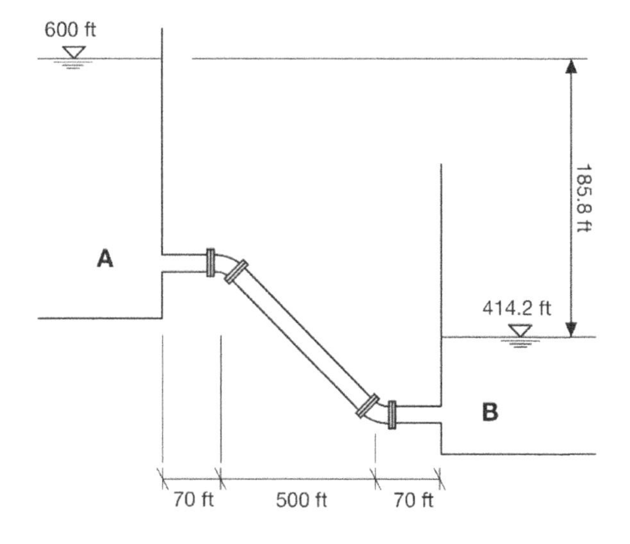

Correct Answer is (A)

SOLUTION 56

The *NCEES Handbook,* Chapter 4 Structural can be used to solve this question.

Equations on deflection can be collected from the relevant Shears, Moments, and Deflections tables provided in the same chapter.

For simply supported, uniformly loaded beam, deflection is calculated as follows:

$$\Delta_{max} = \frac{5wl^4}{384EI}$$

w is the section's self-weight.

I is the gross moment of inertia for the uncracked section – given that the section is uncracked, rebars will not be stressed, and neutral axis has not yet shifted upwards as expected in a cracked section with a positive moment – hence the use of the gross moment of inertia $\boldsymbol{I_{gross}}$ for the concrete section is appropriate in this case.

E is modulus of elasticity of concrete and is calculated using Section 4.3.2.1 as follows:

$$E_c = 33\, w_c^{1.5}\, \sqrt{f_c'}$$

$$= 33 \times 145^{1.5} \times \sqrt{4{,}000}$$

$$= 3.6 \times 10^6\ psi$$

$$w = b \times d \times \gamma_{concrete}$$

$$= 25\ in \times 10\ in \times 160\ \frac{lb}{ft^3} \times \frac{ft^3}{1{,}728\ in^3}$$

$$= 23.1\ lb/in$$

$$I_{gross} = \frac{b \times d^3}{12}$$

$$= \frac{25\ in \times (10\ in)^3}{12}$$

$$= 2{,}083.33\ in^4$$

$$\Delta_{max} = \frac{5wl^4}{384E_c\, I_{gross}}$$

$$= \frac{5 \times 23.1 \frac{lb}{in} \times \left(16\ ft \times \frac{12\ in}{ft}\right)^4}{384 \times 3.6 \times 10^6 \frac{lb}{in^2} \times 2{,}083.33\ in^4}$$

$$= 0.055\ in$$

Correct Answer is (D)

<u>Question extras:</u>

Although not requested in the question and may be beyond the scope of a civil breadth, it is worth checking if the section will crack under its self-weight or shall remain uncrack as assumed in the body of the question.

If the section cracks due to self-weight, then the above assumption by which the question was resolved is wrong and the moment of inertia for the cracked section $\boldsymbol{I_{cracked}}$ shall be calculated, which is beyond the scope of the civil breadth exam.

The maximum strain that concrete sections can withstand before cracking the section is $\varepsilon = 0.0035$.

Based on this, the following equations shall apply:

$$\varepsilon = \frac{\sigma}{E} \quad \& \quad \sigma = \frac{MC}{I}$$

$$M = \frac{wl^2}{8}$$

$$= \frac{23.1\ lb/in \times \left(16\ ft \times \frac{12\ in}{ft}\right)^2}{8}$$

$$= 106{,}445\ lb.in$$

$$\sigma = \frac{MC}{I}$$

$$= \frac{106{,}445\ lb.in \times \left(\frac{10\ in}{2}\right)}{2{,}083.33\ in^4}$$

$$= 255.5\ psi$$

$$\varepsilon = \frac{\sigma}{E}$$

$$= \frac{255.5\ psi}{3.6\times10^{6}\ psi}$$

$$= 0.000071 < 0.0035$$

→ Section is uncracked.

SOLUTION 57

Company revenue:

Contract fees (company revenue)

$$= \$100{,}000{,}000 \times 2.5\%$$

$$= \$2{,}500{,}000$$

Project cost and Gross Margin (GM):

Direct labor (salaries)

$$= \$80{,}000 \times 5 \times 1.3 \times 2.5$$

$$= \$1{,}300{,}000$$

Truck expenses

$$= \$0.25 \times 5 \times 95\ mile \times 240 \times 2.5$$

$$= \$71{,}250$$

Total project expenses

$$= \$1{,}300{,}000 + \$71{,}250$$

$$= \$1{,}371{,}250$$

Gross Margin (GM)

$$= \$2{,}500{,}000 - \$1{,}371{,}250$$

$$= \$1{,}128{,}750$$

Gross Margin (%)

$$= \$1{,}128{,}750/\$2{,}500{,}000$$

$$= 45.2\%$$

Company Operating Income (OI):

$$OI = GM - OH - Capital\ Cost - \ Alloc$$

Overhead cost (OH)

$$= Revenue \times OH\%$$

$$= \$2{,}500{,}000 \times 12.5\%$$

$$= \$312{,}500$$

Company capital costs (trucks in this case)

$$= \$55{,}000 \times 5$$

$$= \$275{,}000$$

Allocations (Alloc)

$$= Revenue \times Alloc\%$$

$$= \$2{,}500{,}000 \times 7.5\%$$

$$= \$187{,}500$$

$$OI = \$1{,}128{,}750 - \$312{,}500 - \$275{,}000 - \$187{,}500$$

$$= \$353{,}750$$

$$OI\ (\%) = \$353{,}750/\$2{,}500{,}000$$

$$= 14.2\%$$

The capital value for trucks is not usually factored into projects' Gross Margins, this is why this value was taken towards the bottom line.

Therefore, in an effort to maintain an Operating Income of :

$$OI = \$2{,}500{,}000 \times 15\% = \$375{,}000$$

Trucks should be salvaged at a minimum of:

$$\$375{,}000 - \$353{,}750 = \$21{,}250$$

$$\$21{,}250/5 = \$4{,}250\ per\ truck$$

Correct Answer is (C)

Solution discussion:

The proposed solution can be implemented using various avenues. However, the solution was presented in a manner consistent with the prevalent practices observed among professional organizations for their financial reporting.

In this context, company revenue (or part of it) is represented by its contract fees of $2,500,000.

The overhead cost $312,500 includes expenditures related to marketing, administration, and other non-chargeable project costs, and is customarily computed. It is reasonable to consider this cost as a percentage of the company's overall revenue in this case 12.5%.

The same principle applies to allocations calculated as $187,500. In certain instances, particularly within large organizations, these costs may be separately itemized, especially in cases where the company operates across multiple regions, maintains various offices, and supports executive functions such as the C-suite and the President's office.

Direct labor accounts for actual staff salaries which averages at $80,000, while the 30% fringe benefits incorporate provisions for annual leave, medical leave, and other contractual benefits granted to employees.

Although trucks' capital cost was not factored into the project cost, it is noteworthy that such capital expenditures reflect a deliberate choice made by the company when alternate options, such as renting trucks, could be available. However, the assumption to include trucks capital cost within the capital expenditure framework has been made as part of the solution. Consequently, such costs do not factor into the project margin calculation.

In other instances, the company would (internally) rent out those trucks (after procuring them) to the project and reduces its project Gross Margin. None of these methods would change the final outcome of the solution.

Truck expenses (such as mileage and operation and maintenance costs = $71,250 over $2.5\,years$) are classified as project-related expenditures, and therefore reduces project's Gross Margin or its profitability.

Operating Income OI in this case $353,750 – which, is also termed as EBITDA (Earnings Before Interest, Taxes, Depreciation, and Amortization) – is derived by subtracting the overhead costs $312,500, allocations $187,500, and any other associated capital investments (exemplified by trucks in this case) $275,000, from the project's Gross Margin $1,128,750.

Considering the scenario at hand, where the company did not achieve the target Operating Income from this project $375,000, selling those trucks and realizing a minimum final sales value of $4,250 per truck could be deemed a viable option, particularly if there are no alternative contracts that necessitate the use of the trucks and given those trucks are at the end of their operating life.

SOLUTION 58

The *NCEES Handbook,* Chapter 6 Water Resources and Environmental, Section 6.5.9.2 Erosion/ Revised Universal Soil Loss Equation is referred to in order to provide context into the solution of this question.

Revised Universal Soil Loss Equation:

$$A = R.K.LS.C.P$$

A is the amount of soil loss due to erosion ($tons\ per\ acre\ per\ year$), ***R*** is the rainfall erosion index or the climatic erosivity, ***K*** is soil erodibility factor, ***LS*** is the topographic factor and is taken from the table provided in page 420 of the handbook's version 1.2. ***C*** represents vegetation and is called the crop and cover management factor, and ***P*** is the erosion control practices factor.

Factors Influencing Soil Loss:

Rainfall erosion index ***R*** considers the intensity, duration, and continuity of rainfall. Rainfall erosivity and its relationship to kinetic energy play a crucial role in erosion. **This makes Statement III true.**

Soil erodibility ***K*** is its susceptibility to erosion. Factors affecting ***K*** include soil aggregation and structure. The more porous the soil, the reduced runoff it shall experience and the lesser effect it would have on its continuous erodibility.
This marks Statement I as incorrect.

In a similar fashion, vegetation cover acts as a barrier against erosion by obstructing water velocity, hence lesser runoff is experienced with more cover.
This makes Statement II true.

Topographic factor ***LS***, specifically the slope length and its steepness, are both proportional to erosion. Which means, more length and more slope causes more erosion.
This marks Statement IV as incorrect, however, it makes Statement V true.

Correct Answer is (B)

SOLUTION 59

The *NCEES Handbook,* Chapter 6 Water Resources and Environmental, Section 6.5.9.2 Erosion/ Revised Universal Soil Loss Equation page 419 is referred to.

$$A = R.K.LS.C.P$$

P is the conversation factor, given all inputs of this equation are provided in the body of the question, ***P*** is calculated as follows:

$$P = \frac{A}{R.K.LS.C}$$

A is the amount of soil loss due to erosion measured in $tons\ per\ acre\ per\ year$:

$$A = 11\ \frac{tons}{hectare.yr} \times \frac{1\ hectare}{2.47\ acre}$$

$$= 4.45\ \frac{tons}{acre.yr}$$

K is the soil erodibility factor taken from the same section of *NCEES Handbook* as $'0.27'$ for Ontario Loam.

LS is the topographic factor taken from the same section of *NCEES Handbook* as $'0.28'$ for a 300 ft land sloped at 2%.

C is the crop and cover management factor taken as $'1.0'$ for bare land.

$$P = \frac{A}{R.K.LS.C}$$

$$= \frac{4.45}{200 \times 0.27 \times 0.28 \times 1.0}$$

$$= 0.29$$

A conservation value of $P = 0.29$ requires strip cropping and contour farming. In a nutshell this requires growing crops in a systematic arrangement of strips along the contours and across a sloping field.

Correct Answer is (A)

SOLUTION 60

The *NCEES Handbook,* Chapter 3 Geotechnical, Section 3.9 Laboratory and Field Compaction is referred to.

Optimum moisture content occurs at the soil's maximum dry density γ_d. In which case, dry density is calculated using the total (moist) density γ_t and moisture content w as follows:

$$\gamma_d = \frac{\gamma_t}{(1+w)}$$

Density is derived from the volume of the mold provided in the *Handbook* as $1/30\ ft^3$, based upon which, compaction curve is created as shown below:

SN	wt_{moist} lb	γ_t lb/ft^3	w %	γ_d lb/ft^3
1	4.32	129.60	11.6%	116.13
2	3.92	117.60	17.0%	100.51
3	3.92	117.60	8.8%	108.09
4	4.36	130.80	14.8%	113.94

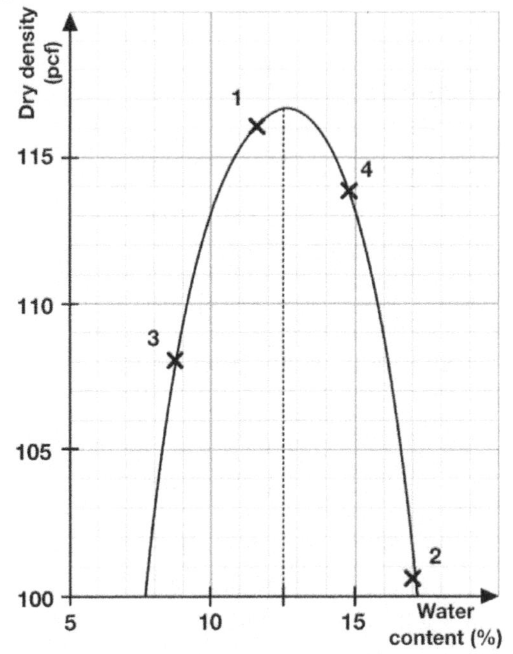

The optimum moisture content as derived from the curve is when dry density is max at 12.5%.

Correct Answer is (C)

SOLUTION 61

The *NCEES Handbook,* Chapter 3 Geotechnical, Section 3.6 Slope Stability is referred to.

The handbook provides two charts for Taylor (1948). The second table is only applicable when friction angle $\emptyset = 0$ and a rock layer has been identified below the slope where the depth factor $D > 1$, which is not the case here.

The first table – copied below for ease of reference – is used in this case.

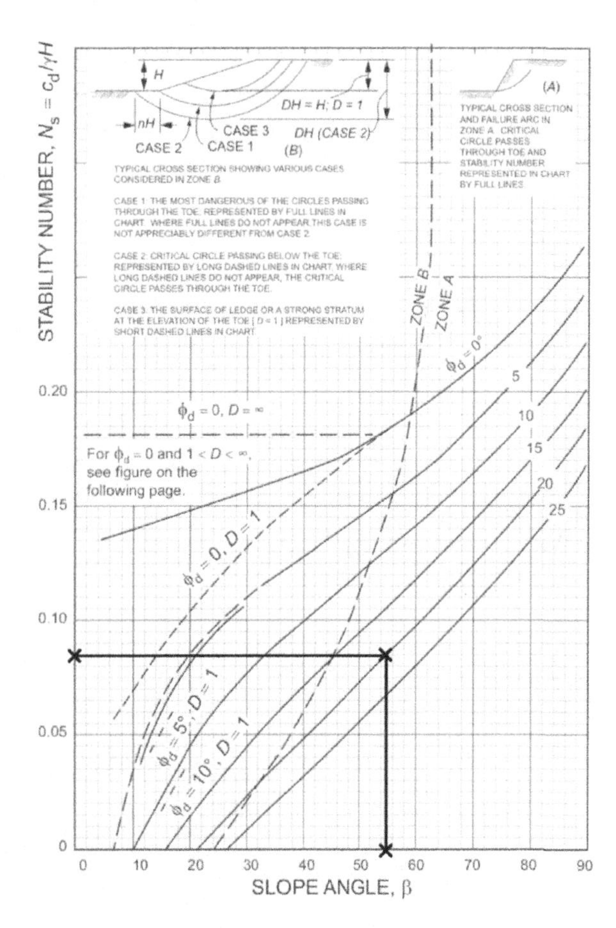

There are two factors that we shall define prior to performing the calculation:

c_d is the developed, or mobilized, cohesion, which is the cohesion that develops at the slip surface upon failure.

ϕ_d is the developed, or mobilized friction angle, which is the friction angle that develops at the slip surface upon failure.

The safety factor requested in this question represents safety against forming a slip, in which case:

$$F.S. = F_c = \frac{c}{c_d}$$

Similarly, the safety factor for friction angle shall be calculated and shall equal to:

$$F.S. = F_\phi = \frac{\tan\phi}{\tan\phi_d}$$

The above process is iterative in nature, and we may have to perform two or more iterations until the following equation is satisfied:

$$F.S. = F_\phi = F_c$$

Iteration 1: assume $\phi = \phi_d = 20^o$

In reference to Taylor's (1948) first chart, using a slope angle $\beta = 55^o$ – first iteration shown on the chart – the stability number $N_s = 0.085$.

$$N_s = \frac{c_d}{\gamma H}$$

$$\rightarrow c_d = \gamma H N_s$$

$$= 95\ \frac{lb}{ft^3} \times 8\ ft \times 0.085$$

$$= 64.6\ psf$$

$$F_c = \frac{c}{c_d} = \frac{58\ psf}{64.6\ psf} = 0.9 = F_\phi$$

$$\phi_d = arctan\left(\frac{\tan\phi}{F_\phi}\right)$$

$$= arctan\left(\frac{\tan 20^o}{0.9}\right)$$

$$= 22^o$$

Iteration 2: assume $\phi_d = 22^o$

In reference to Taylor (1948) chart, using interpolation $\rightarrow N_s = 0.077$.

$$c_d = \gamma H N_s$$

$$= 95\ \frac{lb}{ft^3} \times 8\ ft \times 0.077$$

$$= 58.5\ psf$$

$$F_c = \frac{c}{c_d} = \frac{58\ psf}{58.5\ psf} = 0.99 = F_\phi$$

$$\phi_d = arctan\left(\frac{\tan\phi}{F_\phi}\right)$$

$$= arctan\left(\frac{\tan 20^o}{0.99}\right)$$

$$= 20^o$$

It is obvious at this stage that the safety factor sits somewhere between '0.9' to '1.0'.

Correct Answer is (A)

SOLUTION 62

The *NCEES Handbook,* Chapter 2 Construction, Section 2.6.2 Work Zone and Public Safety – Permissible Noise Exposure (OSHA) is referred to.

The following equation along with the permissible timetable are referred to:

$$D = 100 \times \sum \frac{C_i}{T_i}$$

$\boldsymbol{C_i}$ is time spent at the specified noise pressure. $\boldsymbol{T_i}$ is permissible time taken from the OSHA's table for each exposure specified above represented by $\boldsymbol{i}$.

$$D = 100 \times \left(\frac{2}{32} + \frac{2}{16} + \frac{3}{8} + \frac{1}{4}\right) = 81.25$$

Correct Answer is (D)

SOLUTION 63
The *NCEES Handbook,* Chapter 3 Geotechnical, Section 3.2.1 Normally Consolidated Soils is referred to.

Settlement in a clay layer is calculated using the following equation:

$$S_C = \sum_{1}^{n} \frac{C_c}{1+e_o} H_o \, Log\left(\frac{p_f}{p_o}\right)$$

Where $\boldsymbol{C_c}$ is the coefficient of consolidated and is calculated using the slope of $'Log\,(p) - e\,'$ graph shown below (*). $\boldsymbol{e_o}$ is initial void ratio and can be picked up from the graph by substituting for $Log\,(p_o)$. $\boldsymbol{H_o}$ is the layer thickness, $\boldsymbol{p_f}$ is the final pressure and $\boldsymbol{p_o}$ is the initial/original pressure. $\boldsymbol{n}$ in the equation represents the number of layers, in which case the question did not specify more than *one* layer.

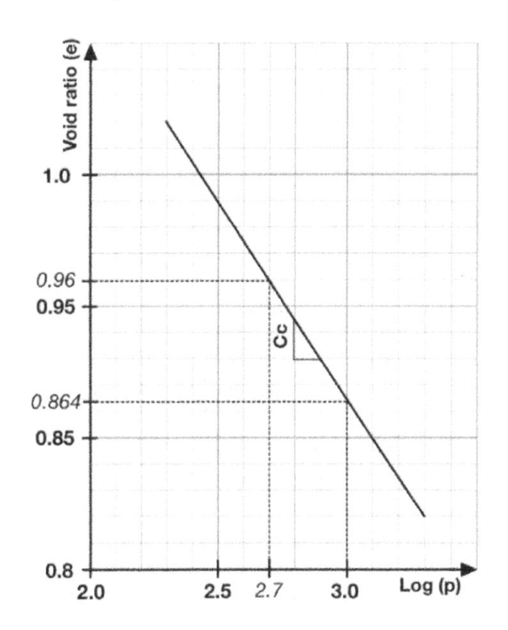

$$C_c = \frac{\Delta e}{\Delta log(p)} \quad (*)$$

$$= \frac{0.96-0.864}{3.0-2.7}$$

$$= 0.32$$

$$e_o = e_{@\,[log(500)\,=\,2.7]} = 0.96$$

$$S_C = \frac{C_c}{1+e_o} H_o \, Log\left(\frac{p_f}{p_o}\right)$$

$$= \frac{0.32}{1+0.96} \times 4\,ft \times Log\left(\frac{1{,}000}{500}\right)$$

$$= 0.2\,ft\,(2.4\,in)$$

Correct Answer is (A)

* For the removal of doubt, C_c is an absolute value and is calculated using the delta of void ratio Δe at the numerator. The graph shown in the *NCEES Handbook version 1.1* might be misleading and one may think that Δe should be at the denominator instead which is wrong, this is corrected in *version 1.2* now.

SOLUTION 64
The *NCEES Handbook,* Chapter 3 Geotechnical, Section 3.10 Trench and Excavation Construction Safety is referred to.

Soil type C should be excavated with a slope of 1 vertical to 1 ½ horizontal as shown below:

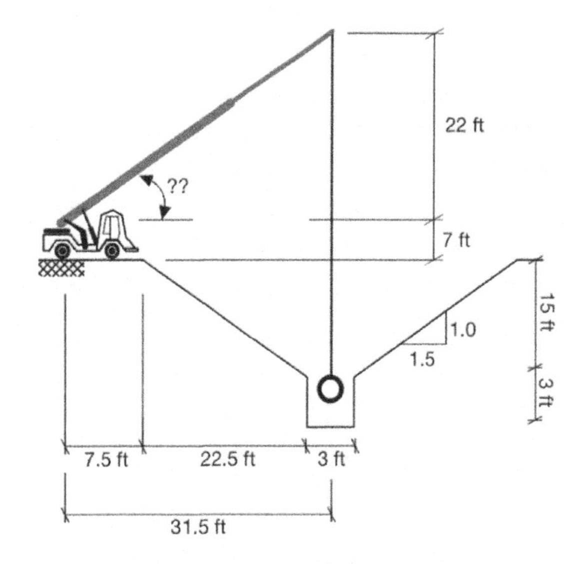

With the above slope configuration, all missing dimensions can be determined, and the required angle can be calculated using trigonometry as follows:

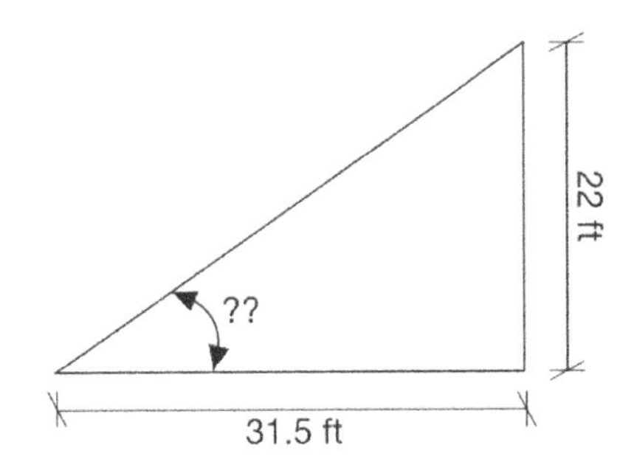

$$\emptyset = arctan\left(\frac{22}{31.5}\right) = 34.9^o$$

Correct Answer is (B)

SOLUTION 65

The *NCEES Handbook,* Chapter 3 Geotechnical, Section 3.19 Pavements is referred to.

The description given in the question is termed as Corrugation. Corrugation is caused by insufficient based stiffness and strength.

Correct Answer is (A)

SOLUTION 66

The solution starts off with calculating service loads for a $1\,ft$ wide strip of the plywood. Deflection (mid and edge) is then calculated and compared with the upper limit in order to determine the maximum allowable distance between joists (L).

$$w_{concrete} = 9\,in \times \frac{1\,ft}{12\,in} \times 145\,\frac{lb}{ft^3}$$

$$= 108.75\,\frac{lb}{ft}/ft$$

$$w_{plywood} = 8\,\frac{lb}{ft^2} = 8\,\frac{lb}{ft}/ft$$

$$w_{live} = 50\,\frac{lb}{ft^2} = 50\,\frac{lb}{ft}/ft$$

$$w_{total} = 108.75 + 8 + 50 = 166.75\,\frac{lb}{ft}/ft$$

$$I_{for\ 1\,ft\ of\ plywood} = \frac{1\,ft \times \left(3/4\,in \times \frac{1\,ft}{12\,in}\right)^3}{12}$$

$$= 2.035 \times 10^{-5}\,ft^4$$

<u>Apply *end of span* deflection equation:</u>

This calculation is not needed provided the question is asking for middle spans only. This calculation is added here for completeness.

$$\Delta_{end\ span} = \frac{wL^4}{148\,EI} \leq \frac{L}{360}$$

$$L \leq \sqrt[3]{\frac{148\,EI}{360\,w}}$$

$$\leq \sqrt[3]{\frac{148 \times 1{,}500{,}000\,\frac{lb}{in^2} \times \frac{144\,in^2}{1\,ft^2} \times 2.035 \times 10^{-5}\,ft^4}{360 \times 166.75\,\frac{lb}{ft}}}$$

$$\leq \sqrt[3]{10.83\,ft^3}$$

$$\leq 2.21\,ft\ (26.5\,in)$$

<u>Apply *mid span* deflection equation:</u>

$$\Delta_{mid\ span} = \frac{wL^4}{1{,}923\,EI} \leq \frac{L}{360}$$

$$L \leq \sqrt[3]{\frac{1{,}923\,EI}{360\,w}}$$

$$\leq \sqrt[3]{\frac{1{,}923 \times 1{,}500{,}000\,\frac{lb}{in^2} \times \frac{144\,in^2}{1\,ft^2} \times 2.035 \times 10^{-5}\,ft^4}{360 \times 166.75\,\frac{lb}{ft}}}$$

$$\leq \sqrt[3]{140.8\,ft^3}$$

$$\leq 5.2\,ft\ (62.4\,in)$$

The middle span requested in this case is $L = 62.4\,in$

Correct Answer is (C)

SOLUTION 67

The *NCEES Handbook,* Chapter 3 Geotechnical, Section 3.7.4 Rock Classification is referred to.

Rock Quality Designation RQD is measured as follows:

$$RQD = \frac{\sum Length\ of\ Sound\ Core\ Pieces > 4\ in}{Total\ Core\ Run\ Legnth}$$

$$= \frac{70\ ft}{100\ ft}$$

$$= 70\%$$

Per the description provided in the FHA Soils and Foundation reference manual, which can be found in the *NCEES Handbook*, an *RQD* of 70% has a fair quality.

Correct Answer is (C)

SOLUTION 68

The *NCEES Handbook,* Chapter 4 Structural, Section 4.3.2 Design Provisions is referred to in which the nominal moment is calculated as follows:

$$M_n = 0.85\ f_c'\ ab\left(d - \frac{a}{2}\right)$$

The depth of the compression block a is calculated as follows:

$$a = \frac{A_s f_y}{0.85\ f_c'\ b}$$

$$= \frac{(4\times1\ in^2)\times60{,}000\ lb/in^2}{0.85\ \times4{,}000\ lb/in^2\times20\ in}$$

$$= 3.5\ in$$

$$M_n = 0.85\ f_c'\ ab\left(d - \frac{a}{2}\right)$$

$$= 0.85 \times 4{,}000\ \frac{lb}{in^2} \times 3.5\ in \times 20\ in \times \left(43.5\ in - \frac{3.5\ in}{2}\right)$$

$$= 9{,}936{,}500\ lb.in\ (828\ kip.ft)$$

Correct Answer is (C)

SOLUTION 69

The point load is broken down into its Y and Z components as follows:

$$P_z = -75\ sin\ (30) = -37.5\ kip$$

$$P_y = -75\ cos\ (30) = -65\ kip$$

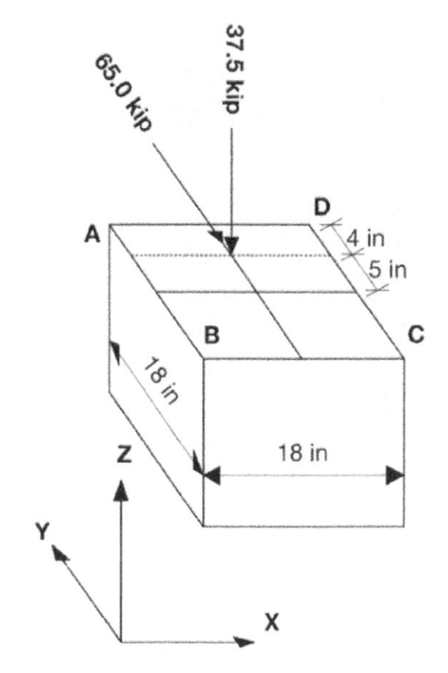

Force P_z is eccentric. Moving this force back to the center of the section creates a moment around X-axis as follows:

$$M = -37.5 \times 5\ in = -187.5\ kip.in$$

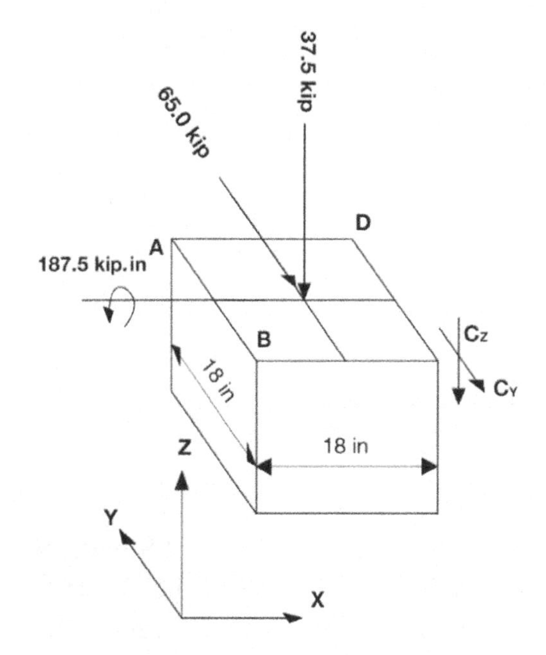

Stresses created at corner 'C' are calculated using equations from the *NCEES Handbook* Chapter 1 General Engineering, Section 1.6.7.2 stresses in beams, as follows:

Stresses due to M_x and P_z:

$$\sigma = \frac{Mc}{I} + \frac{P}{A}$$

$$\sigma_c = \frac{M_x c}{I} + \frac{P_z}{A}$$

$$= \frac{+187.5\ kip.in \times \frac{18\ in}{2}}{\frac{18\ in \times (18\ in)^3}{12}} + \frac{-37.5\ kip}{18\ in \times 18\ in}$$

$$= +0.08\ ksi$$

Stresses due to P_y:

$$\tau = \frac{VQ}{IB}$$

Q is the first moment of area above or below the point where the shear stress is to be determined. It equals to the distance from the area's centroid to the centroid of the shape multiplied by the area itself.

Taking the above definition into account, $Q_{CB} = 0$. Further details on moment of area can be found in the solution of Problem 24.

$$\tau_c = \frac{P_y Q}{IB}$$

$$= \frac{-65\ kip \times 0}{\left(\frac{18\ in \times (18\ in)^3}{12}\right) \times 18\ in}$$

$$= 0$$

Correct Answer is (A)

SOLUTION 70

The *NCEES Handbook,* Chapter 4 Structural, Section 4.2 Fastener Group in Shear is referred to in order to calculate the requested force.

Load P is an eccentric load which shall generate a moment as shown below when the load is moved to the center of the bolts group:

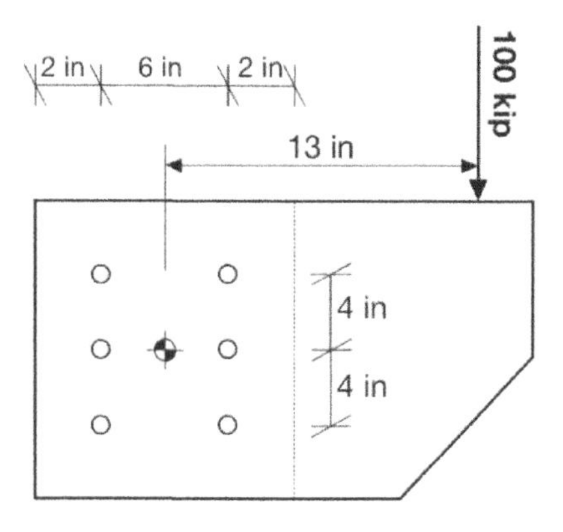

$$M = 100\ kip \times 13\ in = 1{,}300\ kip.in$$

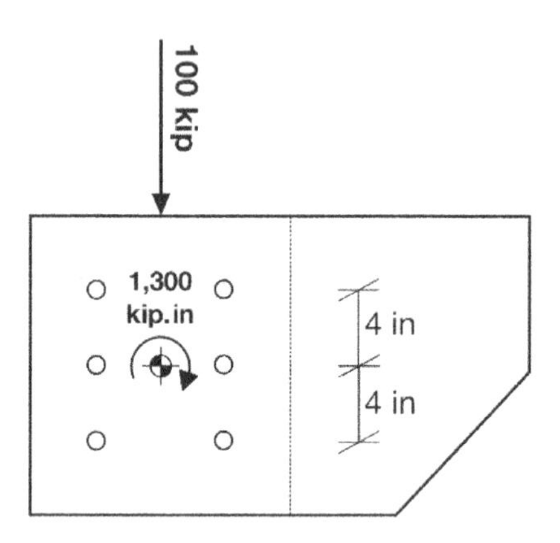

The magnitude of the vertical shear force generated at bolt No. 1 from load P $'F_{P,1}'$ alone is calculated as follows:

$$F_{P,1v} = \frac{100\ kip}{6\ bolts} = 16.7\ kip \downarrow$$

3 in 3 in

16.7 kip

4 in

4 in

The magnitude of the shear force generated at bolt No. 1 from the moment M alone $F_{M,1}$:

$$F_{M,1} = \frac{M \times r_1}{\sum_1^n r_i^2}$$

r_i represents the radius from the centroid of the bolt group to the ith fastener assuming all bolts fall on a perimeter of a circle (*):

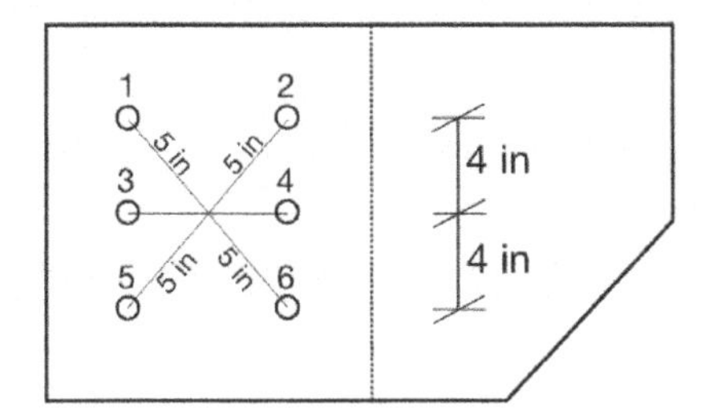

$$F_{M,1} = \frac{1{,}300\ kip.in \times 5\ in}{5^2+5^2+3^2+3^2+5^2+5^2} = 55\ kip$$

This force is perpendicular to the line drawn from the center of the group to the bolt under consideration. See below:

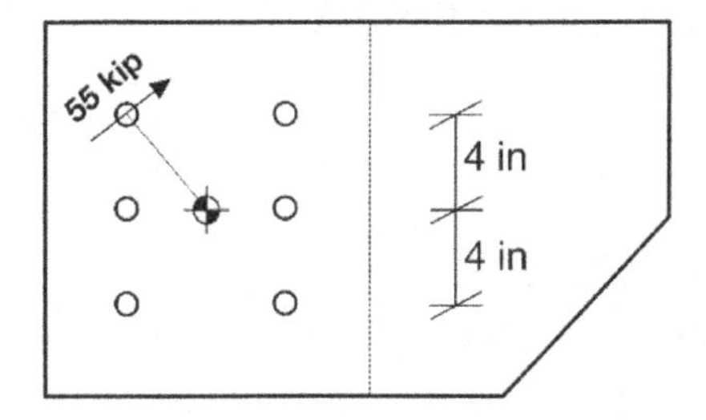

This force is analyzed into its vertical and horizontal components as follows:

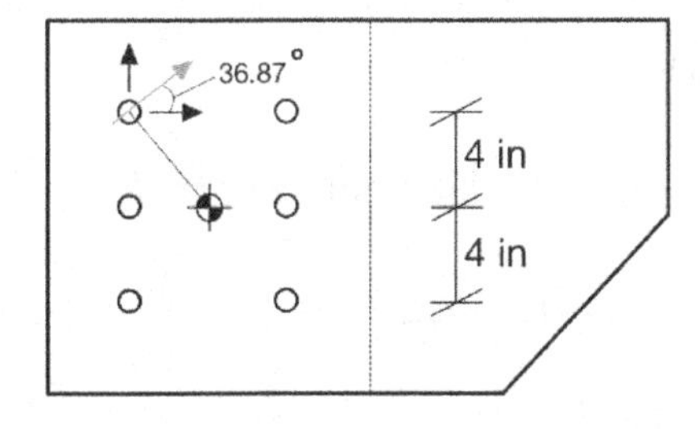

$F_{M,1h} = 55\ kip \times cos(36.87) = 44\ kip \rightarrow$

$F_{M,1v} = 55\ kip \times sin(36.87) = 33\ kip \uparrow$

All forces are shown in the below diagram:

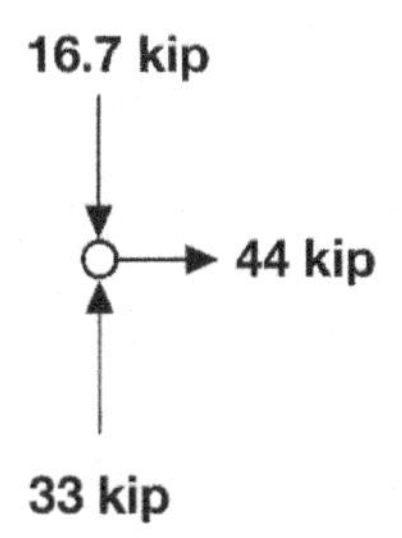

Total vertical and total horizontal shear forces are therefore as follow:

$F_h = F_{M,1h} = 44\ kip \rightarrow$

$F_v = F_{P,1v} + F_{M,1v}$

$= -16.7\ kip + 33\ kip$

$= 16.3\ kip\ \uparrow$

The resultant force is calculated as follows:

$R = \sqrt{F_h{}^2 + F_v{}^2}$

$= \sqrt{44^2 + 16.3^2}$

$= 46.9\ kip$

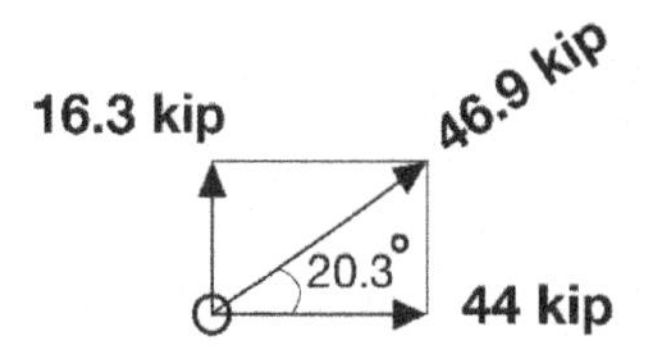

Correct Answer is (C)

* This equation assumes all bolts are on the perimeter of a circle, which does not fully apply given that bolts No. 3 are 4 are not. The above solution is a good approximation though since the difference is slight.

SOLUTION 71

Concrete volume is initially calculated as follows:

$$V = (5 \times 10\,ft + 1) \times (3 \times 9\,ft + 1) \times \frac{9}{12}\,ft$$
$$= 1{,}071\,ft^3$$

Total weight for the concrete slab, live load and plywood is calculated as follow:

$$W_{Total} = 1{,}071\,ft^3 \times 145\,\frac{lb}{ft^3} + 50\,\frac{lb}{ft^2} \times (28\,ft) \times (51\,ft) + 8\,\frac{lb}{ft^2} \times (28\,ft) \times (51\,ft)$$

$$= 238{,}119\,lb\ (238\,kip)$$

Number of props required to support the slab using the service total load is calculated as follows:

$$\#props = \frac{238\,kip}{7\,kip} \cong 34$$

Calculate the number of props using spacing as criteria and use the highest figure:

Number of bays in the long direction:

$$= \frac{(5\times10ft+1ft)\times12\frac{in}{ft}}{80\,in}$$
$$= 7.65\,bays$$

$(use\ 8\ bays\ to\ make\ a\ spacing\ of\ 76.5\ in\ the\ long\ direction)$

Number of bays in the short direction:

$$= \frac{(3\times9ft+1ft)\times12\frac{in}{ft}}{80\,in}$$
$$= 4.2\,bays$$

$(use\ 5\ bays\ to\ make\ a\ spacing\ of\ 67.2\ in\ the\ short\ direction)$

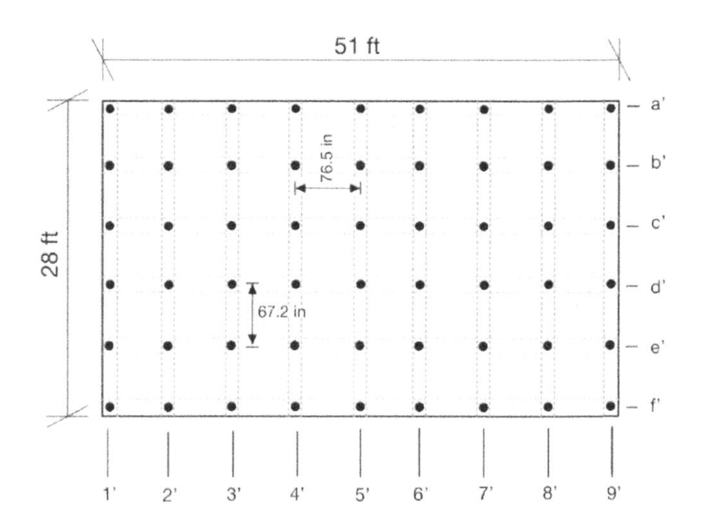

Number of props is calculated using the number of axes as shown in the figure above as follows:

$$\#props = 9 \times 6 = 56\ (*)$$

Use the largest number of props as follows:

$$= max\,(56, 34) = 56\,props$$

Correct Answer is (D)

(*) Dotted lines in the figure represent joists and stringers.

SOLUTION 72

There are few methods that can be used to solve this question, below is one.

Start with calculating the surface area for one wall combinations as follows:

$$A = 2 \times \left[\left(32 - \frac{14}{12}\right) \times \frac{14}{12}\right] + 30 \times \frac{14}{12}$$
$$= 106.9\,ft^2$$

Rate of concrete placement in ft^3/hr:

$$\alpha = 70\,yard^3/hr \times \frac{27\,ft^3}{yard^3}$$
$$= 1{,}890\,ft^3/hr$$

In *four* hours, the volume of pumped concrete should be as follows:

$$V = 1{,}890\ ft^3/hr \times 4\ hrs$$
$$= 7{,}560\ ft^3$$

Height of concrete poured in *four* hours:

$$h = \frac{7{,}560\ ft^3}{106.9\ ft^2} = 70.7\ ft$$

Divide this by each wall combination height to determine how many walls could be poured in full:

$$No.of\ walls = \frac{70.7\ ft}{11\ ft} = 6.43\ wall$$

This answer represents *six* walls fully poured and 43% of the seventh wall being poured which equals to $43\% \times 11 = 4.73\ ft$

Correct Answer is (C)

SOLUTION 73

Start by rearranging all speeds from least to greatest as follows:

SN	Speed *mph*
1	12
2	21
3	25
4	32
5	45
6	45
7	45
8	46
9	55
10	56
11	59
12	59
13	69
14	72
15	73

The 85th percentile speed is the speed at or below which 85% of all vehicles are observed travelling past a monitoring point.

The position where 85% of the dataset fall under equals to $0.85 \times 15 = 12.75$. Since this is not a whole number, the whole number right after this position will be used, in which case $'13'$. In other means, 85% of the speeds observed are below the speed at position 13 which is $69\ mph$ - this speed represents the 85th percentile of this data set.

Correct Answer is (A)

SOLUTION 74

Safety Incidence Rate IR equation can be picked up from the *NCEES Handbook* Section 2.6.1.1.

$$IR = \frac{N \times 200{,}000}{T}$$

$$IR_{1st\ project} = \frac{250 \times 200{,}000}{0.85 \times 2 \times 12 \times 260 \times 150}$$
$$= 62.85$$

$$N_{2nd\ project} = \frac{IR_{first\ project,improved} \times T}{200{,}000}$$
$$= \frac{(0.75 \times 62.85) \times 0.85 \times 1.5 \times 12 \times 260 \times 75}{200{,}000}$$
$$= 70$$

Correct Answer is (B)

SOLUTION 75

The *NCEES Handbook* Chapter 4 Structural, Section 4.1.7 Moment, Shear, and Deflection Diagrams is generally referred to in this solution to determine reactions for simply supported and continuous beams.

The following figure presented below is numbered for ease of reference:

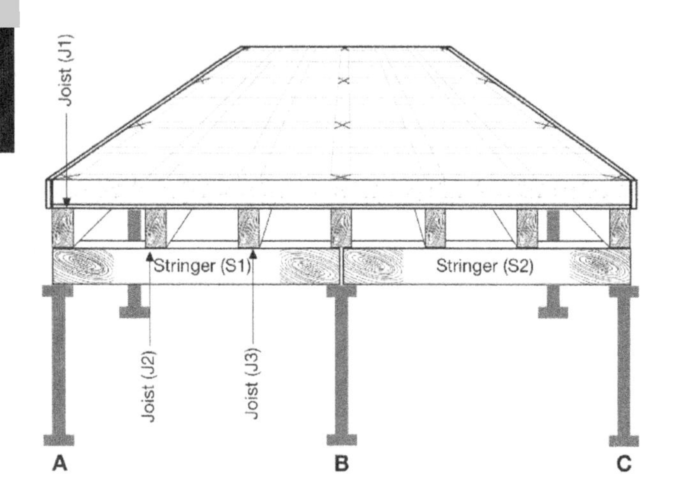

Determine load per square foot as follows:

$$w = 145\ lb/ft^3 \times 9\ in \times \frac{1\ ft}{12\ in} + 50\ lb/ft^2$$
$$= 158.75\ lb/ft^2$$

Load applied on Joist (J1):

$$w_{J1} = 158.75\ lb/ft^2 \times \frac{\left(60\ in \times \frac{1\ ft}{12\ in}\right)}{2}$$
$$= 396.9\ lb/ft$$

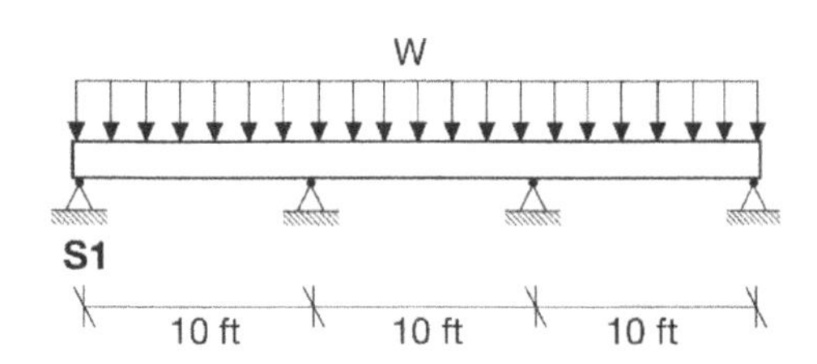

Using diagrams of the *NCEES Handbook version 1.2,* Section 4.1.7*;* reaction (R) at Stringer (S1) due to load on Joist (J1) – using diagram 39 page 259 of *version 1.2* – is calculated as follows:

$$R_{J1} = 0.4\ w_{J1} \times l$$
$$= 0.4 \times 396.9\ lb/ft \times 10\ ft$$
$$= 1{,}587.6\ lb\ (1.59\ kip)$$

Load applied on Joist (J2 and J3):

$$w_{J2,3} = 158.75\ lb/ft^2 \times \left(60\ in \times \frac{1\ ft}{12\ in}\right)$$
$$= 793.8\ lb/ft$$

$$R_{J2\ or\ 3} = 0.4\ w_{J2\ or\ 3} \times l$$
$$= 0.4 \times 793.8\ lb/ft \times 10\ ft$$
$$= 3{,}175.2\ lb\ (3.18\ kip)$$

Load applied on Stringer (S1):

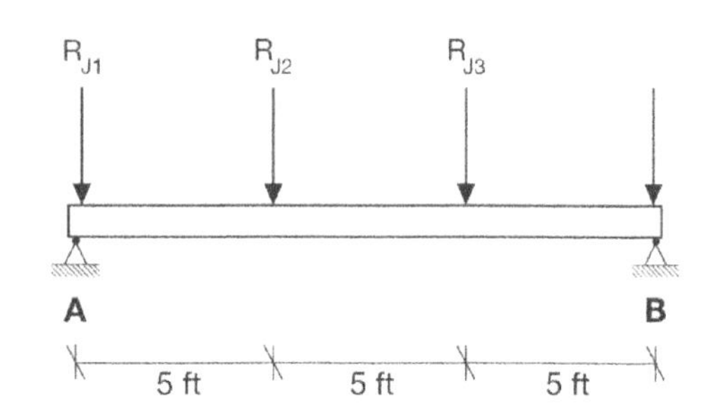

$$R_A = R_{J1} + \frac{R_{J2} + R_{J3}}{2}$$
$$= 1.59\ kip + \frac{3.18\ kip + 3.18\ kip}{2}$$
$$= 4.77\ kip\ (*)$$

Correct Answer is (A)

(*) This reaction is missing the quantity of concrete from the center of the edge joist (J1) to the edge of the slab. Assuming a joist width of 2 *in*, this quantity would increase the load on prop A by a negligible 20 *lb* (i.e., 0.02 *kip*).

SOLUTION 76

The *NCEES Handbook,* Chapter 4 Geotechnical, Section 3.4.2.1 Bearing Capacity for Concentrically Loaded Square or Rectangular Footings, is referred to in order to provide a solution for this question.

First start with calculating the ultimate bearing capacity for the conditions provided in the question, then have it divided by the acting pressure from the column loading in order to determine the safety factor.

$$q_{ult} = c(N_c)s_c + q(N_q)s_q + 0.5\gamma(B_f)(N_\gamma)s_\gamma$$

The bearing capacity factors $\boldsymbol{N_c}$, $\boldsymbol{N_q}$ and $\boldsymbol{N_\gamma}$ are collected from the table provided in the *NCEES Handbook* as 46.1, 33.3 and 48 respectively.

The shape correction factors $\boldsymbol{s_c}$, $\boldsymbol{s_q}$ and $\boldsymbol{s_\gamma}$ are calculated using the following equations when $\emptyset > 0$:

$$s_c = 1 + \left(\frac{B_f}{L_f}\right)\left(\frac{N_q}{N_c}\right)$$

$$= 1 + \left(\frac{6}{6}\right)\left(\frac{33.33}{46.1}\right)$$

$$= 1.72$$

$$s_q = 1 + \left(\frac{B_f}{L_f}\ tan\emptyset\right)$$

$$= 1 + \left(\frac{6}{6}\ tan35\right)$$

$$= 1.7$$

$$s_\gamma = 1 - 0.4\left(\frac{B_f}{L_f}\right)$$

$$= 1 - 0.4\left(\frac{6}{6}\right)$$

$$= 0.6$$

Given there is no surcharge load:

$$q = \gamma\ D_f$$

$$= 130\ pcf \times 5\ ft$$

$$= 650\ psf$$

It is also important to remember that the density γ in the bearing capacity equation represents soil at the bottom of the footing, in which case the buoyant one will be used given that the bottom of the footing is submerged. It can either be calculated using the effective stress method as γ' by deducting pore/water pressure from it, or a correction factor of 0.5 can be applied to it which can be collected from the Correction Factor table presented in the same chapter.

$$\gamma' = 130\ pcf - 62.4\ pcf$$

$$= 67.6\ pcf$$

$$q_{ult} = c(N_c)s_c + q(N_q)s_q + 0.5\gamma'(B_f)(N_\gamma)s_\gamma$$

$$= 450(46.1) \times 1.72 + 650(33.3) \times 1.7$$

$$+ 0.5 \times 67.6(6)(48) \times 0.6$$

$$= 78{,}318.54\ psf$$

$$q_{actual} = \frac{650{,}000\ lb}{6\ ft \times 6\ ft}$$

$$= 18{,}055.56\ psf$$

$$S.F. = \frac{q_{ult}}{q_{actual}}$$

$$= \frac{78{,}318.54\ psf}{18{,}055.56\ psf}$$

$$= 4.34$$

Correct Answer is (A)

SOLUTION 77

The *NCEES Handbook,* Chapter 4 Geotechnical, Section 3.4 Bearing Capacity can be referred in this question.

Bearing capacity equations of Section 3.4.2 for strip footings – copied below for ease of reference – which are fundamentally similar to bearing capacity equations for other footing types, has the total surcharge pressure at the base of the footing q as part of, and a major contributor to, the bearing capacity of the soil beneath. See below:

$$q_{ult} = c(N_c) + q(N_q) + 0.5\gamma(B_f)(N_\gamma)$$

$$q = q_{app} + \gamma_a D_f$$

Where q_{app} is the surcharge pressure at surface, which also has a positive impact on bearing capacities. γ_a is density of soil above the base of the footing, and D_f is the depth of the footing.

In conclusion, the removal of soil, or any other surcharge load on top of embedded foundations, by either excavation or scour, can substantially reduce the ultimate bearing capacity and may cause a catastrophic shear failure.

Correct Answer is (B)

SOLUTION 78

Project expenses include the cost of material along with the cost of borrowing, in which case you will be the one borrowing from the bank on the behalf of the contractor to supply them with the material.

$$Cost\ of\ borrowing = 6\% \times \$100{,}000 = \$6{,}000$$

$$Cost\ of\ material = \$100{,}000$$

$$Total\ expenses = \$100{,}000 + \$6{,}000 = \$106{,}000$$

$$Overall\ sales\ inclusive\ of\ profit\ on\ all\ expeses$$

$$= \$106{,}000 \times 1.2 = \$127{,}200$$

$$Monthly\ payment = \frac{127{,}200}{12} = \$10{,}600$$

Correct Answer is (A)

SOLUTION 79

The tributary area method is explained briefly in the *NCEES Handbook* Chapter 5 Geotechnical, Section 3.18.3 Anchor Loads.

Using the traditional method explained in Solution 36 for lateral load calculation may give you indicative, not very accurate, results. The pressure behind retaining walls with anchors is complex and is best described per the below diagram (*).

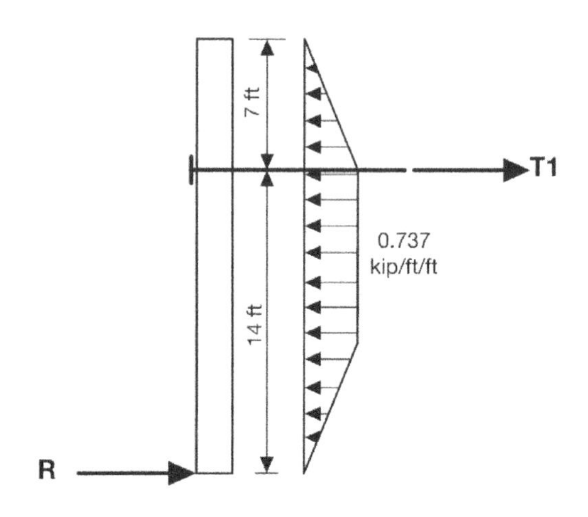

$$k_a = tan^2\left(45 - \frac{\phi'}{2}\right) = tan^2\left(45 - \frac{35^o}{2}\right) = 0.27$$

Maximum ordinate (p) – although given in the question – is available from the FHWA reference mentioned in the *NCEES Handbook*, and is calculated as follows:

$$p = k_a\, H\, \gamma_{soil} = 0.27 \times 21\ ft \times 0.13\ kcf = 0.737\ kip/ft/ft$$

Using the tributary area method explained in the *NCEES Handbook* Section 3.18.3 (*):

$$T1 = load\ over\ \left(H1 + \frac{H2}{2}\right)$$
$$= 0.5 \times 0.737 \times 7 + 0.5 \times (14 \times 0.737)$$
$$= 7.74\ kip/ft$$

Given that anchors are placed at $3\,ft$ intervals, each anchor should have a capacity of $\mathbf{23.2\ kip}$.

A capacity of $23.2\,kip$ per anchor is considered significant. Another layer or two may be required in this case.

Correct Answer is (C)

(*) We assumed that the top and bottom triangular sections of the trapezoidal shaped pressure are $7\,ft$ high, which is a fair assumption for such a simple question.

The accurate method for estimating the pressure profile given in FHWA reference found in the *NCEES Handbook* is explained below briefly.

The trapezoidal loading per FHWA can be reconstructed in reference to the following diagram:

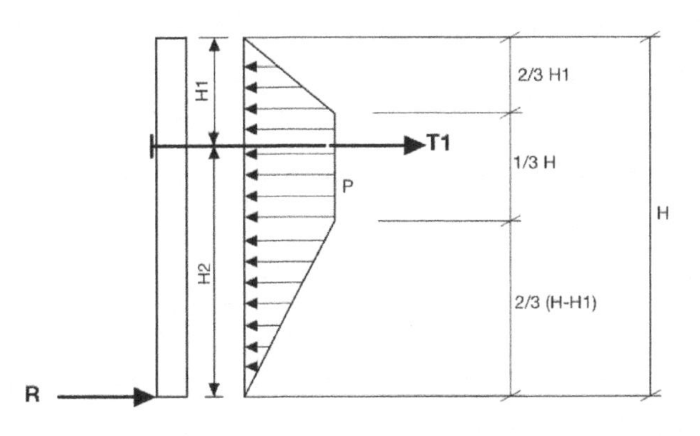

Based on the above, the following diagram is constructed to represent the pressure profile for the question in hand:

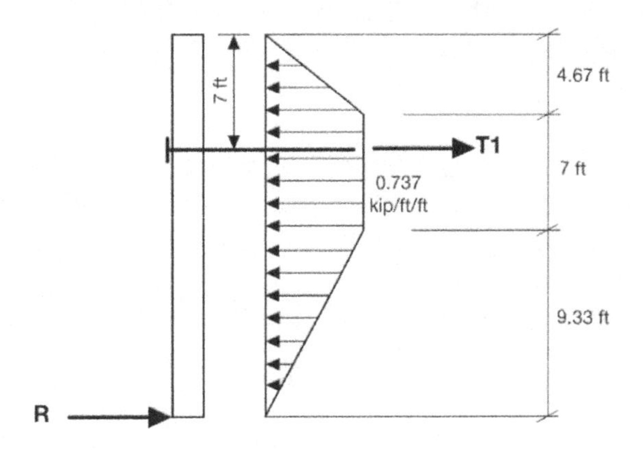

The tributary area contributing to the anchor loading in this case is:

$$T1 = load\ over\ \left(H1 + \frac{H2}{2}\right)$$

This area is better understood using the following diagram:

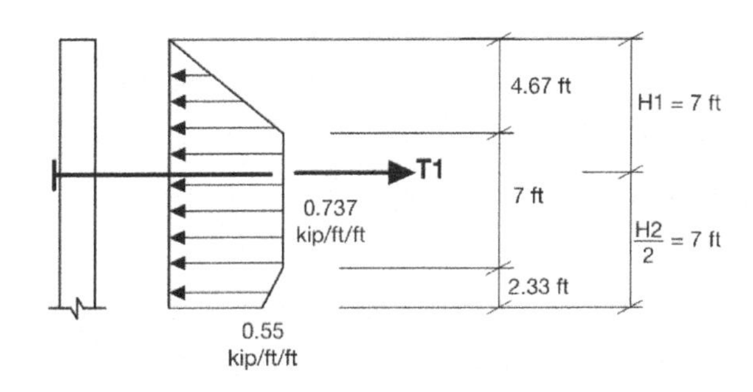

Load $T1$ is therefore calculated as follows:

$$T1 = \frac{0.737 \times 4.67}{2} + 0.737 \times 7 + \frac{(0.55 + 0.737)}{2} \times 2.33$$
$$= 8.37\ kip/ft$$

$$Load\ per\ Anchor = 8.37 \times 3 = 25.1\ kip$$

Reference to the above is as follows:

Federal Highway Administration, June 1999. Ground Anchors and Anchored Systems, Geotechnical Engineering Circular No.4. *U.S. Department of Transportation.*

SOLUTION 80

Determine: (1) skin friction with an unknown depth of h along with (2) the bearing capacity of soil, both shall equal to 2.5 times the required loading.

$$Skin\ friction = 2\pi r \times h \times 150\ psf$$
$$= 2\pi \times (½\ ft) \times h \times 150\ psf$$
$$= 150\pi h\ lb$$

$$Bearing\ resistance = \pi r^2 \times 200\ psf$$
$$= \pi \times (½\ ft)^2 \times 200\ psf$$
$$= 50\pi\ lb$$

$$Total\ resistance = (150\pi h + 50\pi)\ lb$$

$$S.F. = \frac{Resistance}{Actual\ Loading}$$

$$2.5 = \frac{150\pi h + 50\pi}{2000\ lb}$$

$$h = 10.3\ ft$$

Correct Answer is (B)

PART II

STRUCTURAL DEPTH

SECTION 1

Structural Analysis

Problems & Solutions

PROBLEM 1.1 ***Two Fixed Ends***

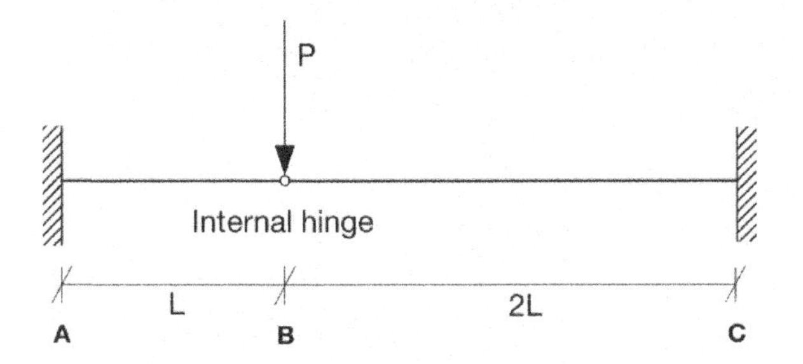

Moment at 'A' is:

(A) $\frac{1}{3}$ PL

(B) $\frac{8}{9}$ PL

(C) $\frac{1}{2}$ PL

(D) $\frac{6}{9}$ PL

PROBLEM 1.2 ***A Strand with a Cantilever Beam***

Ignoring the beam and the strand's self-weight. There is a concentrated service live load of 3 *kip* applied to the end of the *W* section shown. The *W* section is connected to a 0.6 *in* steel strand with a pin connection.

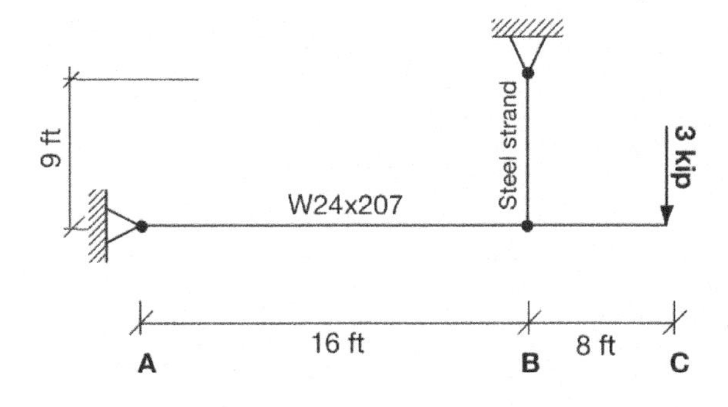

Knowing that the structure will stay within the elastic range, deflection at point 'C' due to the concentrated service load is most nearly:

(A) 0.13 *in*

(B) 0.02 *in*

(C) 0.10 *in*

(D) 0.17 *in*

PROBLEM 1.3 3D ***Truss Reactions***

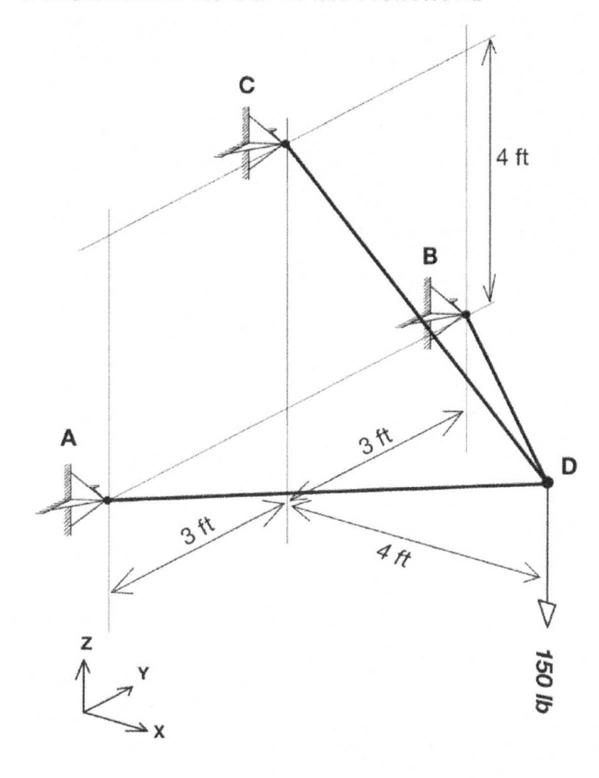

Reactions (*) at pinned support 'A' in *lb* are as follows:

(A) $R_X = -(75)$
$R_Y = -(56)$
$R_Z =$ Zero

(B) $R_X = +75$
$R_Y = +56$
$R_Z =$ Zero

(C) $R_X = +56$
$R_Y = +75$
$R_Z =$ Zero

(D) $R_X = +75$
$R_Y = +56$
$R_Z = +75$

* For the removal of doubt, points 'A', 'B' and 'D' fall on the same horizontal XY-plane.

PROBLEM 1.4 ***Structural Changes***

The below 30 ft long simply supported beam represents an actual beam in one of the projects with four joists each carrying a load of 5.5 kip as shown.

Due to changes on the project, the second joist from the right moved 3 ft to the left, reduced its load to 30 kip, and its original location replaced with a 9.0 $kip.ft$ concentrated moment a shown:

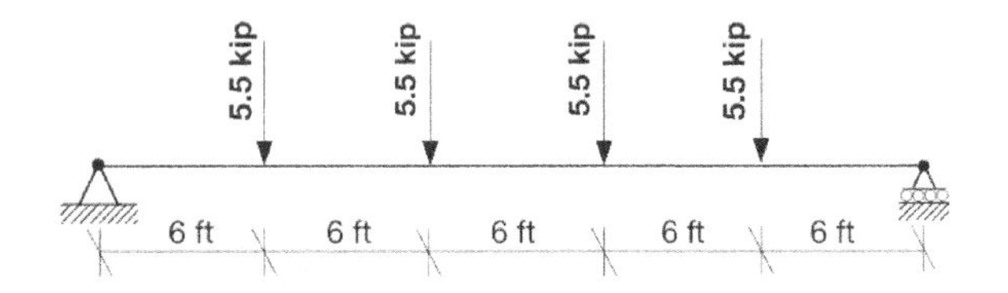

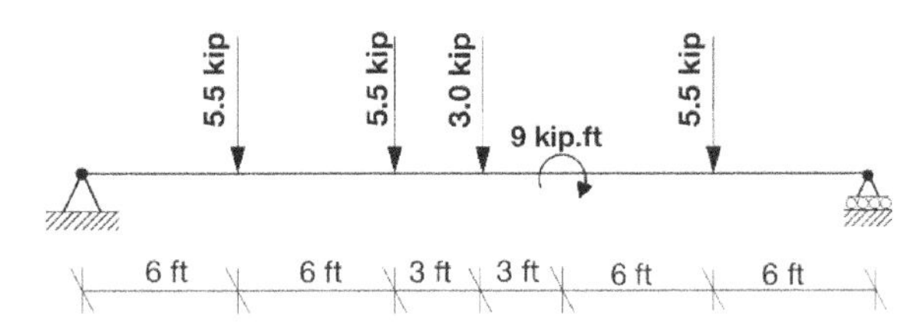

The net increase/decrease in $kip.ft$ in maximum moment for the main beam, regardless of the location of maximums before and after the change, is:

(A) Decrease of 18 $kip.ft$

(B) Increase of 18 $kip.ft$

(C) Decrease of 21.6 $kip.ft$

(D) Increase of 21.6 $kip.ft$

PROBLEM 1.5 ***Moment Distribution Method***

Consider a consistent cross-section across the below three spans along with a consistent modulus of elasticity as well.

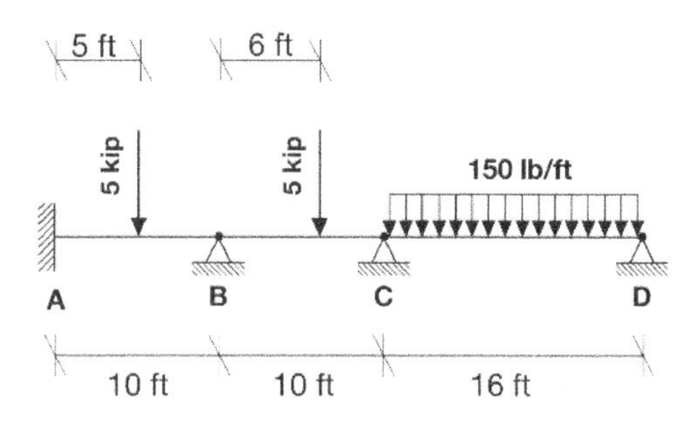

Using the moment distribution method, the unbalanced moment at 'C' is most nearly:

(A) 4.0 $kip.ft$ clockwise

(B) 2.4 $kip.ft$ counterclockwise

(C) 4.0 $kip.ft$ counterclockwise

(D) 2.4 $kip.ft$ clockwise

PROBLEM 1.6 ***Moving Load on a Truss***

The below point load moves on the upper chord of the truss shown below:

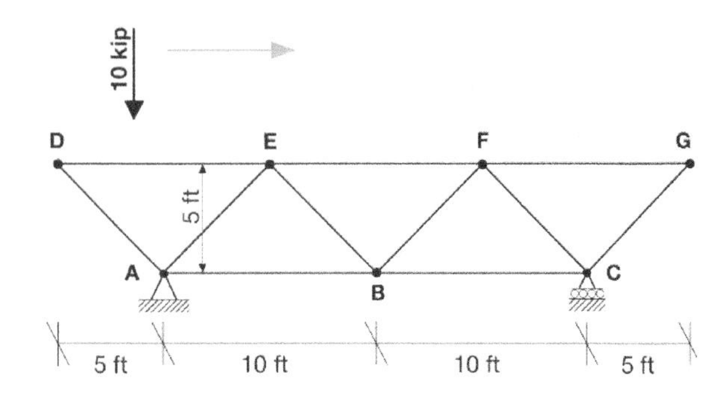

The shear influence line for panel AE and the maximum internal force for AE are as follows:

(A) $AE_{max} = 10.6\ kip$ in compression

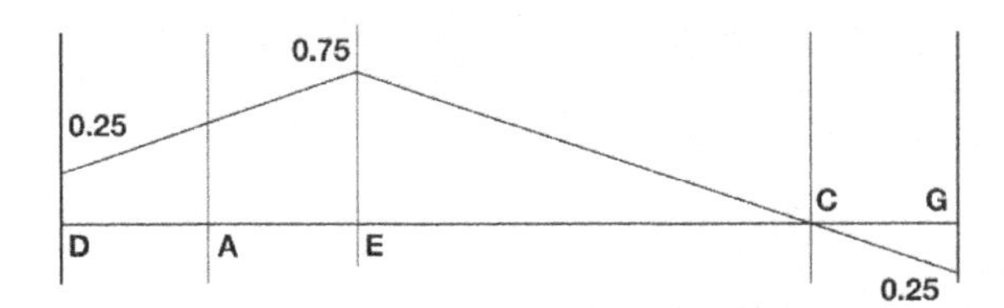

(B) $AE_{max} = 8.5\ kip$ in compression

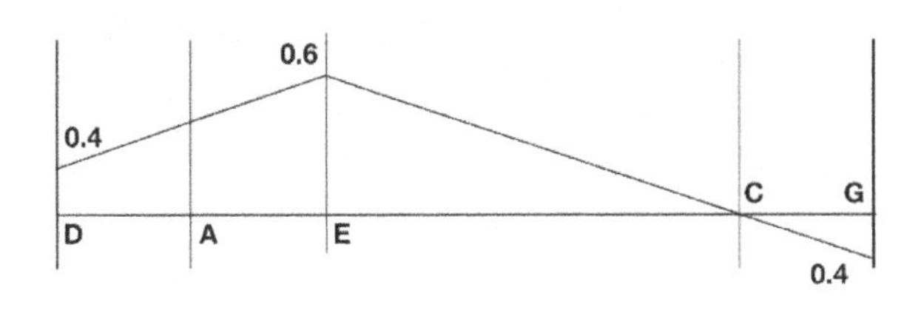

(C) $AE_{max} = 10.6\ kip$ in compression

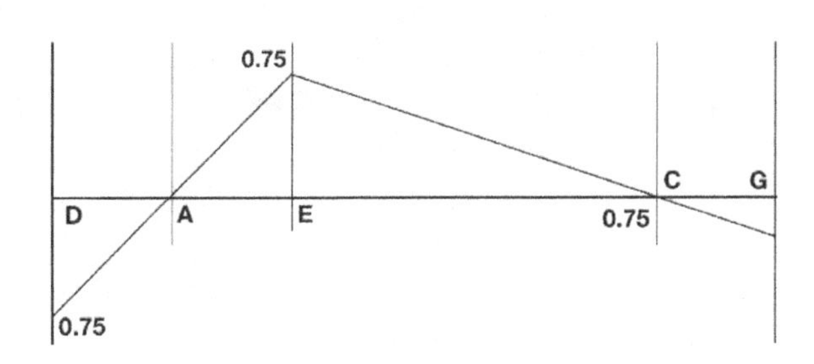

(D) $AE_{max} = 8.5\ kip$ in compression

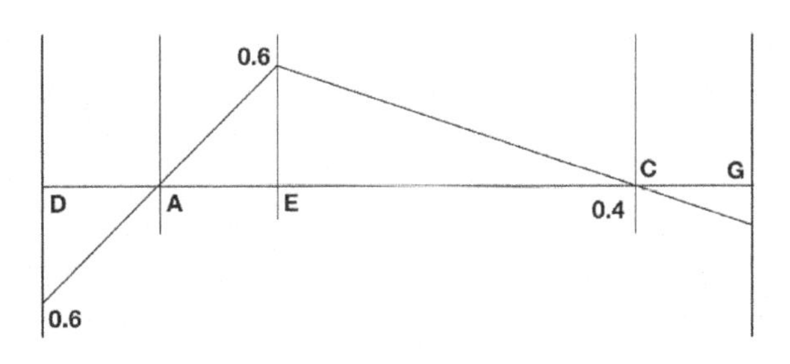

PROBLEM 1.7 ***Moving Load on a Continuous Beam***

The below three points load represents an AASHTO HS20 truck moving on a short three span $120\,ft$ bridge with four hinges/rollers at point 'A', 'B', 'C' and 'D' along with two internal hinges at points 'E' and 'G'.

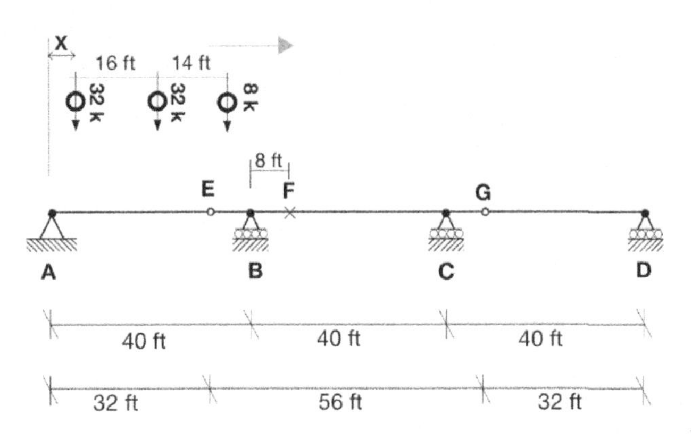

The largest shear that occurs at point 'F' and the location of the truck that produces this shear (distance 'x') are:

(A) $42.0\ kip\ @\ x = 48\,ft$

(B) $42.0\ kip\ @\ x = 40\,ft$

(C) $35.6\ kip\ @\ x = 32\,ft$

(D) $39.0\ kip\ @\ x = 48\,ft$

PROBLEM 1.8 ***3D Truss Deflection***

The below is a four equal legged 3D truss with a horizontal point load applied at joint 'E' in direction Y as shown:

$EA = 187 \times 10^3\ kip$ for all members

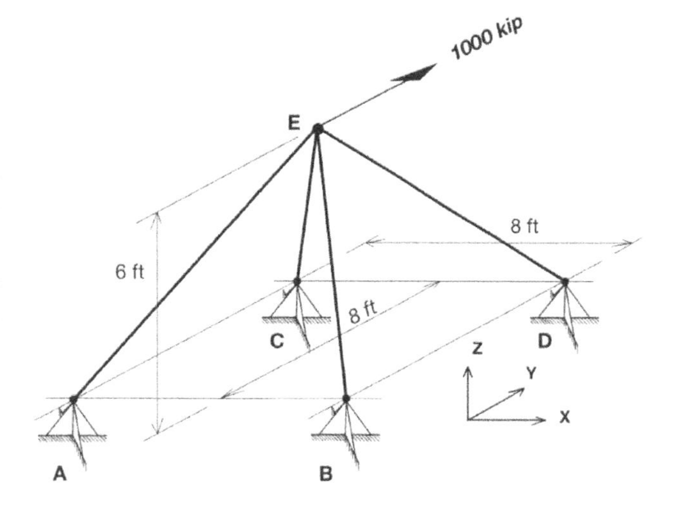

Joint E's lateral displacement in the Y-axis is most nearly:

(A) + 0.43 in

(B) + 0.14 in

(C) + 0.27 in

(D) + 0.56 in

PROBLEM 1.9 ***Frame Deflection***

The frame shown below has the following cross section properties for both members AB and BC:

$E = 29{,}000\ ksi$

$I = 1{,}000\ in^4$

The frame has a fixed support at 'A' and a rigid connection at 'B' with a free end at 'C'.

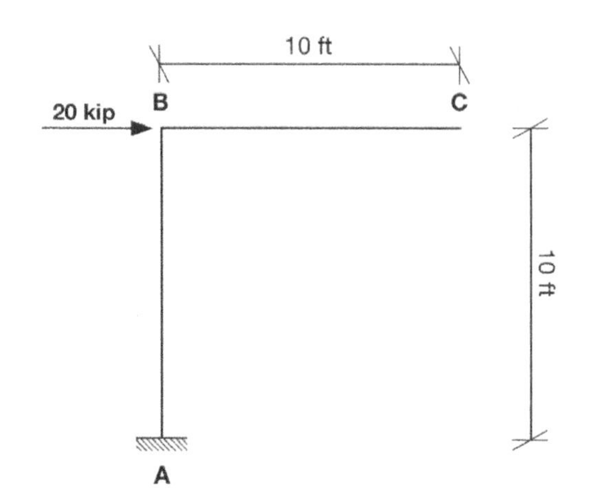

Horizonal and vertical displacement at joint 'C' are most nearly:

(A) $5 \times 10^{-3}\ in\ \downarrow, 0.4\ in\ \rightarrow$

(B) $0.4\ in\ \downarrow, 0.6\ in\ \rightarrow$

(C) $0.6\ in\ \downarrow, 0.4\ in\ \rightarrow$

(D) $5 \times 10^{-3}\ \downarrow, 0.6\ in\ \rightarrow$

PROBLEM 1.10 3D ***Frame Deflection and Torque***

The below is a 3D steel pipe with a 7 in outer diameter, 0.75 in thick pipe wall, fixed end at 'A', rigid connection at 'B', and a free end at 'C'.

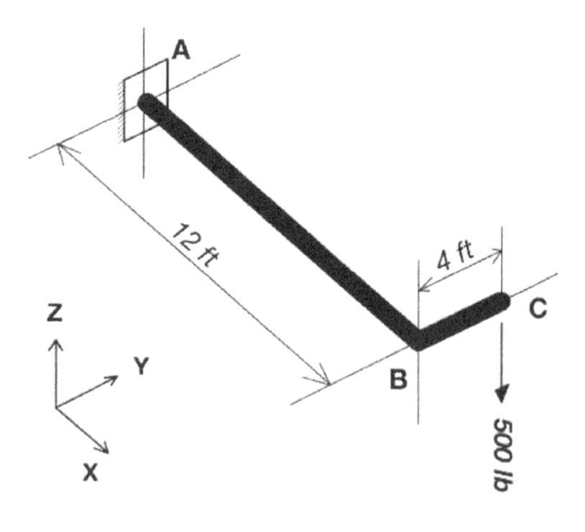

Vertical displacement at 'C' is most nearly:

(A) 0.42 in

(B) 0.58 in

(C) 0.18 in

(D) 0.75 in

PROBLEM 1.11 ***Frame Moment Diagram***

Which of the following diagrams most nearly represents the moment generated from the lateral load shown below:

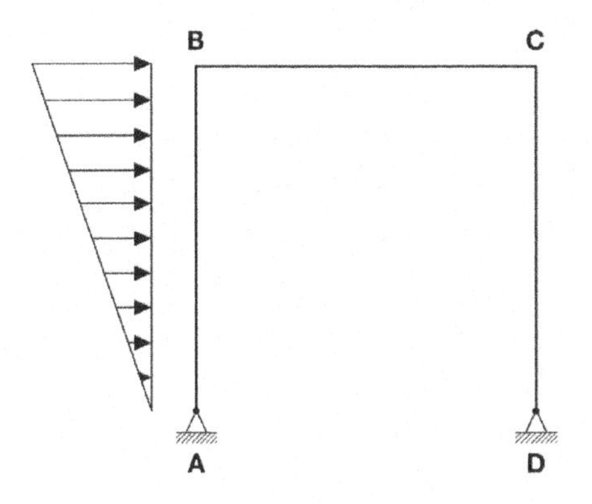

(A) Diagram A

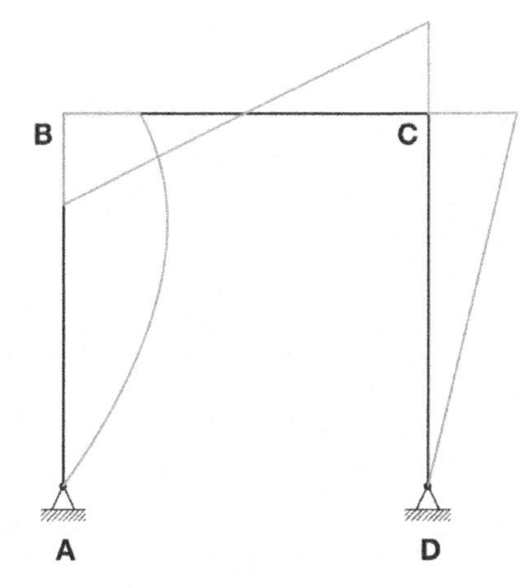

(B) Diagram B

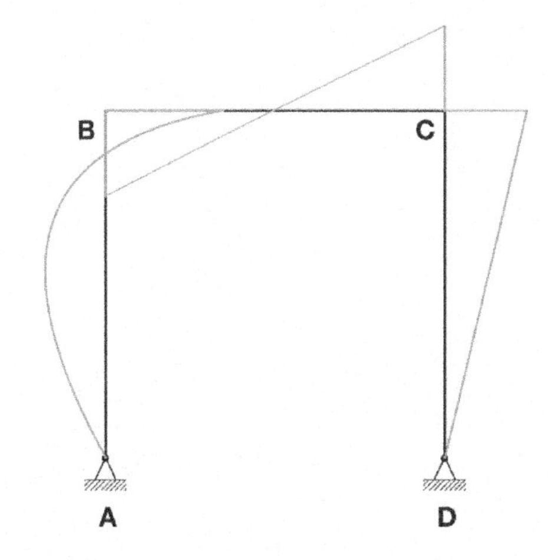

(C) Diagram C

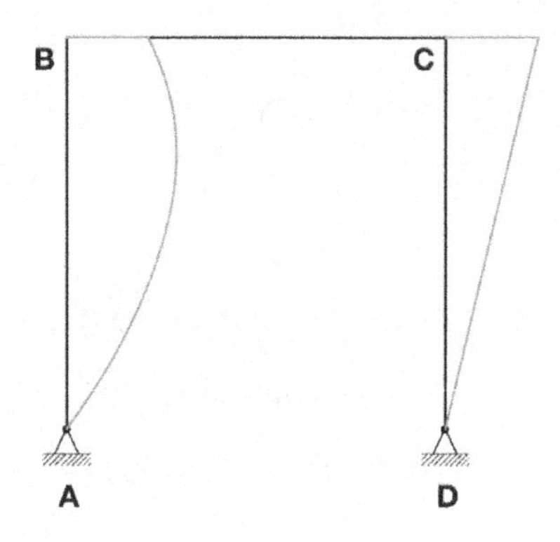

(D) Diagram D

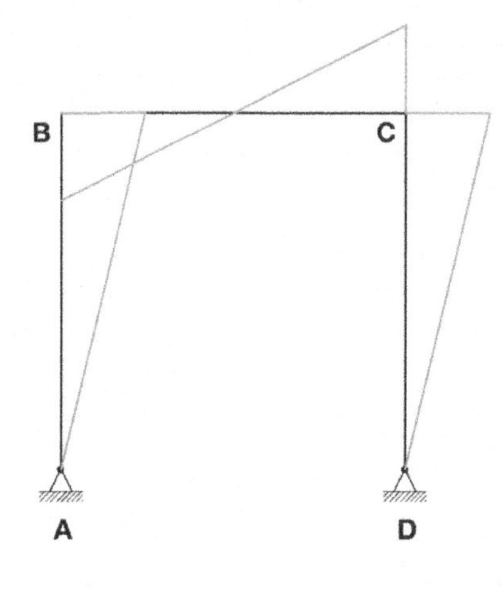

PROBLEM 1.12 ***Moment Distribution Method***

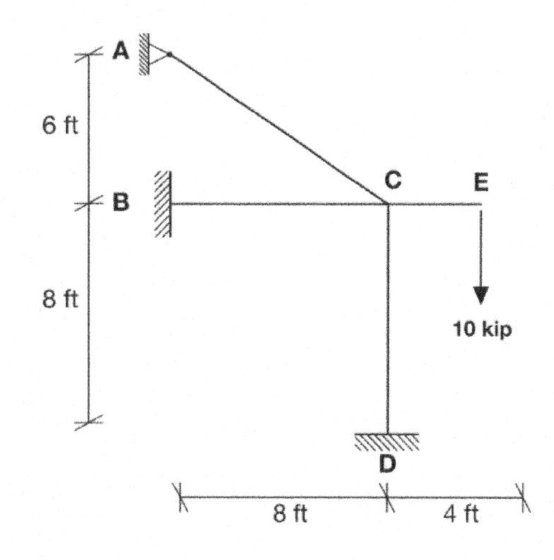

The above structure is a frame with a rigid join at ‘C’.

Moment generated at fixed end ‘D’ due to loading at ‘E’ is most nearly:

(A) 7.2 $kip.ft$ clockwise

(B) 10 $kip.ft$ counterclockwise

(C) 14.4 $kip.ft$ counterclockwise

(D) 7.2 $kip.ft$ counterclockwise

SOLUTION 1.1

The beam can be split into two at the hinge location. Each side of the beam resists a portion of the load P based on its stiffness. Moment can be calculated accordingly.

$M_A = L \times P_L$

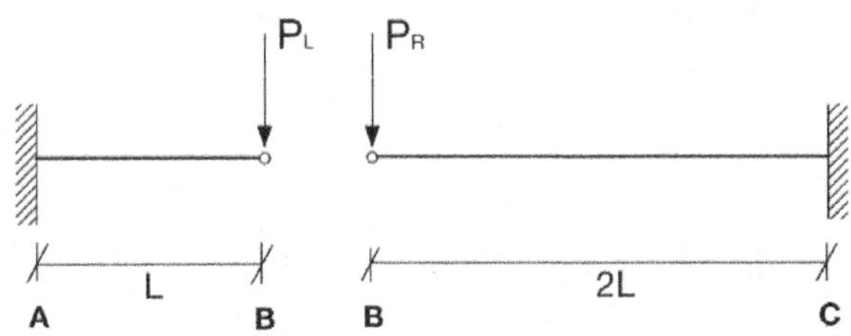

Vertical deflection at hinge 'B' is the same for the two load portions. Based upon which the following equations can be formulated and resolved simultaneously:

<u>Equation 1:</u>

$P = P_L + P_R$

$P_R = P - P_L$

<u>Equation 2:</u>

Calculate deflection at 'B', once using P_L, and another using P_R and equate the two resulting equations. Deflection equations are provided by the *NCEES handbook.*

$$\delta_{B,P_L} = \frac{P_L(L)^3}{3EI}$$

$$\delta_{B,P_R} = \frac{P_R(2L)^3}{3EI}$$

$$\frac{P_L(L)^3}{3EI} = \frac{P_R(2L)^3}{3EI}$$

$P_L = 8\,P_R$

<u>Substitute the first equation into the second equation:</u>

$P_L = \frac{8}{9}P$

$M_A = \frac{8}{9}PL$

Correct Answer is (B)

SOLUTION 1.2

The cross-section properties are as follows:

$A_{n,0.6\ in\ dia\ strand} = 0.217\ in^2$

$I_{W24\times207} = 6{,}820\ in^4$

$E_{steel} = 29{,}000\ ksi$

Determine tension force in the strand P_{strand}:

$\sum M_A = 0$

$16 \times P_{strand} = 3 \times 24$

$P_{strand} = 4.5\ kip$

This force shall determine the elongation at the strand in a later stage.

Knowing that the structure is in the elastic range, the principle of superposition applies. Dissecting the structural deflections as follows shall help determine the final vertical deflection at 'C':

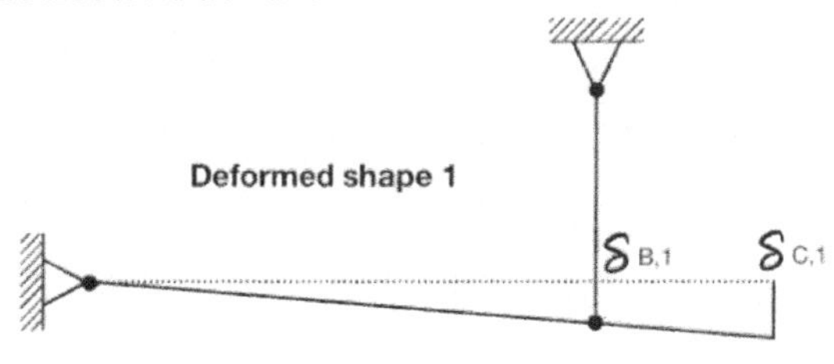

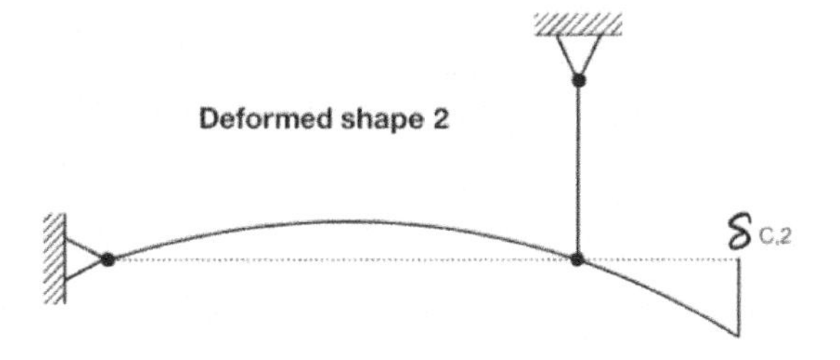

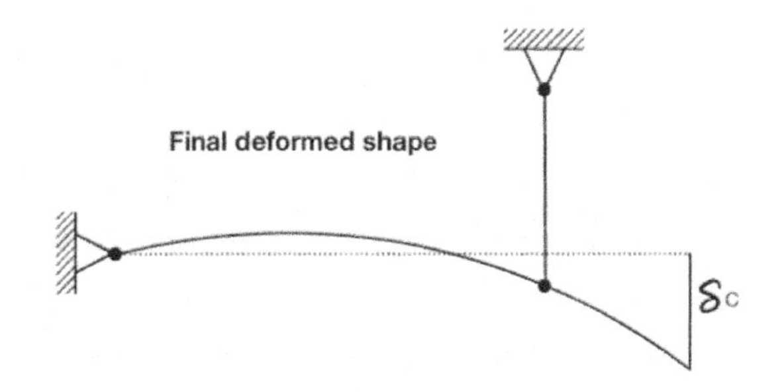

The final deformed shape is the combination of:

Deformed shape (1) + Deformed shape (2)

$$\delta_{C,1} \quad + \quad \delta_{C,2}$$

Deformed shape (1) is influenced by the elongation of the steel strand alone ignoring stresses or deformations in the W section.

Deformed shape (2) is the deflection of the beam alone ignoring the tension force in the steel strand.

$$\delta_{B,1} = \frac{PL}{AE}$$

$$= \frac{4.5\ kip \times 9\ ft \times \left(12 \frac{in}{ft}\right)}{0.217\ in^2 \times 29{,}000\ ksi}$$

$$= 0.077\ in$$

Using triangles from the first shape:

$$\delta_{C,1} = \frac{24 \times \delta_{B,1}}{16}$$

$$= 0.116\ in$$

Using the *NCEES handbook* Chapter 4 *Moment, Shear and Deflections*:

$$\delta_{C,2} = \frac{P\,a^2}{3\,EI}(L + a)$$

$$= \frac{3.0\ kip \times \left(8\ ft \times \frac{12\ in}{ft}\right)^2}{3 \times 29{,}000\ ksi \times 6{,}820\ in^4} \times \left(24\ ft \times \frac{12\ in}{ft}\right)$$

$$= 0.013\ in$$

$$\delta_C = \delta_{C,1} + \delta_{C,2}$$

$$= 0.116 + 0.013$$

$$= 0.129\ in$$

Correct Answer is (A)

SOLUTION 1.3

This is a 3D truss with three members, three supports with nine reactions and four joints. A test of determinacy is carried out to determine which method to be used to solve this problem.

A determinate stable 3D truss should satisfy the following:

$$reactions + members - 3 \times joints = 0$$

$$3 \times 3 \quad + \quad 3 \quad - \quad 3 \times 4 = 0$$

→ *determinate, stable*

Hence, truss can be analyzed using traditional methods.

One simplistic method is to take each plane and analyze it separately and then carry over reactions from one plane to another.

A more laborious method is to generate three equations per joint and have them balanced out.

Using the simplistic method, each plane will be looked at separately, and reactions from one plane will be carried over to the other.

The first plane to be dealt with is the XZ-plane as shown in the following figure. A pseudo support (AB') is created as shown, and the virtual member (AB')D shall carry the compression generated from the external downward load, then carried over to the XY-plane for further analysis.

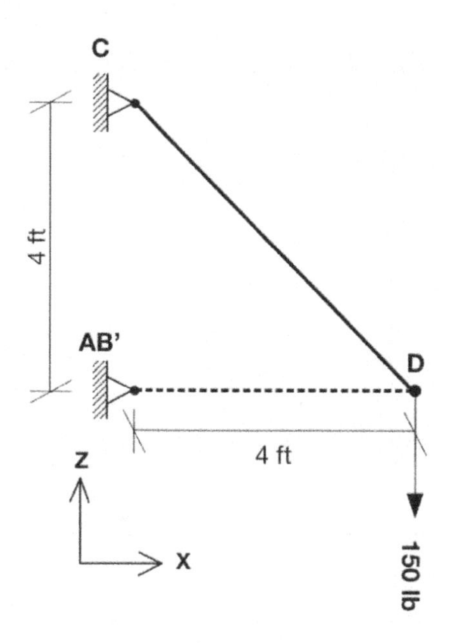

Balancing joint 'D':

$\sum Z = 0$

$CD \times \sin(45) - 150 = 0$

$CD = +212.13\ lb\ (Tension)$

$\sum X = 0$

$(AB')D - CD \times \cos(45) = 0$

$(AB')D = 150\ lb\ (Compression)$

This load will be carried over to the XY-plane as shown:

$\sum Y = 0$

$AD \times \sin(36.87) = BD \times \sin(36.87)$

$\rightarrow AD = BD$

$\sum X = 0$

$(-150) + 2 \times (AD \times \cos(36.87)) = 0$

$AD = BD = 93.75\ lb\ (compression)$

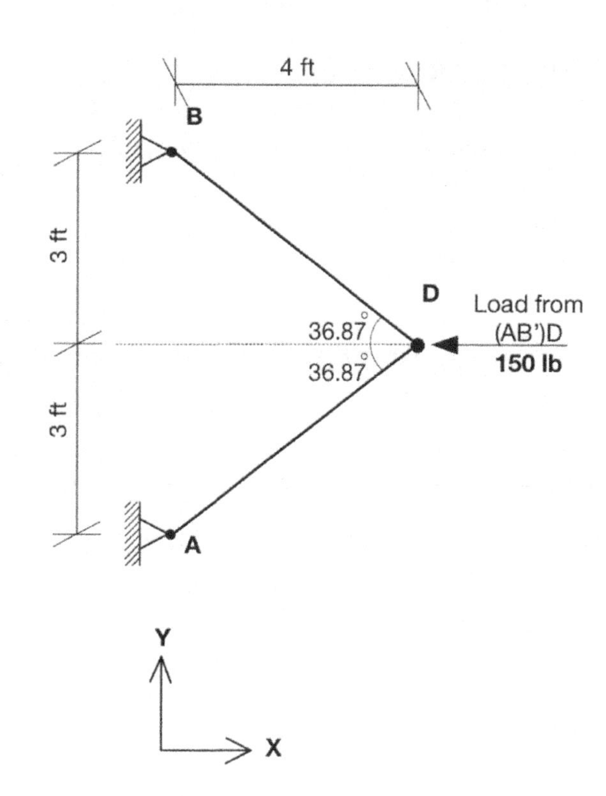

Force in AD is then transferred to support 'A' as follows:

$R_{A,x} = AD \times \cos(36.87) = 75\ lb \rightarrow$

$R_{A,y} = AD \times \sin(36.87) = 56.25\ lb \uparrow$

$R_{A,z} = 0$

Correct Answer is (B)

SOLUTION 1.4

This is a common problem and a common practice in projects when joists must move to create an opening in an adjacent slab, the new arrangement could create extra moment forces due the addition of equipment or other members.

To simplify the solution of this problem, and as the question is only looking for the differences between first case and second case maximums, joists that did not move will be ignored in the solution.

Before moving the load:
The original arrangement generates the following moment and moment diagram:

$$M_{max,before} = \frac{Pab}{L}$$

$$= \frac{5.5 \times 18 \times 12}{30}$$

$$= 39.6\ kip.ft$$

5.5 kip

18 ft 12 ft

36.9 kip.ft

Moment diagram due to load P @ x=18ft

Y
X

After moving the load:
Moving this load 3 ft to the left and adding a concentrated moment of 9 $kip.ft$ clockwise creates the following moments:

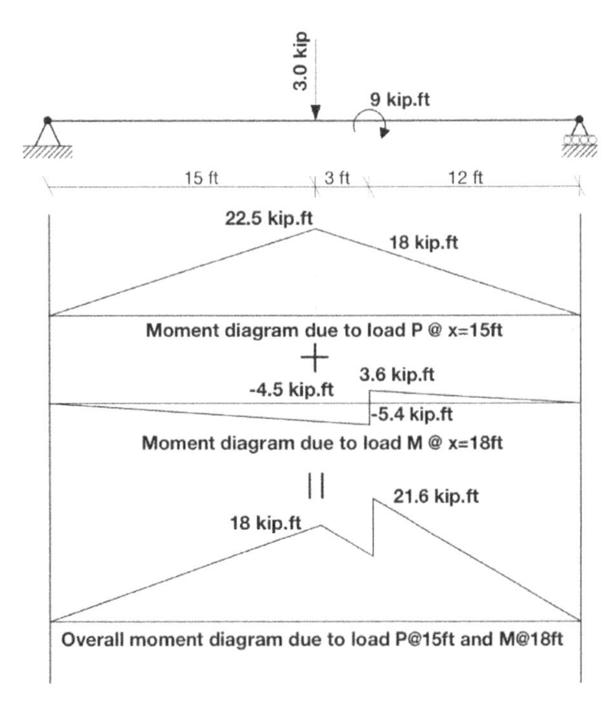

Moment due to concentrated load $P = 3.0\ kip$ alone:

$$M_{p,mid} = \frac{PL}{4}$$

$$= \frac{3 \times 30}{4}$$

$$= 22.5\ kip.ft\ @\ x = 15\ ft$$

$$M_{P@18ft} = \left(\frac{12}{15}\right) \times 22.5\ kip.ft$$

$$= 18\ kip.ft\ @\ x = 18\ ft$$

The concentrated 9 $kip.ft$ moment generates the following:

$$R_{left} = (9/30)\ lb \downarrow$$

$$M_{M,mid} = -\left(\frac{9}{30}\right) \times 15\ kip.ft$$

$$= -4.5\ kip.ft\ @\ x = 15\ ft$$

$$M_{M@18ft} = -\left(\frac{9}{30}\right) \times 18\ kip.ft$$

$$= -5.4\ kip.ft$$

Final (total) moments:

$$M_{mid} = M_{p,mid} + M_{M,mid}$$

$$= 22.5 - 4.5 = 18\ kip.ft$$

$$M_{@18ft} = M_{P@18ft} + M_{M@18ft} + M_{external\ @18ft}$$

$$= 18 - 5.4 + 9 = 21.6\ kip.ft$$

$$\rightarrow M_{max,after} = 21.6\ kip.ft$$

Differences in maximum moments:

$$M_{max,before} - M_{max,after} = 39.6 - 21.6$$

$$= 18\ kip.ft$$

$$(in\ reduction)$$

Correct Answer is (A)

SOLUTION 1.5

Fixed End Moments (FEMs) to be determined for member BC at 'C', and member CD at 'C', both deducted from each other taking sign into account.

Reference is made to the *NCEES handbook*.

Member BC is two fixed ends while CD is a one fixed end only and the other end is pinned.

$$FEM_{BC@C} = \frac{Pa^2b}{l^2} = \frac{5\ kip \times (6\ ft)^2 \times 4\ ft}{(10\ ft)^2}$$

$$= 7.2\ kip.ft\ \ Clockwise$$

$$FEM_{CD@C} = \frac{wl^2}{8} = \frac{0.15\ kip/ft \times (16\ ft)^2}{8}$$

$$= 4.8\ kip.ft\ \ Counter\ clock$$

<u>Unbalance moment at C:</u>

$$FEM_{BC@C} - FEM_{CD@C}$$

$$= 7.2\ kip.ft - 4.8\ kip.ft$$

$$= 2.4\ kip.ft\ \ Clockwise$$

Correct Answer is (D)

SOLUTION 1.6

In a similar fashion to a beam' influence diagram, an influence line can be constructed for trusses. Inclined truss members are represented by shear influence diagrams and horizontal members are represented by moment influence diagrams.

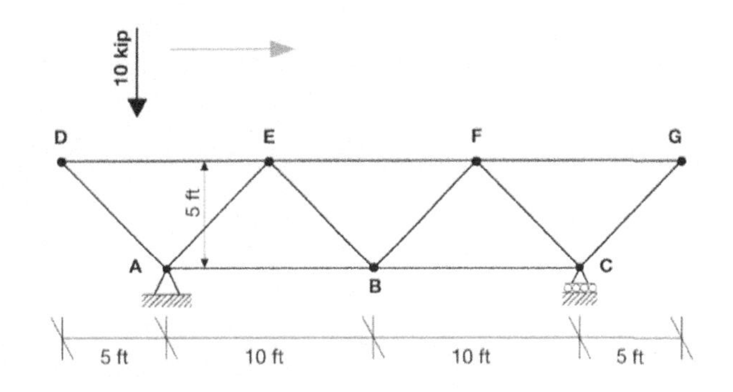

Joint 'E' is cut vertically throughout the entire section – imagining this truss as an overhanging beam that runs from 'D' to 'G' – the right beam section from the cut point 'E' is lifted upwards, and the left section is pushed downwards.

Section G - E should be parallel to beam section E' - D, and the gap created equals to 1.0 unit as shown in the following diagrams.

Because this is a truss, section D - E' - E is neglected, and the influence line travels from point 'D' to point 'E' as shown. The rest of the ordinates are determined through simple trigonometry.

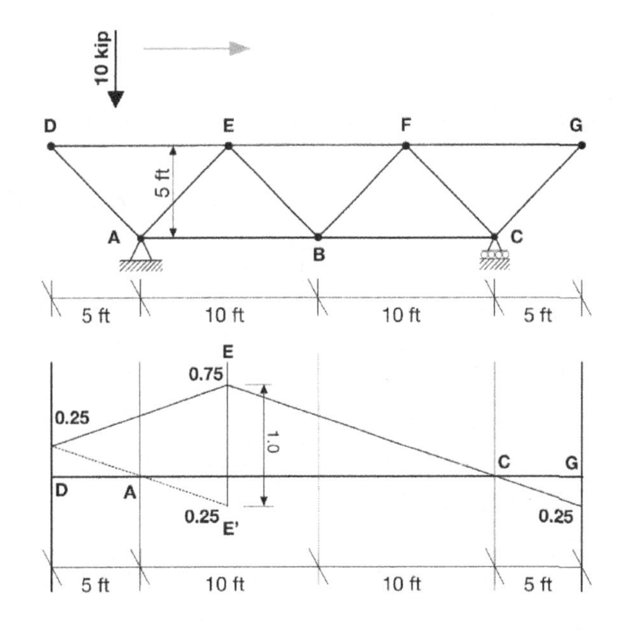

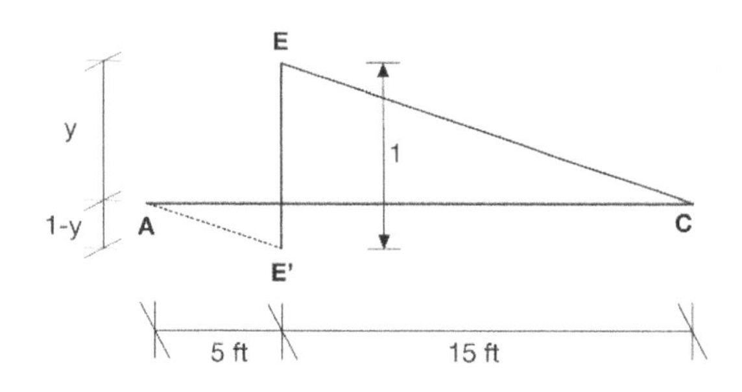

$\frac{y}{15} = \frac{1-y}{5}$ → $y = 0.75$

Therefore, maximum force in AE occurs when the point load is at joint 'E' and is determined by assuming a shear force that equals to $0.75 \times 10\ kip = 7.5\ kip$ is applied upwards throughout the entire section as shown below:

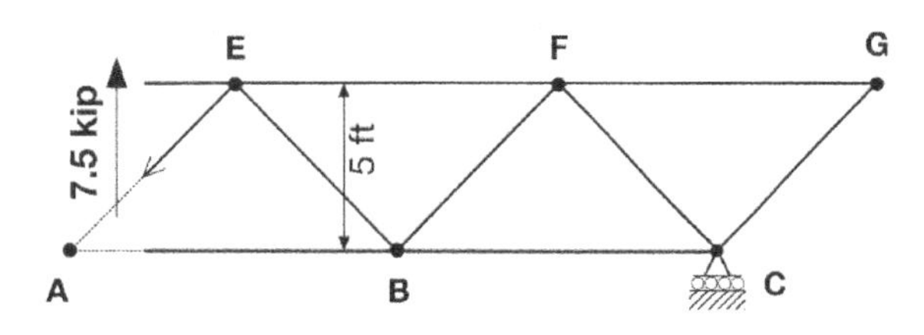

$AE \times \sin(45) = 7.5\ kip$

$\rightarrow AE = 10.6\ kip\ Compression$

Correct Answer is (A)

SOLUTION 1.7

A shear influence diagram is the fastest and most efficient method to determine both the position of wheels and the maximum shear generated from those wheels at point 'F'.

The shear influence diagram for a specific point on the beam can be drawn by cutting that section of the beam, pushing the right side section upwards from the cut section, and the left side downwards, keeping all internal hinges and supports in place. The generated deflected shape of this beam is the influence diagram being sought.

There are two considerations though: First, the vertical gap created at the point of interest, which is 'F' in this case, should equal to unity. Second, the left side beam portion (E - B - F) should be parallel to the right side one (F - C - G). The rest of the diagram ordinates can be determined using trigonometry.

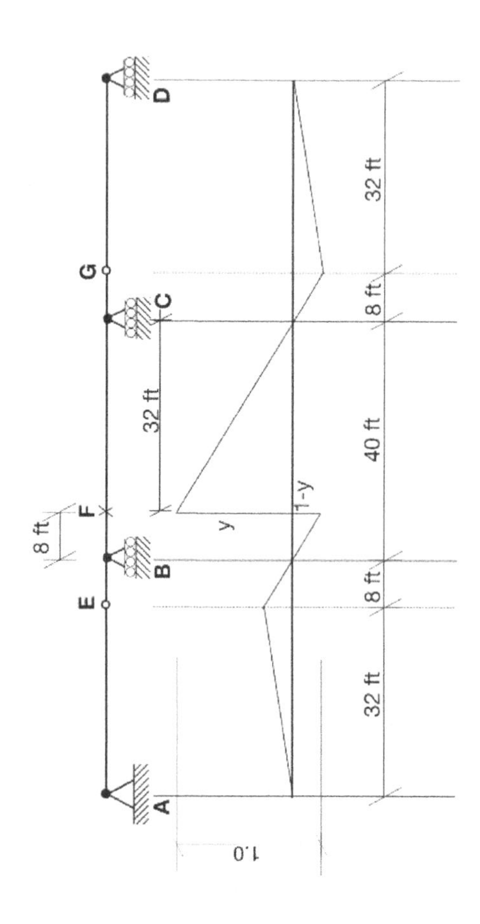

Trigonometry is used as follows to determine $'y'$:

$\frac{y}{32} = \frac{1-y}{8}$ → $y = 0.8$

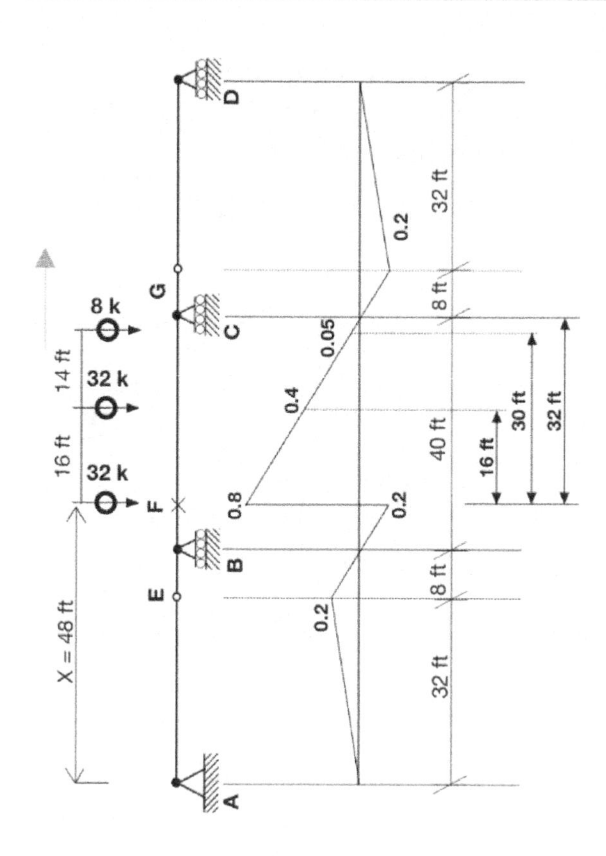

The location of the wheels that maximizes shear at 'F' occurs when they are placed at the highest positive ordinates of the diagram. Each of those ordinates is then multiplied by the load above to determine the resultant maximum shear as follows:

$$V_{max} = 0.8 \times 32 + 0.4 \times 32 + 0.05 \times 8$$

$$= 38.8\ kip\ @\ x = 48\ ft$$

Correct Answer is (D)

SOLUTION 1.8

The unit load method is used to solve this question. The unit load method is briefed in the *NCEES handbook*.

Deflection per member due to the applied external load must be determined first. A unit load is then applied to the direction of the wanted displacement and the following equation applies:

$$\Delta_{joint} = \sum f_i(\delta)_i$$

f_i is the force in each member due to unit load applied to the direction of the wanted displacement.

δ is the deflection per member from the applied external load which has been determined $= \frac{PL}{AE}$

Internal force P can be determined as follows:

$P_{BE} = P_{AE}$ both are in tension

$P_{DE} = P_{CE}$ both are in compression

Force per member is equivalent to the resultant of its hinged support:

$$P_{BE} = \sqrt{R^2_{(B\ or\ A),x} + R^2_{(B\ or\ A),y} + R^2_{(B\ or\ A),z}}$$

$$P_{AE} = P_{BE}$$

$$P_{DE} = \sqrt{R^2_{(D\ or\ C),x} + R^2_{(D\ or\ C),y} + R^2_{(D\ or\ C),z}}$$

$$P_{CE} = P_{DE}$$

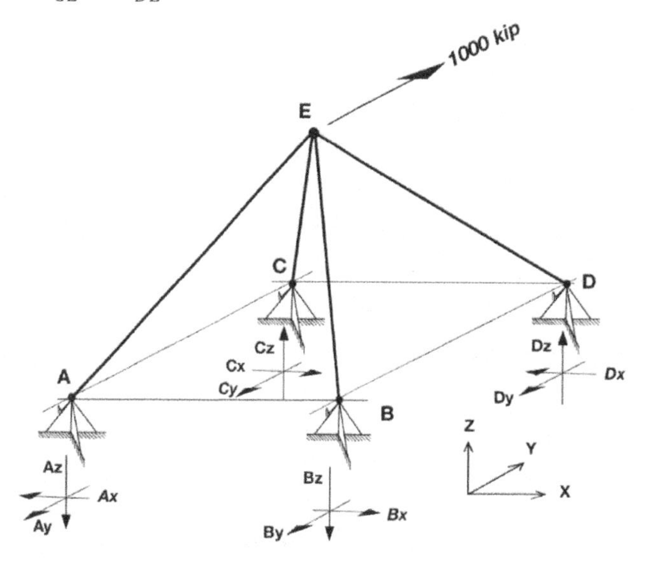

Moment sum is taken around each axis do determine all reactions:

$$\sum M_{D\&C\ about\ 'x'} = 0$$

$$6 \times 1{,}000 = 8 \times B_Z + 8 \times A_Z$$

$$Also\ \ B_Z = A_Z$$

$B_Z = A_Z = 375\ kip \downarrow$

$\sum M_{A\&B\ about\ 'x'} = 0$

$6 \times 1{,}000 = 8 \times D_Z + 8 \times C_Z$

$Also\ \ D_Z = C_Z$

$D_Z = C_Z = 375\ kip \uparrow$

Other (horizontal) reactions such as *By* and *Bx* are determined by analyzing each member separately. Below is an example on member *BE*.

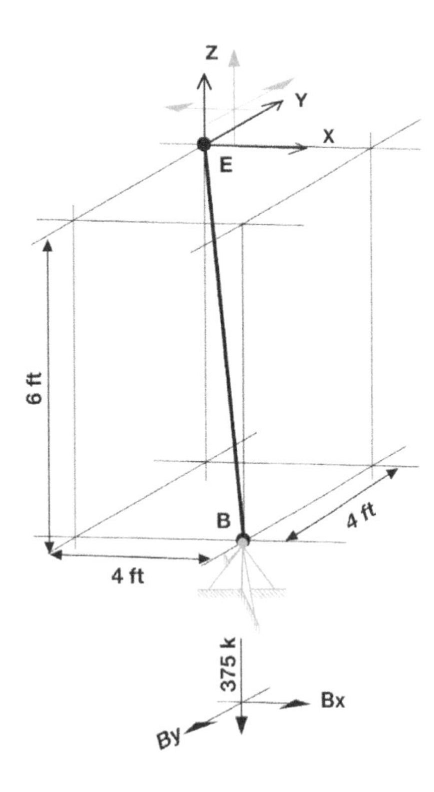

$\sum M_{E\ about\ 'x'} = 0$

$6 \times B_Y = 4 \times 375$

$\rightarrow B_Y = 250\ kip\ \ \ backwards$

$\sum M_{E\ about\ 'y'} = 0$

$6 \times B_X = 4 \times 375$

$\rightarrow B_X = 250\ kip\ \ \ to\ the\ right$

The rest of the reactions are identical in value and opposite in direction and are represented on the below figure:

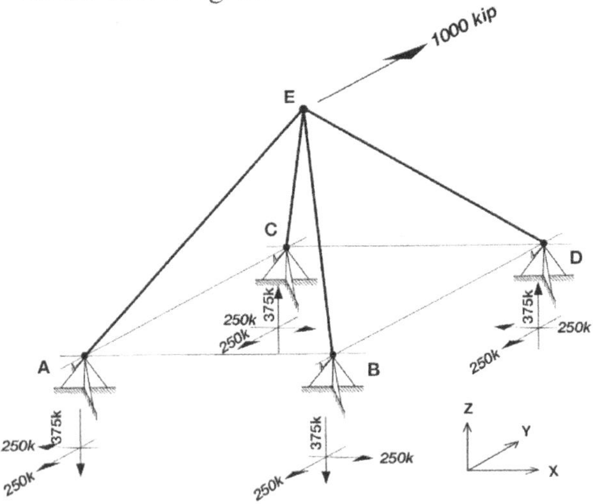

The two tension members:

$$P_{BE} = \sqrt{R^2_{(B\ or\ A),x} + R^2_{(B\ or\ A),y} + R^2_{(B\ or\ A),z}}$$

$$= \sqrt{250^2 + 250^2 + 375^2}$$

$$= 515\ kip\ \ \ Tension$$

$$\delta_{AE} = \delta_{BE} = \frac{P_{AE\ or\ BE} \times L_{AE\ or\ BE}}{AE}$$

$$= \frac{+515\ kip \times 8.25\ ft \times (12\frac{in}{ft})}{187 \times 10^3\ kip}$$

$$= +0.27\ in\ \ \ Longer$$

Applying a unit in the same external load direction:

$f_{BE} = f_{AE} = +0.515\ \ \ Tension$

The two compression members:

$$P_{DE} = \sqrt{R^2_{(D\ or\ C),x} + R^2_{(D\ or\ C),y} + R^2_{(D\ or\ C),z}}$$

$$= \sqrt{250^2 + 250^2 + 375^2}$$

$$= -515\ kip\ \ \ Compression$$

$$\delta_{DE} = \delta_{CE} = \frac{P_{DE\ or\ CE} \times L_{DE\ or\ CE}}{AE}$$

$$= \frac{-515\ kip \times 8.25\ ft \times (12\frac{in}{ft})}{187\times10^3\ kip}$$

$$= -0.27\ in \quad Shorter$$

Applying a unit in the same external load direction:

$$f_{DE} = f_{CE} = -0.515 \quad Compression$$

Apply the unit load method equation:

$$\Delta_{joint} = \sum f_i(\delta)_i$$

$$\Delta_{E,y} = 2 \times \begin{bmatrix} (f_{(A\ or\ B)E} \cdot \delta_{(A\ or\ B)E}) \\ + \\ (f_{(C\ or\ D)E} \cdot \delta_{(C\ or\ D)E}) \end{bmatrix}$$

$$= 2 \times \begin{bmatrix} (0.515 \times 0.27) \\ + \\ (-0.515 \times -0.27) \end{bmatrix}$$

$$= 0.56\ in$$

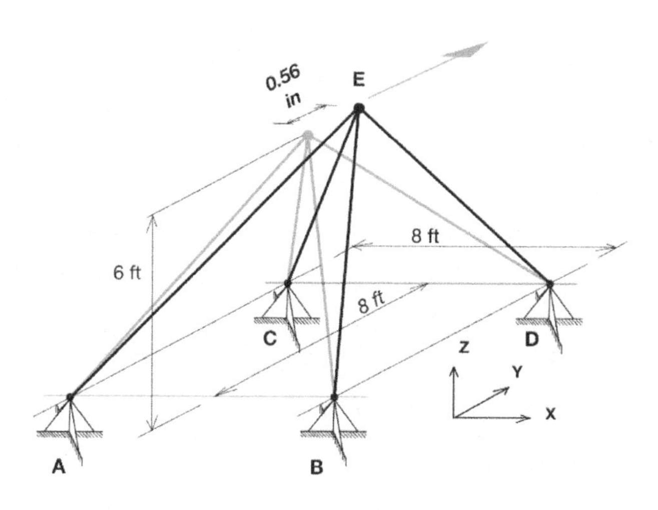

Correct Answer is (D)

SOLUTION 1.9

Member AB alone is a fixed beam and its horizonal deflection at 'B' is as follows:

$$\delta_{B,Horizonatl} = \frac{Pl^3}{3EI}$$

$$= \frac{20\ kip \times \left(10\ ft \times 12\frac{in}{ft}\right)^3}{3 \times 29{,}000\ ksi \times 1{,}000\ in^4}$$

$$= 0.40\ in$$

Horizontal displacement at 'B' is carried over to vertical and horizontal displacements at 'C' using trigonometry upon determining the slope at 'B', and this is possible because joint at 'B' is rigid and there are no loads on the BC portion of the frame.

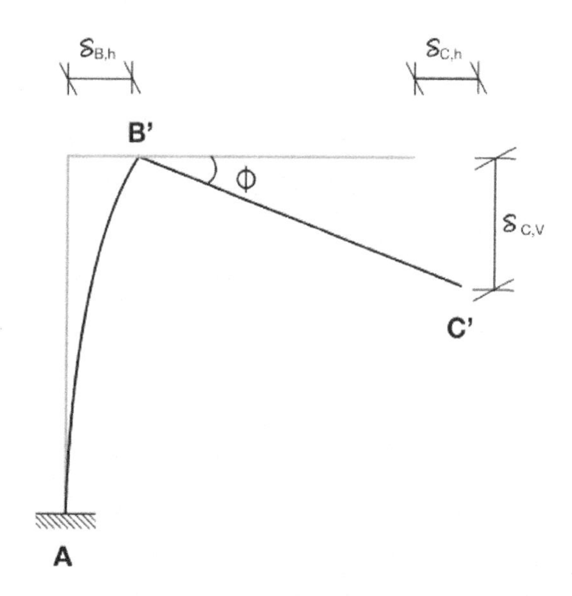

Slope at 'B' can be determined using the conjugate beam method (*). When loading the conjugate beam with the (M/EI) diagram, the slope at 'B' is the shear resultant of the (M/EI) loading at support B.

PART II Structural Depth
Section 1 Structural Analysis

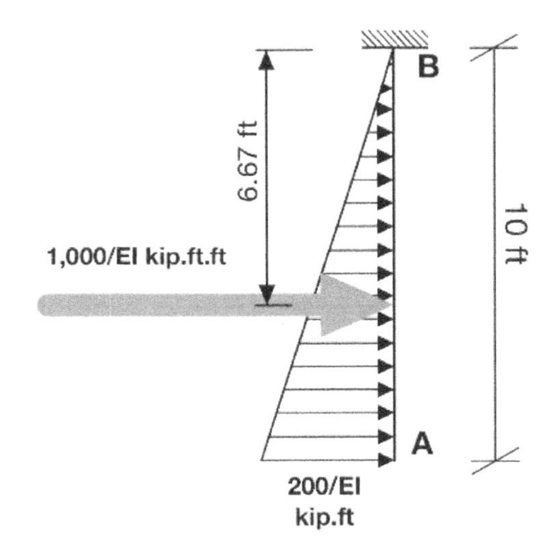

$$\emptyset_B = area\ under\ \frac{M}{EI}$$

$$= \frac{½ \times [\,200\ kip.ft \times (12in/ft) \times 10\ ft \times (12in/ft)]}{29{,}000\ ksi \times 1{,}000\ in^4}$$

$$= 5.0 \times 10^{-3}\ rad$$

Deflection can be calculated using the same method, and deflection in this case equals to the moment at support 'B' divided by EI:

$$\delta_{B,hor} = \frac{1{,}000\ kip.ft.ft \times 144 \frac{in^2}{ft^2} \times 6.67\ ft \times 12 \frac{in}{ft}}{29{,}000\ ksi \times 1{,}000\ in^4}$$

$$= 0.40\ in$$

Using trigonometry as follows:

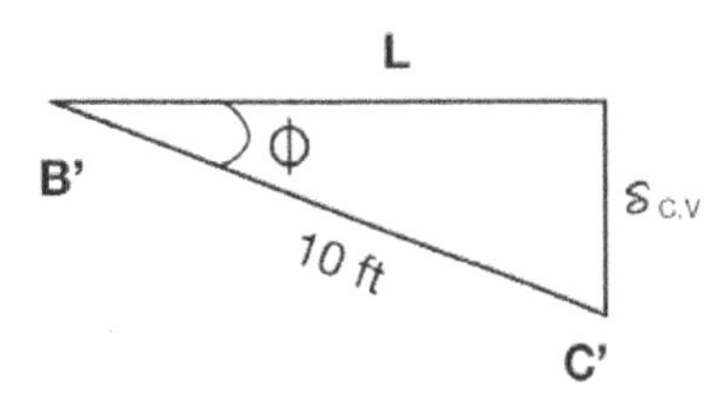

$$\delta_{C,Ver} = 10\ ft \times 12 \frac{in}{ft} \times sin(5.0 \times 10^{-3})$$

$$= 0.6\ in$$

Correct Answer is (C)

* Conjugate beam method can only be used with beams and cannot be used for frames

SOLUTION 1.10

Flexural deflection at AB is determined separate from torsional deflection at AB and both effects are added together at point 'C' assuming all stresses are acting within the elastic range.

Due to its short length, flexural deflection at member BC can be neglected.

Member properties for the *7 in* Dia steel pipe are as follows:

$$I = \frac{\pi(r_o^4 - r_i^4)}{4} = \frac{\pi \times (3.5^4 - 3.125^4)}{4} = 43\ in^4$$

$$J = \frac{\pi(r_o^4 - r_i^4)}{2} = \frac{\pi \times (3.5^4 - 3.125^4)}{2} = 86\ in^4$$

$$E_{steel} = 29{,}000\ ksi$$

$$G_{steel} = 11{,}200\ ksi$$

$$\delta_{B,Flexure} = \frac{Pl^3}{3EI}$$

$$= \frac{0.5\ kip \times (12\ ft \times 12\ in/ft)^3}{3 \times 29{,}000\ ksi \times 43\ in^4}$$

$$= 0.4\ \text{in}$$

$$Torsion\ (T) = 0.5\ kip \times 4\ ft = 2.0\ kip.ft$$

$$\emptyset_B = \frac{TL}{GJ} = \frac{2.0\ kip.ft \times 12\ in/ft \times (12\ ft \times 12\ in/ft)}{11{,}200\ ksi \times 86\ in^4}$$

$$= 3.6 \times 10^{-3}\ rad$$

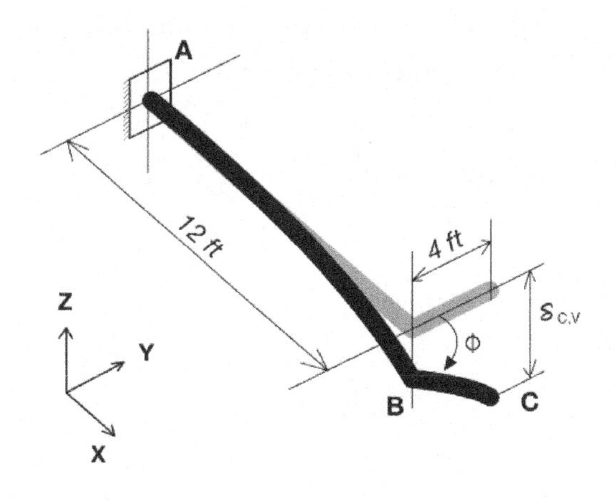

$$\delta_{C,Tor.} = 4\,ft \times 12\frac{in}{ft} \times sin(0.0036\,rad)$$

$$= 0.17\,in$$

$$\delta_{C,Flexure} = \frac{Pl^3}{3EI}$$

$$= \frac{0.5\,kip \times \left(4\,ft \times 12\frac{in}{ft}\right)^3}{3 \times 29{,}000\,ksi \times 43\,in^4} = 0.01\,in$$

$$\delta_C = \delta_{C,Flexure} + \delta_{C,Torsion} + \delta_{B,Flexure}$$

$$= 0.01 + 0.17 + 0.4 = 0.58\,in$$

Correct Answer is (B)

SOLUTION 1.11

The following suggested steps can be followed to visually construct a moment diagram:

Step 1: Determine external reactions:

The structure is indeterminate because it has four unknown reactions and only three equilibrium equations are available to solve them. Because of this, it is sufficient to understand the directions of those reactions which are as follows:

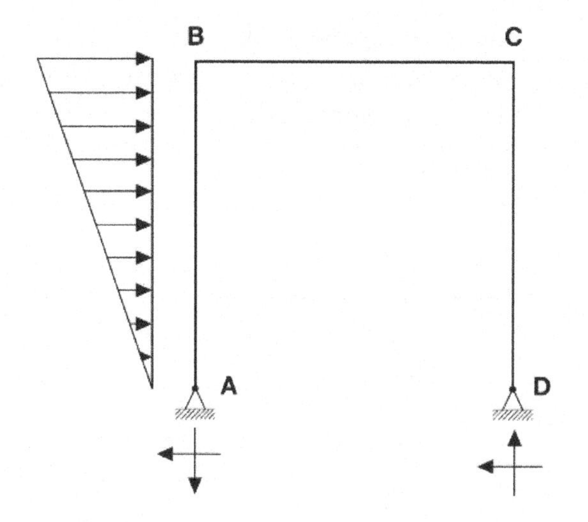

Step 2: Determine members' internal forces:

Dismantle all members and understand the directions of internal forces acting at member ends.

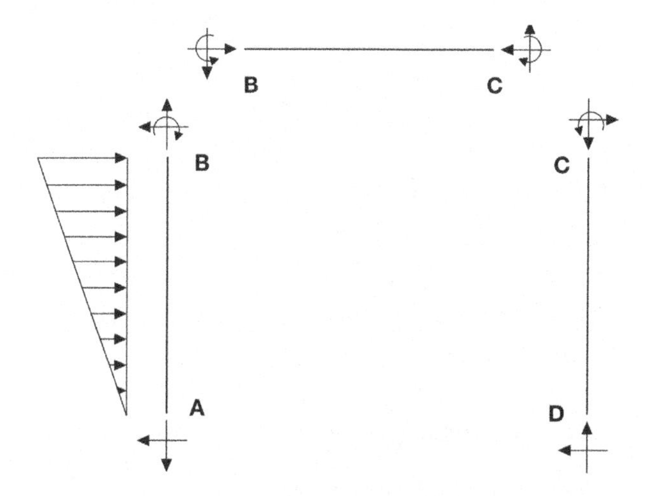

Step 3: Establish the shear diagram:

At this stage the requested moment diagram can be computed, however, it is always a good strategy to start with the shear diagram first to:

- Determine the location of maximum moment as it occurs at zero shear locations.
- Determine whether there is a point of inflection in the moment diagram as it occurs when shear changes sign.
- Determine the shape function of the moment diagram. i.e., the moment

diagram represents the area under the shear diagram and its shape function is the integral of that of the shear diagram.

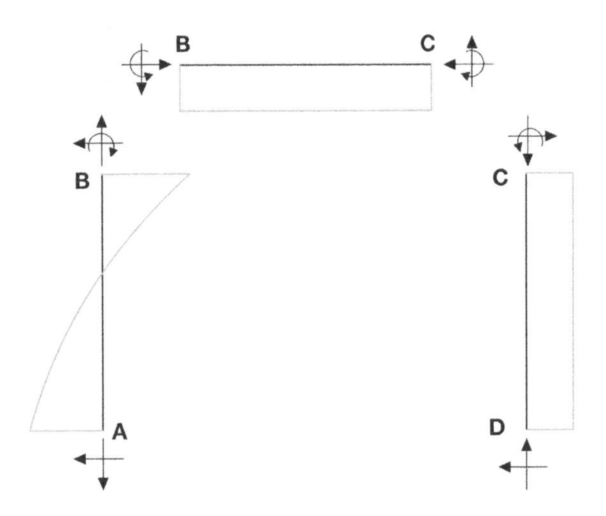

It is also important to observe that the shear diagram for member AB has a parabolic shape because the shear diagram represents the area under the load diagram, this indicates that its shape function is the integral of that of the load diagram. The load diagram in this case is represented by a linear equation (first degree equation), consequently, the shear diagram would be represented using a second-degree equation (parabolic function).

Step 4: Establish the moment diagram:
The moment diagram is the integral of the shear function. In which case:

- Member AB: the shear diagram has a parabolic shape; moment diagram in this case would best be represented by a cubic equation/function. Moreover, maximum moment occurs at the location where shear equals to zero.
- Members BC & CD: Shear diagram is represented by a constant function; hence, moment diagram would be represented by a first-degree function/linear equation.

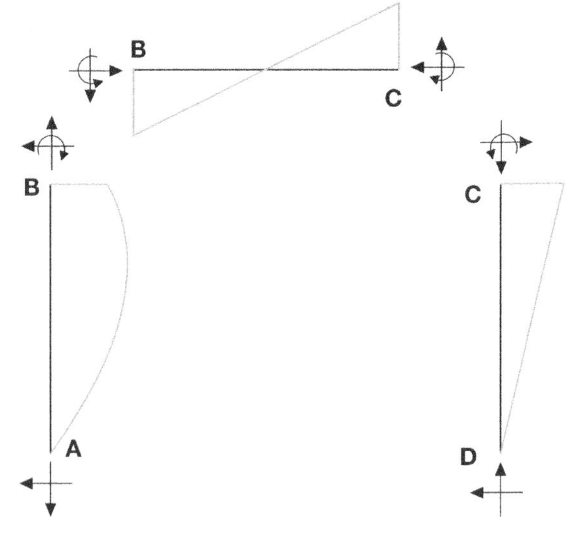

Finally:

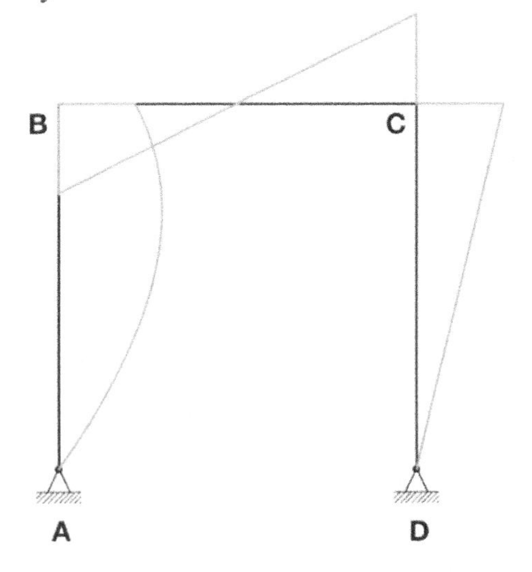

Correct Answer is (A)

<u>Question discussion:</u>
One may think that diagram B could be a good representation for the requested moment diagram given that joint B of diagram B has an internal moment that is similar to diagram A, and joint A has no moment due to the hinged support.

For the removal of doubt, it is best to draw the deflected shape and detect if there could be a point of inflection in member AB – i.e., double curvature. A double curvature in the deflected shape indicates that moment is changing signs. Also, moment equals zero at the point of inflection of the deflected shape.

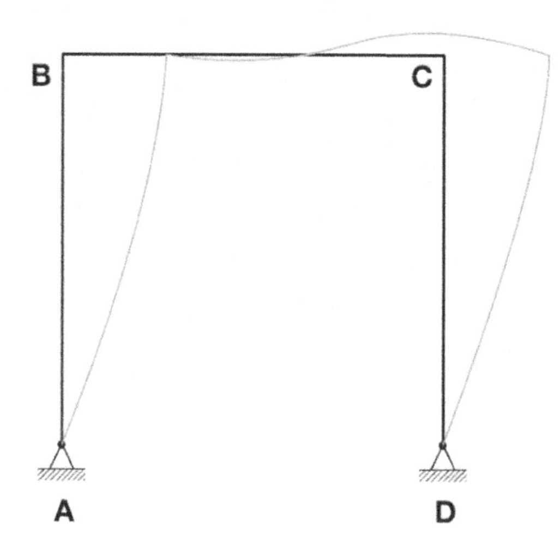

Given that joint A is a hinged support, member AB shall rotate freely around this joint. Joint B on the other hand is not restrained from horizontal movement.

Those two conditions shall prevent the formation of a double curvature in member AB and further confirms that moment in this member will not change signs.

The above argument verifies that diagram A is the best representation for the requested moment diagram.

SOLUTION 1.12

Moment distribution method is used to solve this question. Also, Chapter 4, Structural, of the *NCEEES Handbook* is referred to determine carryover and stiffness factors '*k*'.

Determine moment at joint C as follows:

$$M_C = 10\ kip \times 4\ ft = 40\ kip.ft$$

Beam Stiffness is calculated as follows:

$$k = \frac{4EI}{L}$$

$$k_{CA} = \frac{4EI}{10} = 0.4\ EI$$

$$k_{CB} = \frac{4EI}{8} = 0.5\ EI$$

$$k_{CD} = \frac{4EI}{8} = 0.5\ EI$$

Distribution Factor is calculated as follows:

$$DF = \frac{k}{\sum k}$$

$$DF_{CA} = \frac{0.4\ EI}{1.4\ EI} = 0.28$$

$$DF_{CB} = \frac{0.5\ EI}{1.4\ EI} = 0.36$$

$$DF_{CD} = \frac{0.5\ EI}{1.4\ EI} = 0.36$$

Distribution factors for fixed ends are zeros as they receive moments (as carryovers) and they keep it – i.e., these moments are not redistributed after being carried over:

$$DF_{DC} = 0$$

$$DF_{BC} = 0$$

Distribute moments at 'C' using the moment distribution factors as follows:

$$M_{CD} = 40\ kip.ft \times 0.36 = 14.4\ kip.ft$$

$$M_{CB} = 40\ kip.ft \times 0.36 = 14.4\ kip.ft$$

$$M_{CA} = 40\ kip.ft \times 0.28 = 11.2\ kip.ft$$

Half the moment at member CD is carried over from joint C to joint D as follows:

$$M_{DC} = 0.5 \times M_{CD} = 7.2\ kip.ft$$

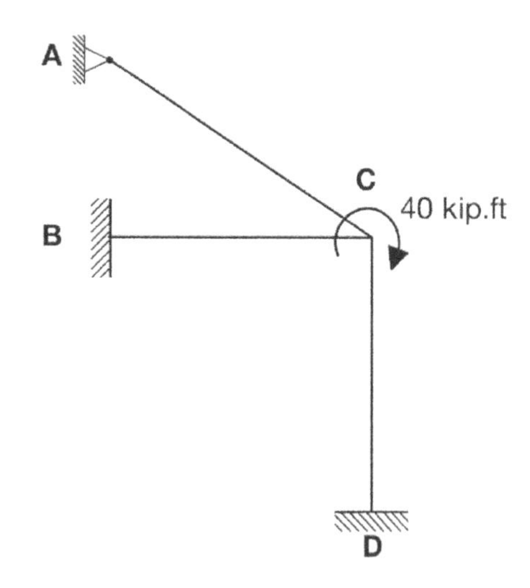

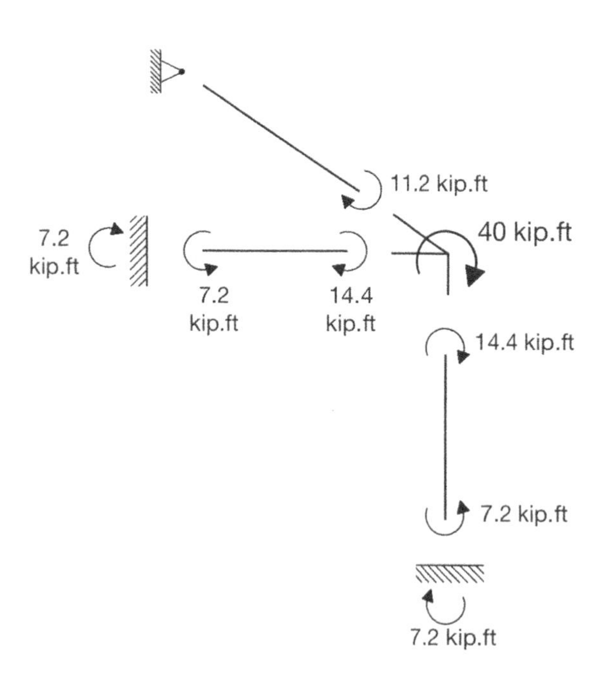

Correct Answer is (A)

SECTION 2

Cross Section Analysis

Problems & Solutions

PROBLEM 2.1 ***C Section Internal Stresses***
The below cross section is for a simply supported beam spanning 10 ft. A 1.5 kip point load is applied at the centerline of the section's centroidal y-axis as shown:

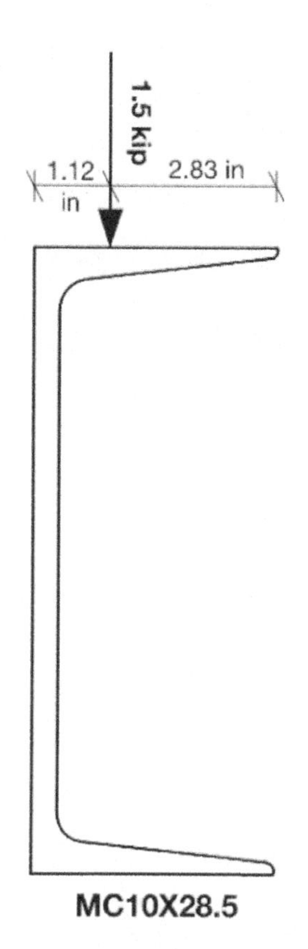

Ignoring the beam's self-weight, internal forces experienced at section mid-span are:

(A) Moment = 3.75 $kip.ft$
Shear = 0.75 kip
Torsion = 0.29 $kip.ft$

(B) Moment = 3.75 $kip.ft$
Shear = 0.75 kip
Torsion = 0

(C) Moment = 3.75 $kip.ft$
Shear = 0.75 kip
Torsion = 0.14 $kip.ft$

(D) Moment = 3.75 $kip.ft$
Shear = 1.5 kip
Torsion = 0.14 $kip.ft$

PROBLEM 2.2 ***Shear Center***
Of the following, only two statements describing shear center of a section are true:

(A) Shear center for a double angle steel section coincides with its centroid.

(B) Shear center of a shape coincides with its center of gravity.

(C) Shear center of a section is where moment and torsion are decoupled.

(D) Only when lateral loads and transverse loads passes through the shear center simultaneously, member will be subjected to moment with no torsion.

(E) Forces passing through a shape's centroid do not generate twisting forces.

PROBLEM 2.3 ***Section Unsymmetrical Loading***
The below is a vertical rectangular 18 in × 12 in section subjected to the below:

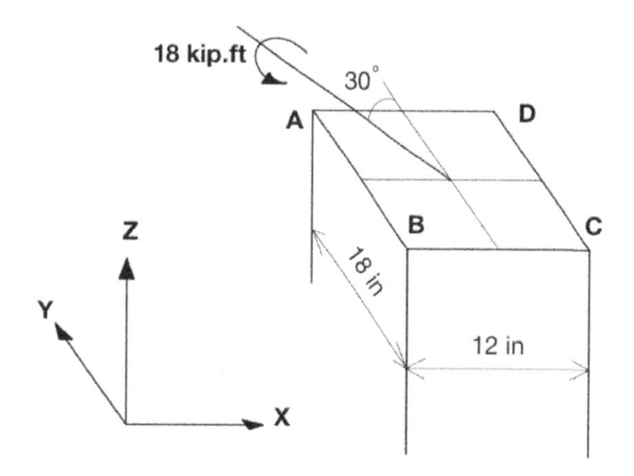

Stress at edge 'B' and 'D' due to this unsymmetric loading in ksi is most nearly:

(A) B = 0.60 (compression)
D = 0.60 (tension)

(B) B = 0.26 (compression)
D = 0.60 (tension)

(C) B = 0.60 (compression)
D = 0.26 (tension)

(D) B = 0.43 (compression)
D = 0.17 (tension)

PROBLEM 2.4 ***Shear Stress Calculation***

The ultimate shear load applied to this section is 200 kip.

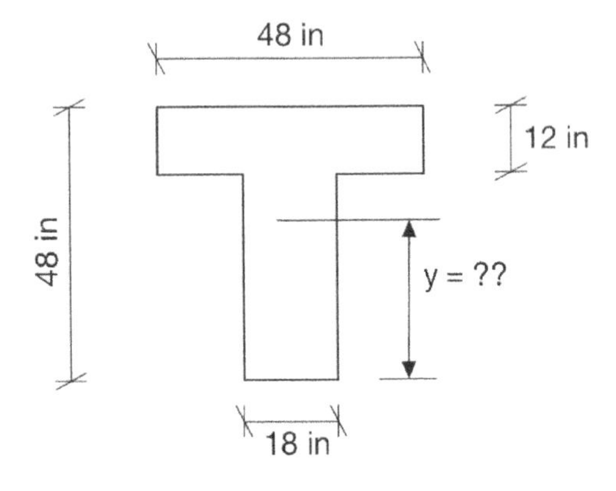

The distance $'y'$ to the maximum shear stress along with the maximum shear stress are as follows:

$y =$ ______ in

$\tau_{\max} =$ ______ psi

PROBLEM 2.5 ***Maximizing Moment of Inertia***

The below section located at the top of the figure is a fictitious cross section of a building's footprint constructed with connected shear walls.

The stiffness properties for the entire plan cross-section for the cross-section at the top are:

$I_x = 16{,}500\ ft^4$

$I_y = 14{,}750\ ft^4$

$I_{xy} = 7{,}775\ ft^4$

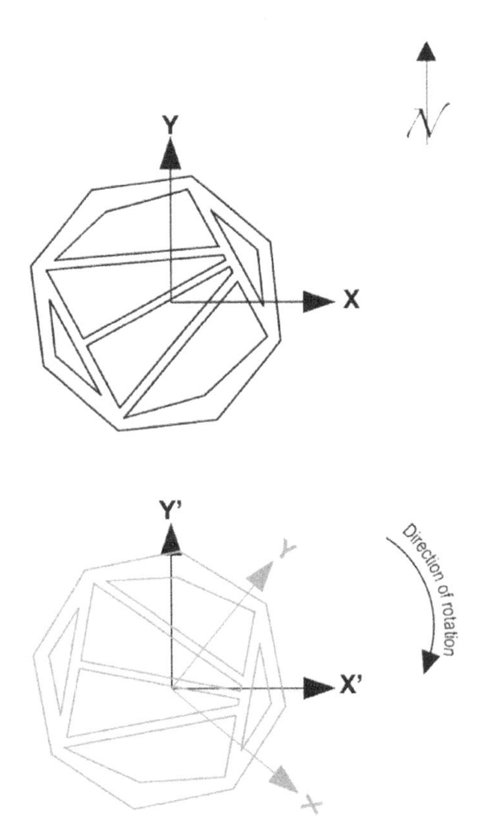

The clockwise orientation in degrees that produces the largest flexural stiffness for the building to resist a lateral load on the East-West direction, along with the resultant maximum moment of inertia for this orientation about the North direction in ft^4 are:

(A) Orientation = 42^0
$I_{max} = 23{,}449\ ft^4$

(B) Orientation = 84^0
$I_{max} = 23{,}449\ ft^4$

(C) Orientation = 42^0
$I_{max} = 31{,}250\ ft^4$

(D) Orientation = 84^0
$I_{max} = 31{,}250\ ft^4$

PROBLEM 2.6 ***Composite Prestressed Beam Analysis***
The below section is for a girder that supports a short bridge. The girder is an AASHTO type I section with properties noted below.

Top slab is an in-situ concrete slab with effective width $b_e = 3\ ft$ and $f_c' = 3{,}500\ psi$.

The slab and girder are made of normal weight concrete.

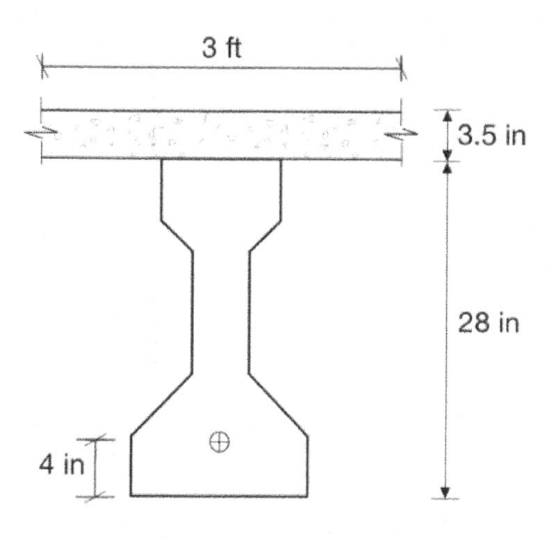

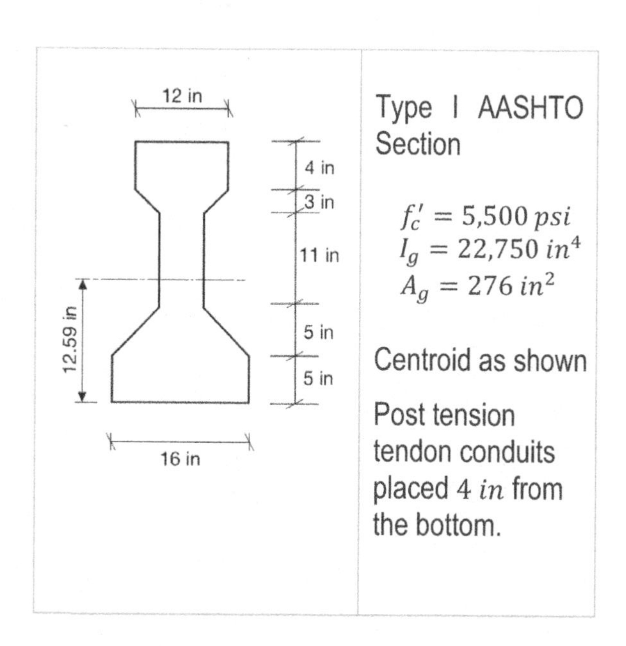

The top slab and the girder act as one unit before applying the prestressing tension force.

Ignoring self-weight, stress at the topmost of the concrete slab after applying $150\ kip$ pretension force at the tendons' location is most nearly:

(A) $0.20\ ksi$

(B) $0.50\ ksi$

(C) $0.60\ ksi$

(D) $0.70\ ksi$

PROBLEM 2.7 ***Moment of Inertia Calculation***

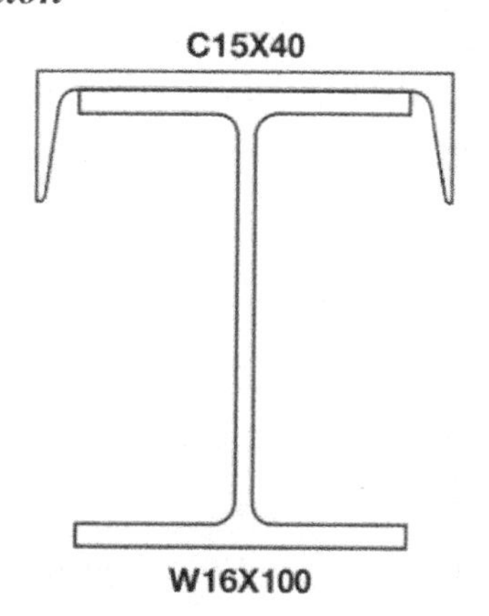

The moment of inertia in in^4 of the strong access for the above two sections, assuming the two sections are riveted properly and act as one, is most nearly:

(A) 1,500

(B) 2,100

(C) 2,500

(D) 1,980

PROBLEM 2.8 ***Shear Flow Determination***
The below is a $24\ ft$ long $W30 \times 292$ steel beam subjected to a service total shear of $50\ kip$. The beam section is strengthened with a $20\ in \times 1.5\ in$ steel plate bolted at its

top flange at two rows as shown below. Bolts in the figure are not to scale.

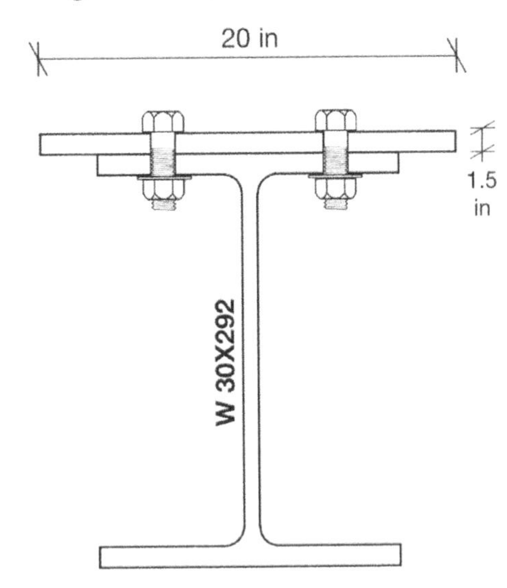

Using $A307$ $5/8\ in$ diameter bolts and considering ASD method, the minimum number of bolts requires per row is:

(A) 11

(B) 58

(C) 31

(D) 22

PROBLEM 2.9 *Shear Flow Profile*

The below section is subjected to a shear load.

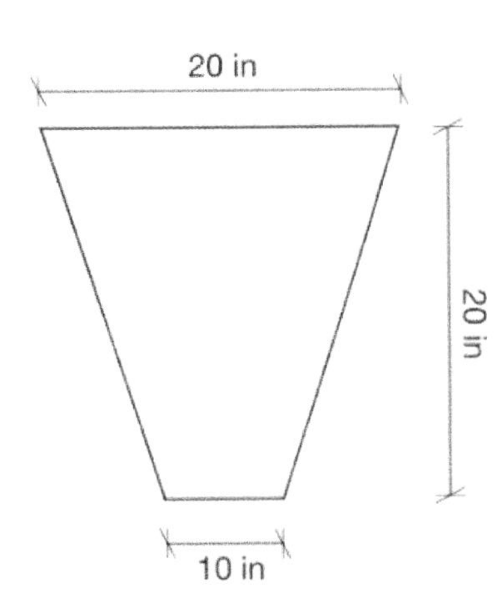

Shear flow profile for this cross-section is best represented by:

(A) Profile A

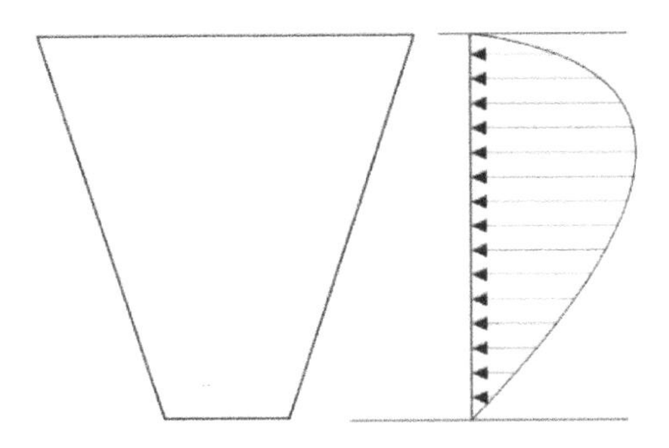

(B) Profile B

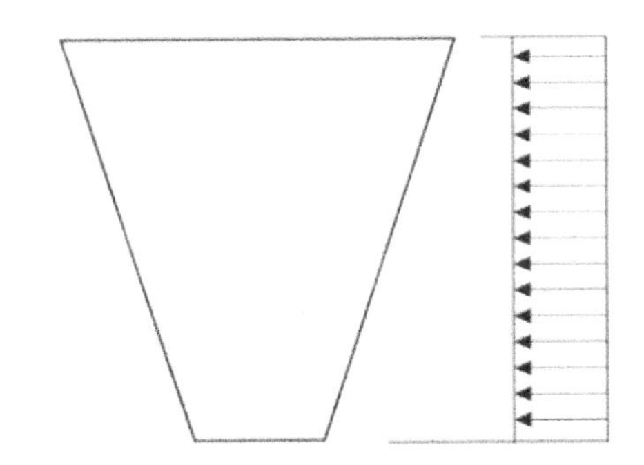

(C) Profile C

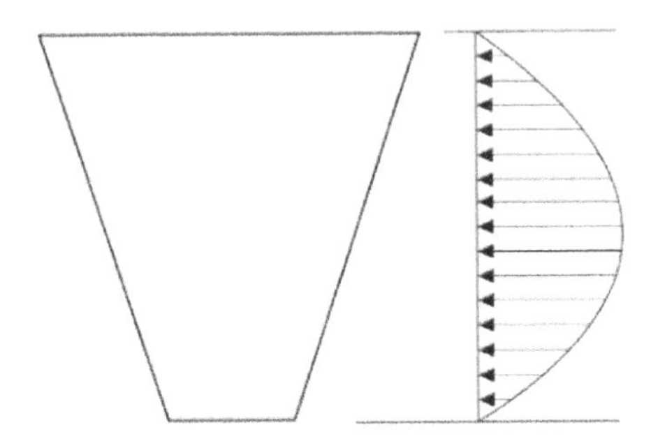

(D) Profile D

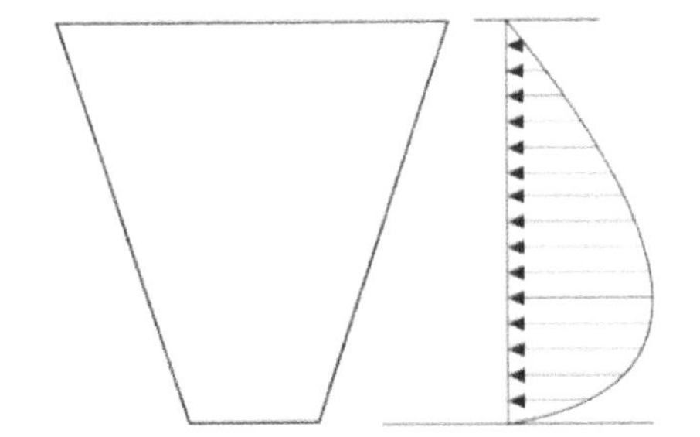

PROBLEM 2.10 ***Shear Stress Profile***
The below section is subjected to a shear load.

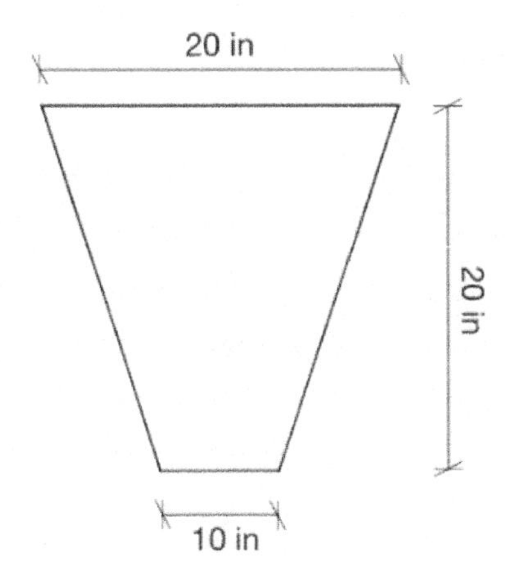

Based on this load, the profile that best represents the shear stress profile across this section is:

(A) Profile A

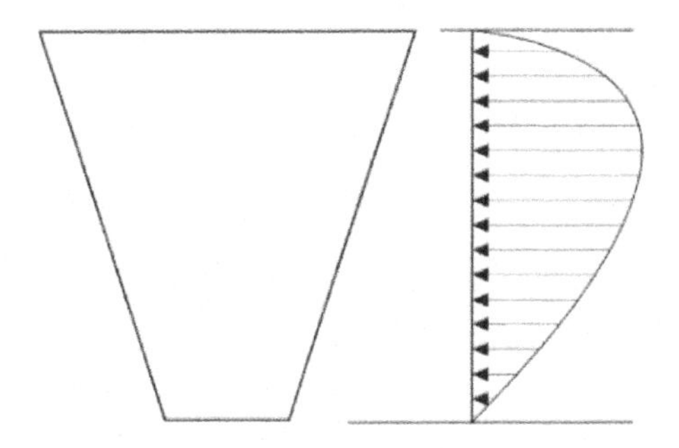

(B) Profile B

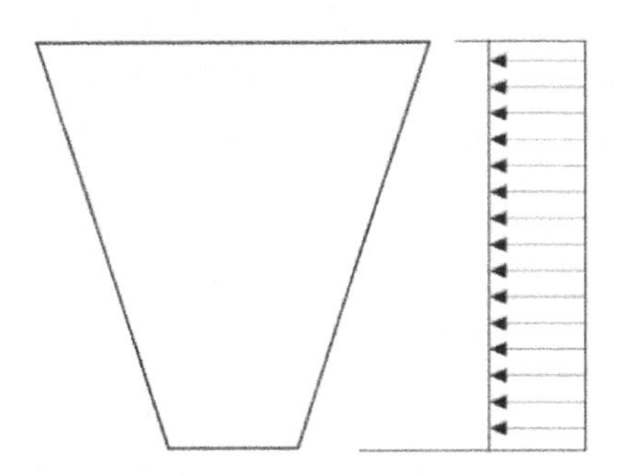

(C) Profile C

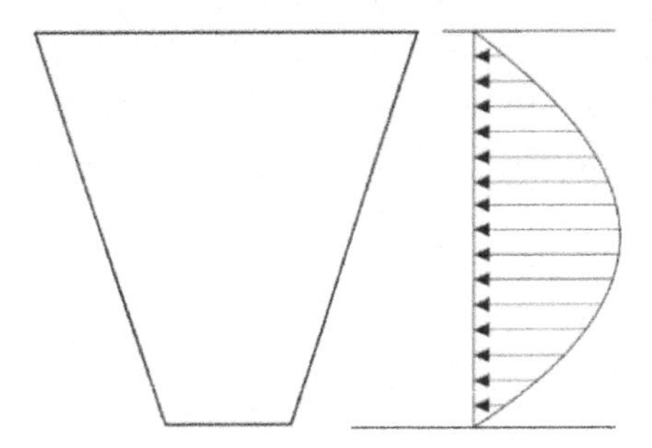

(D) Profile D

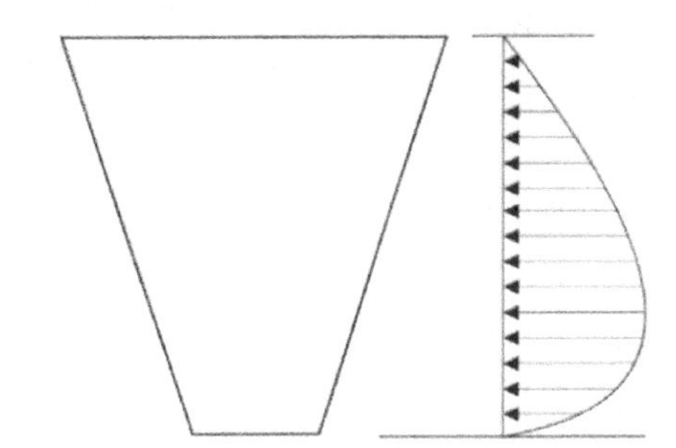

SOLUTION 2.1

From the *AISC Steel Construction Manual* the shear center for $MC10 \times 28.5$ is located 1.21 in from the left side of the section's web as shown below.

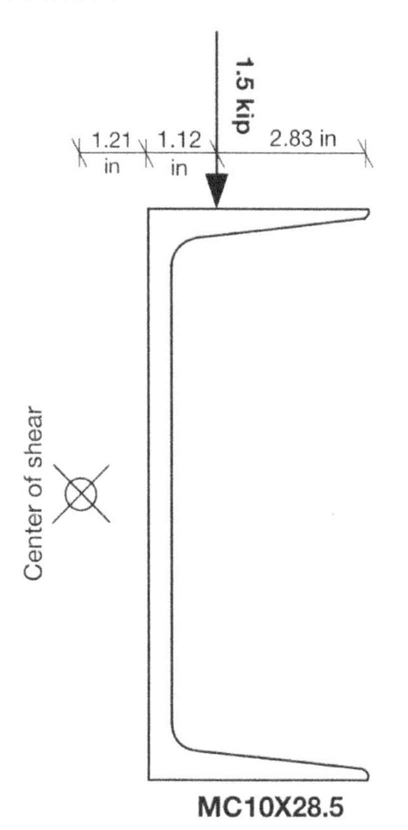

$$T_{mid} = 1.5\ kip \times (1.12 + 1.21)in \times \frac{1ft}{12\ in}$$
$$= 0.29\ kip.ft$$

$$M_{mid} = \frac{PL}{4} = \frac{1.5\ kip \times 10\ ft}{4} = 3.75\ kip.ft$$

$$V_{mid} = \frac{P}{2} = \frac{1.5\ kip}{2} = 0.75\ kip$$

Correct Answer is (A)

SOLUTION 2.2

Shear center is the center of the shape that does not produce twisting forces when a force is applied to it, be it lateral or transverse – this does not apply to normal forces, however. It can also be defined as the center of the shear force system.

Shear center does not coincide with the centroid of the shape, or its center of gravity, the latter being the point that corresponds to the geometric center of the shape, or the centroid of the area.

Unlike the center of gravity, which can be determined using simple calculus, the shear center involves the calculation of the first and the second moment of inertia and sometimes requires the use of more complex numerical equations depending on the complexity of the shape under consideration. Regardless of which, sections with symmetrical axis have their shear centers coincide with their center of gravity.

Based on the brief description above, the following can be concluded for each of the statements of this problem:

Item	Statement	Justification
A	Shear center for a double angle steel section coincides with its centroid.	True – a double angle is a symmetrical shape and because of this its centroid coincides with its shear center.
B	Shear center of a shape coincides with its center of gravity.	Wrong – shear center could only coincide with shapes' centroid for shapes with axis of symmetry.
C	Shear center of a section is where moment and torsion are decoupled.	True – when lateral or transverse forces pass through the shear center of a section, only bending moments are produced with no torsion.
D	Only when lateral loads and transverse loads passes through the shear center simultaneously, member will be subjected to moment with no torsion.	Wrong – lateral and transverse forces do not have to be simultaneous in order to cancel torsion.

E	Forces passing through a shape's centroid do not generate twisting forces.	Wrong – this could only be true for shapes with axis of symmetry where the shear center of a shape coincides with its centroid.

Correct Answers are (A & C)

SOLUTION 2.3

$M_{x-x} = 18 \times \sin(30)$

$= 9.0\ kip.ft \quad counterclockwise$

$M_{y-y} = 18 \times \cos(30)$

$= 15.6\ kip.ft \quad counterclockwise$

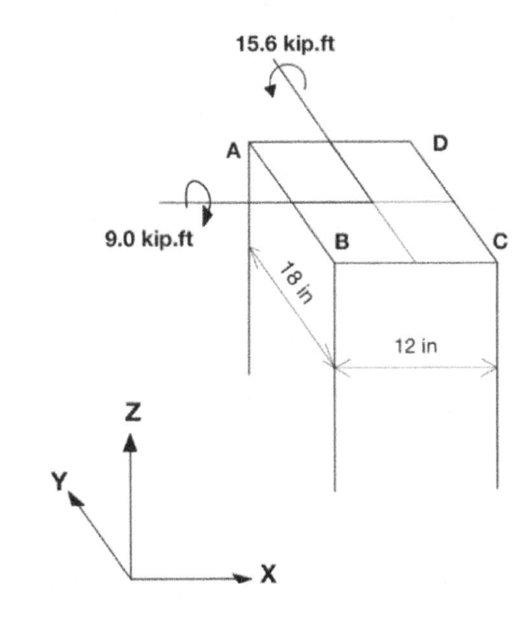

$I_{x-x} = \frac{12 \times 18^3}{12} = 5{,}832\ in^4$

$I_{y-y} = \frac{18 \times 12^3}{12} = 2{,}592\ in^4$

Use the below stress equation to determine stress at each edge then stress at each corner:

$\sigma = \frac{MC}{I}$

Stresses around the y-y axis:

$\sigma_{A-B} = \frac{-15.6\ kip.ft \times 12 \frac{in}{ft} \times 6\ in}{2{,}592\ in^4}$

$= -\ 0.43\ ksi\ Compression$

$\sigma_{D-C} = \frac{+15.6\ kip.ft \times 12 \frac{in}{ft} \times 6\ in}{2{,}592\ in^4}$

$= +\ 0.43\ ksi\ \ Tension$

Stresses around the x-x axis

$\sigma_{A-D} = \frac{+9.0\ kip.ft \times 12 \frac{in}{ft} \times 9\ in}{5{,}832\ in^4}$

$= +\ 0.17\ ksi\ Tension$

$\sigma_{B-C} = \frac{-9.0\ kip.ft \times 12 \frac{in}{ft} \times 9\ in}{5{,}832\ in^4}$

$= -\ 0.17\ ksi\ Compression$

Stresses at each corner

$\sigma_A = \sigma_{A-B} + \sigma_{A-D}$

$= -0.43 + 0.17$

$= -\ 0.26\ ksi$

$\sigma_B = \sigma_{A-B} + \sigma_{B-C}$

$= -0.43 - 0.17$

$= -\ 0.60\ ksi$

$\sigma_C = \sigma_{B-C} + \sigma_{D-C}$

$= -0.17 + 0.43$

$= +\ 0.26\ ksi$

$\sigma_D = \sigma_{A-D} + \sigma_{D-C}$

$= +0.17 + 0.43$

$= +0.60\ ksi$

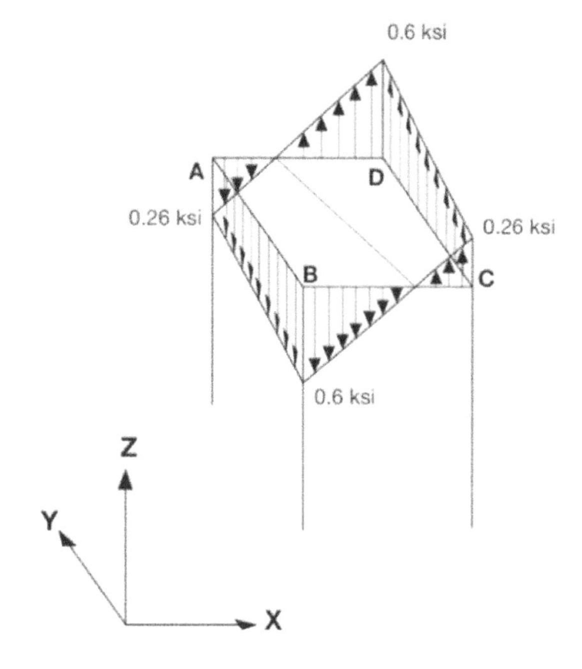

Correct Answer is (A)

SOLUTION 2.4

Shear stress is calculated as follows:

$$\tau = \frac{VQ}{IB}$$

V is the shear load in which case is a constant $200\ kip$. $\boldsymbol{I}$ is the moment of inertia for the overall section.

Centerline for the entire shape measured from the bottom:

$$= \frac{18\times48\times24+12\times30\times42}{18\times48+12\times30}$$

$$= 29.3\ in$$

$$I = \frac{18\times48^3}{12} + 18\times48\times(29.3-24)^2$$

$$+\frac{30\times12^3}{12} + 12\times30\times(42-29.3)^2$$

$$= 252{,}542\ in^4$$

The two items that are left in the shear stress equation that can determine the location of maximum stress is $\boldsymbol{Q}$ and $\boldsymbol{B}$.

$\boldsymbol{Q}$ is the first moment of area above or below the point where shear stress is to be determined and it must be maximum in this case. $\boldsymbol{B}$ is the width of the section and must be minimum to maximize the shear stress.

$\boldsymbol{Q}$ maximum occurs at a plane where the area above it or below it is maximum. This occurs at a plane that divides the shape into two equivalent halves. See below figure for more elaboration.

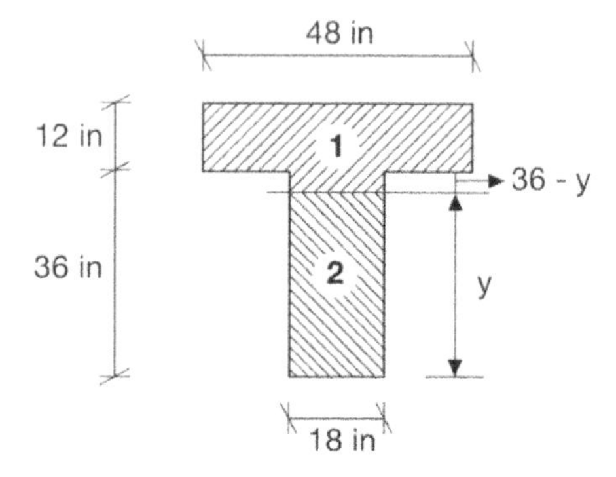

Equalizing the top part with the bottom part and solving for $\boldsymbol{y}$ determines the location of maximum $\boldsymbol{Q}$ is at $y = 34\ in$, which has the minimum $\boldsymbol{B}$ of the section as well.

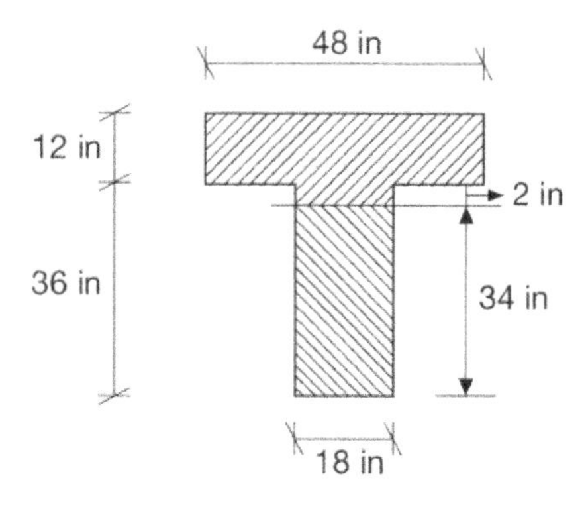

$\boldsymbol{Q}$ now can be calculated using the area at the top or the bottom of the dividing plane. Using the area at the top will add one layer of calculation, it is always good to use both methods and compare notes as an extra layer of quality check.

Using the top area to calculate Q:

Centerline of the area above the plane with maximum shear calculated from its bottom:

$= \frac{12\times48\times8+2\times18\times1}{12\times48+2\times18}$

$= 7.6\ in$

Distance from this centroid to the overall shape's centroid:

$= (48 - 29.3) - (14 - 7.6)$

$= 12.3\ in$

$Q = (12 \times 48 + 18 \times 2) \times 12.3$

$= 7{,}527.6\ in^3$

Using the bottom area to calculate Q:
Centerline of the area below the plane with maximum shear calculated from its bottom:

$= 34/2 = 17\ in$

Distance from this centroid to the overall shape's centroid:

$= 29.3 - 17 = 12.3\ in$

$Q = (34 \times 18) \times 12.3 = 7{,}527.6\ in^3$

$\tau = \frac{VQ}{IB} = \frac{200{,}000\ lb \times 7{,}527.6\ in^3}{252{,}542\ in^4 \times 18\ in} = 331.2\ psi$

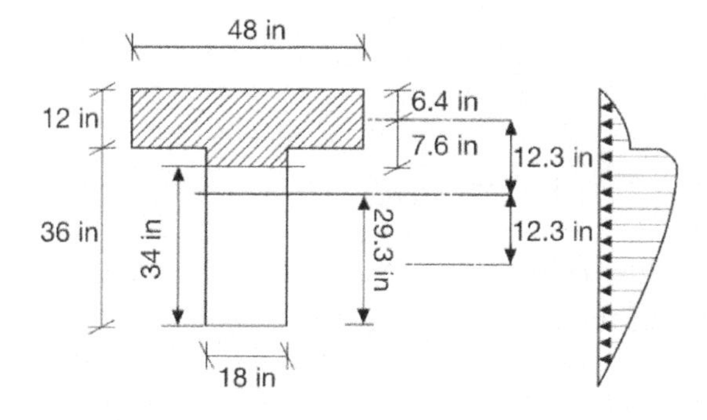

Correct Answers

$(y = 34\ in, \tau_{max} = 331.2\ psi)$

SOLUTION 2.5

The maximums and minimums for moment of inertia with respect to a principal axis can be determined in a similar fashion for normal stresses and shear stresses and a 2D *Mohr Circle* can be constructed for this purpose.

An inertia *Mohr Circle* can be constructed by replacing normal stresses with inertias, and shear stresses with the product of inertia of the cross-section.

The center of the circle is 'C' and radius R, and both can be determined as follows:

$C = \frac{I_x+I_y}{2}$

$= \frac{16{,}500 + 14{,}750}{2}$

$= 15{,}625\ ft^4$

$R = \sqrt{\left(\frac{I_x - I_y}{2}\right)^2 + {I_{xy}}^2}$

$= \sqrt{\left(\frac{16{,}500 - 14{,}750}{2}\right)^2 + 7{,}775^2}$

$= 7{,}824\ ft^4$

$I_{max} = C + R$

$= 15{,}625 + 7{,}824$

$= 23{,}449\ ft^4$

$I_{min} = C - R$

$= 15{,}625 - 7{,}824$

$= 7{,}801\ ft^4$

Based on the above inertia attributes an inertia *Mohr Circle* can be constructed as follows:

PART II Structural Depth

Section 2 X-Section Analysis

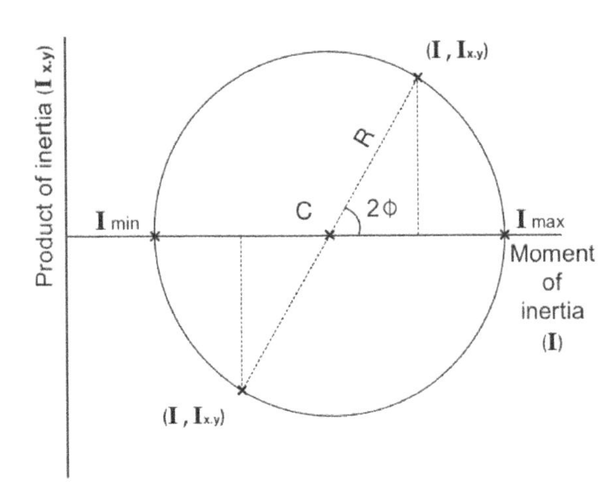

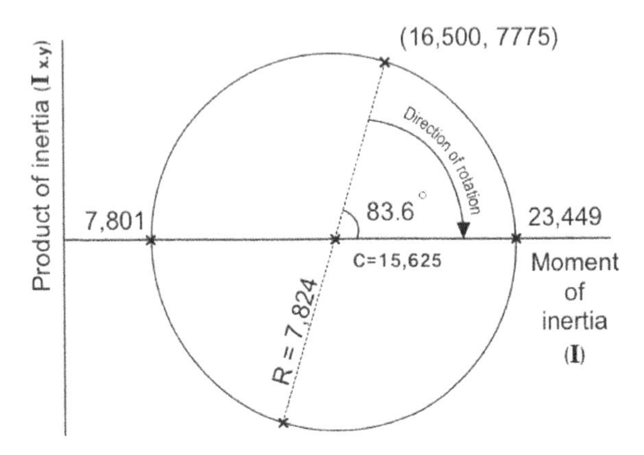

The angle which the maximum inertia occurs at is "ϕ" and can be determined using basic trigonometry as follows:

$$2\phi = tan^{-1}\left(\frac{7{,}775}{16{,}500-15{,}625}\right)$$

$$= 83.6^{o}$$

$$\phi = 41.8^{o}$$

Correct Answer is (A)

SOLUTION 2.6

This is a composite section and it requires its centroid to be recalculated.

Calculate modulus of elasticity for the two materials using ACI 318-14 section 19.2.2.1(b):

$$E_{c,girder} = 57{,}000\sqrt{5{,}500}$$

$$= 4{,}227{,}233\ psi$$

$$E_{c,slab} = 57{,}000\sqrt{3{,}500}$$

$$= 3{,}372{,}165\ psi$$

$$Modular\ ratio\ (n) = \frac{E_{c,slab}}{E_{c,girder}} = 0.8$$

Modular ratio n is used to reduce the effective width of the slab prior to determining the new location of the centroid. It shall reduce the stress on the slab section as well.

$$b_{e,new} = 36\ in \times 0.8 = 28.8\ in$$

$$\bar{y}_{new} = \frac{(\bar{y}\times A_g)_{girder} + (\bar{y}\times A_g)_{slab}}{A_{g,total}}$$

$$= \frac{12.59\times276 + 29.75\times(28.8\times3.5)}{276 + 28.8\times3.5} = 17.2\ in$$

$$Eccentricity\ (e) = 17.2 - 4.0 = 13.2\ in$$

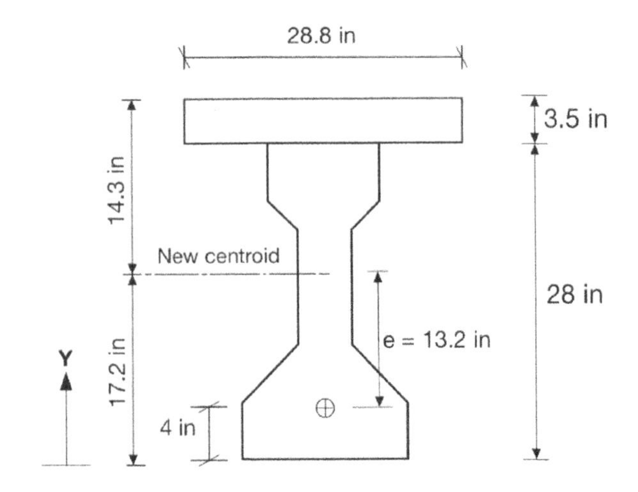

$$I_{g,new} = (I_g + A_g \times d^2)_{Girder} + (I_g + A_g \times d^2)_{Slab}$$

d *is* the distance from the centroid of each area to the centroid of the new shape:

$$d_{girder} = 17.2 - 12.59 = 4.61\ in$$

$$d_{slab} = \left(28 + \frac{3.5}{2}\right) - 17.2 = 12.55\ in$$

$$I_{g,slab} = \frac{28.8 \times 3.5^3}{12} = 102.9\ in^4$$

$$I_{g,new} = (22{,}750 + 276 \times 4.61^2)$$
$$+ (102.9 + (28.8 \times 3.5) \times 12.55^2)$$
$$= 44{,}595\ in^4$$

$$\sigma_{top\ slab} = n \times \frac{(pe)C}{I}$$
$$= 0.8 \times \frac{(150\ kip \times 13.2\ in) \times 14.3\ in}{44{,}595\ in^4}$$
$$= 0.5\ ksi\ Tension$$

Correct Answer is (B)

SOLUTION 2.7

The new centroid for the overall section shall be calculated first. The centroidal moment of inertia for each section is then transferred to the new centroid.

The following dimensions for the two shapes were collected from the *AISC Steel Construction Manual.*

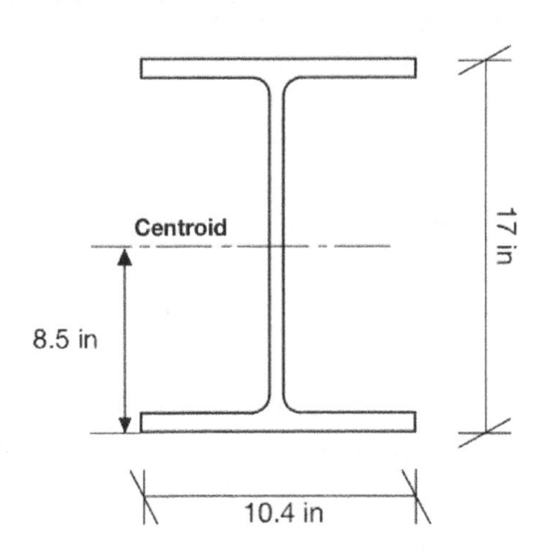

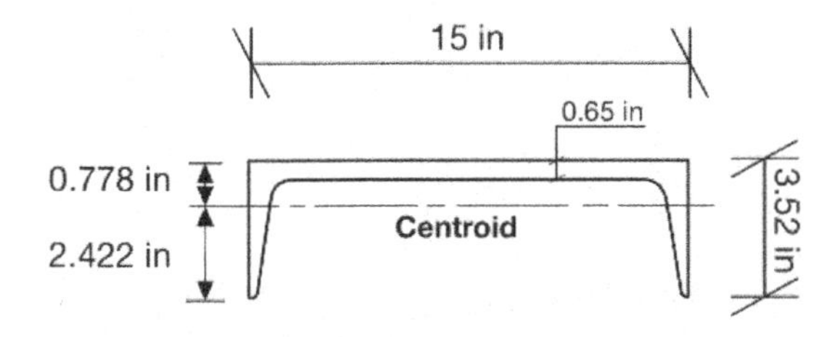

The following are the two sections' attributes:

$$A_{g,W16\times100} = 29.4\ in^2$$
$$I_{g,W16\times100} = 1{,}490\ in^4$$
$$A_{g,C15\times40} = 11.8\ in^2$$
$$I_{g,C15\times40} = 9.17\ in^4$$

Taking datum at the bottom of the two combined sections, the new location of the centroid is:

$$\bar{y} = \frac{29.4 \times 8.5 + 11.8 \times 16.872}{29.4 + 11.8} = 10.9\ in$$

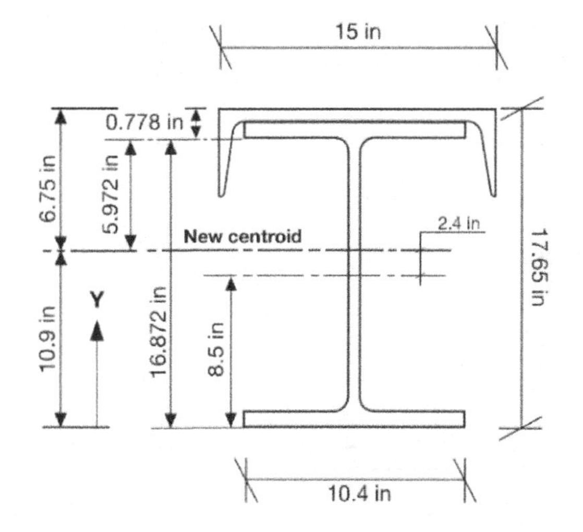

Based on the new centroid location, the new moment of inertia is:

$$I_{new} = \left(I + A_g \times d^2\right)_{W16\times100} + (I + A_g \times d^2)_{C15\times40}$$

d *is* the distance from the centroid of each area to the centroid of the new shape

$$d_{W16\times100} = 10.9 - 8.5 = 2.4\ in$$

$$d_{C15\times40} = 17.0 + 0.65 - 10.9 - 0.778$$
$$= 5.972\ in$$

$$I_{new} = (1{,}490 + 29.4 \times 2.4^2)$$
$$+ (9.17 + 11.8 \times 5.972^2)$$
$$= 2{,}089\ in^4$$

Correct Answer is (B)

SOLUTION 2.8

Shear flow q is to be determined at the interaction between the beam section and the plate, based upon which, the minimum number of bolts can be determined.

$$q = \frac{VQ}{I}$$

Q is the first moment of area above or below the point where shear stress is to be determined and it equals to the distance from the area's centroid A (in which case the steel plate cross section) to the neutral axis $\bar{y}$'.

The new location of centroid for the newly created cross-section is calculated as follows:

Taking datum at the bottom of the new shape:

$$A_{g,W30\times292} = 86\ in^2$$

$$A_{g,plate} = 30\ in^2$$

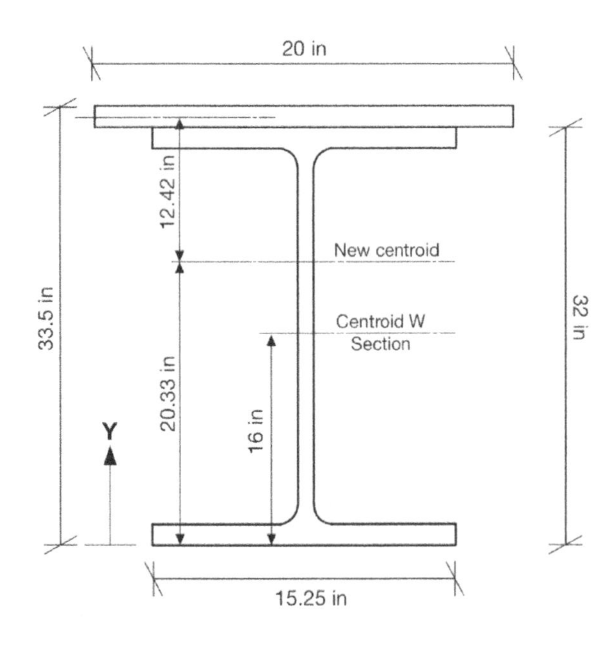

$$(\bar{y} \times A_g)_{plate} + (\bar{y} \times A_g)_{W30\times292} = \bar{y}_{new} \times A_{g\ total}$$

$$32.75 \times 30 + 16 \times 86 = \bar{y}_{new} \times (30 + 86)$$

$$\bar{y}_{new} = 20.33\ in$$

$$Q_{plate} = 30 \times 12.42 = 372.6\ in^3$$

$$I_{plate@centroid} = \frac{bh^3}{12} = \frac{20\times1.5^3}{12} = 5.63\ in^4$$

$$I_{W30\times292@centroid} = 14{,}900\ in^4$$

$$I_{new} = \left(I + A_g \times d^2\right)_{plate} + (I + A_g \times d^2)_{W30\times292}$$

d is the distance from the centroid of each area to the centroid of the new shape:

$$d_{plate} = 33.5 - 0.75 - 20.33$$
$$= 12.42\ in$$

$$d_{W30\times292} = 20.33 - 16$$
$$= 4.33\ in$$

$$I_{new} = (5.63 + 30 \times 12.42^2) + (14{,}900 + 86 \times 4.33^2)$$
$$= 21{,}145.7\ in^4$$

$$q = \frac{VQ}{I} = \frac{50\ kip \times 372.6\ in^3}{21{,}145.7\ in^4} = 0.88\ kip/in$$

The horizontal force F_H acting at the interface is as follows:

$$F_{H,two\ rows} = 0.88\frac{kip}{in} \times 24\ ft \times \frac{12\ in}{ft}$$
$$= 253.7\ kip$$

$$F_{H,one\ row} = \frac{253.7}{2} = 126.9\ kip$$

Per the *AISC Steel Construction Manual*, the available shear strength for one single loaded A307 bolt 5/8 *in* is:

$$\frac{r_n}{\Omega} = 4.14\ kip$$

Number of required bolts per single row:

$$\frac{126.9\ kip}{4.14\ kip} = 30.6\ bolts \cong 31\ bolts\ per\ row$$

Correct Answer is (C)

SOLUTION 2.9

Shear flow is calculated using the following equation:

$$q = \frac{VQ}{I}$$

V is the shear load, and ***I*** is the moment of inertia for the entire section, and both are constant and will not contribute to how the final shear profile will look like. This indicates that the first moment of area ***Q*** controls the requested profile.

Q equals to: (1) the distance from the overall shape's centroid to the area's centroid directly above or below the plane under consideration which is bounded by the plane where shear is to be calculated, multiplied by, (2) the area of that part.

(1) Distance from shape's centroid to area's centroid that is bounded by the plane where shear is to be calculated:

The centroid of the overall shape is constant and does not change, whereas the centroid of each of the selected areas (above or below the plane under consideration) are ever changing as the plane under consideration moves up or down. The distance therefore from this centroid to the centroid of the entire shape changes, it is however almost mirrored across the neutral axis. See figure:

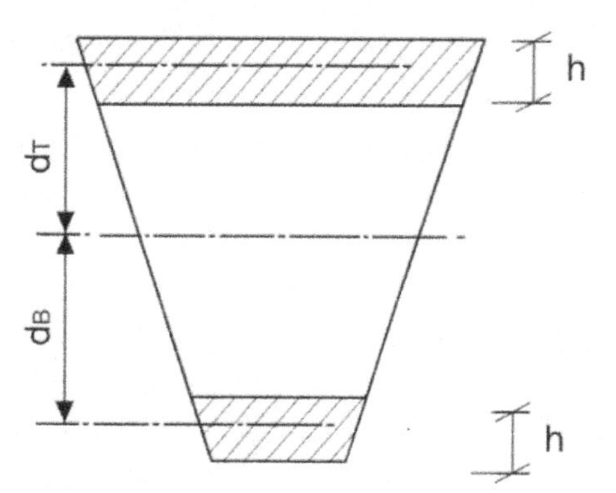

I.e., an area of height *h* measured from the very top of the section downwards, will have almost a similar distance from its centroid to the centroid of the entire shape, to that of an area's centroid of a similar height measured from the very bottom of the entire shape to the centroid of the shape. Hence this distance will have minor contribution to the final shear flow profile.

$$\rightarrow d_B \approx d_T$$

(2) The area above or below the plane under consideration:

The area above or below the plane under consideration tends to be larger when starting from the top of the depicted section and it grows smaller on its way downwards compared to when it starts at the bottom of the overall shape moving upwards.

Hence it is expected that the shear flow profile will be skewed upwards (profile A) compared to downwards (profile D). Simply put, because the contribution from ***Q*** as the section start from top to bottom is higher than from bottom to top.

Correct Answer is (A)

Question extras:

For the removal of doubt, the peak shear flow, or at least distance $'y'$ can be calculated by solving a quadratic equation as shown in the following figure.

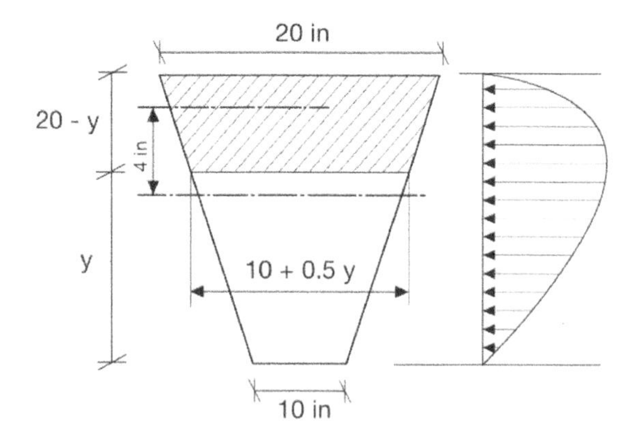

The shaded area that sits at the top section, where the maximum shear flow is expected to occur, is equivalent to the remainder of the area at the bottom of this plane, and this represents the peak or maximum $\boldsymbol{Q_{max}}$.

$$\frac{(20+(10+0.5y))}{2} \times (20-y) = \frac{(10+(10+0.5y))}{2} \times (y)$$

$$y^2 + 40y - 600 = 0$$

$$y = 11.65\ in$$

This proves that shear flow will be skewed upwards for this shape.

The distance from this centroid to the centroid of the entire shape is 4 in and the moment of inertia of the entire shape is 9,629 in^4. This information, although not requested in the question, can now be used to calculate the shear flow at the depicted maximum location if needed.

SOLUTION 2.10

Reference is made to Solution 2.9 in this book. Question 2.9 and its solution refer to the maximum shear flow location and shear flow profile. The equation of shear flow contains only one variable being the moment of area $\boldsymbol{Q}$. This very variable skews the shear flow profile upwards for this section for reasons explained in Solution 2.9 narrative.

The shear stress equation on the other hand, below, has one more variable which is the width $\boldsymbol{B}$ of the section where the stress is being measured.

$$\tau = \frac{VQ}{IB}$$

The width of this section reduces as it moves downwards. Thereby, should increase the shear stress and skews the profile downwards. However, Solution 2.9 of this book confirmed that the moment of area $\boldsymbol{Q}$ would skew the profile upwards. The inverse relation with the width $\boldsymbol{B}$ of the section will neutralize the skew generated by the moment of area $\boldsymbol{Q}$ and therefore produces a final shape with no skew (neither upwards nor downwards).

This makes Profile C the most accurate profile that represents the shear stress profile for this section.

Correct Answer is (C)

SECTION 3

Loads and Load Analysis

Problems & Solutions

PROBLEM 3.1 ***Base Shear for a Tall Building***

A six-story retail building located in Albuquerque, New Mexico (*) with equal story heights of 9 ft each. The building is founded on soil type C and is built using reinforced concrete ordinary moment resisting frames. All floors are equal in mass and each floor weighs 700 kip.

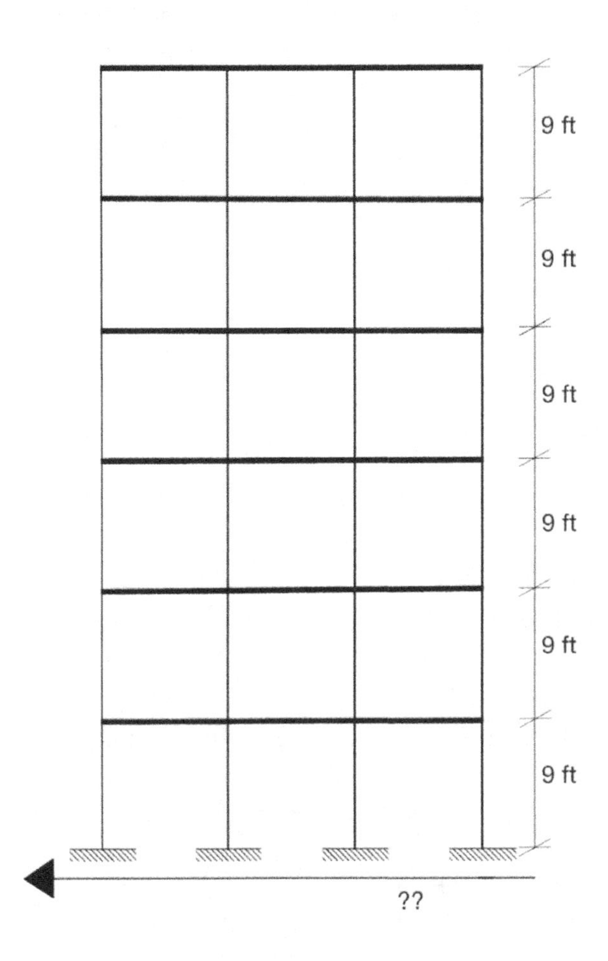

Assume that the effective seismic weight equals to the weight of all floors plus an extra 25% to account for other code requirements.

The Seismic Design Category and the Seismic Base Shear nearly are:

(A) Seismic Design Category C
Base shear = 390 kip

(B) Seismic Design Category C
Base shear = 580 kip

(C) Seismic Design Category B
Base shear = 390 kip

(D) Seismic Design Category B
Base shear = 580 kip

* Use the following spectral parameters for Albuquerque, New Mexico:

$S_S = 0.43$

$S_1 = 0.12$

$T_L = 6$ Sec

PROBLEM 3.2 ***Vertical Seismic Load Distribution***

A four-story hospital located in Miami (*) with equal story heights of 14 ft each. The building is founded on soil type B and is built using reinforced concrete ordinary moment resisting frames. All floors are equal in mass and the portion of dead weight per story is 850 kip.

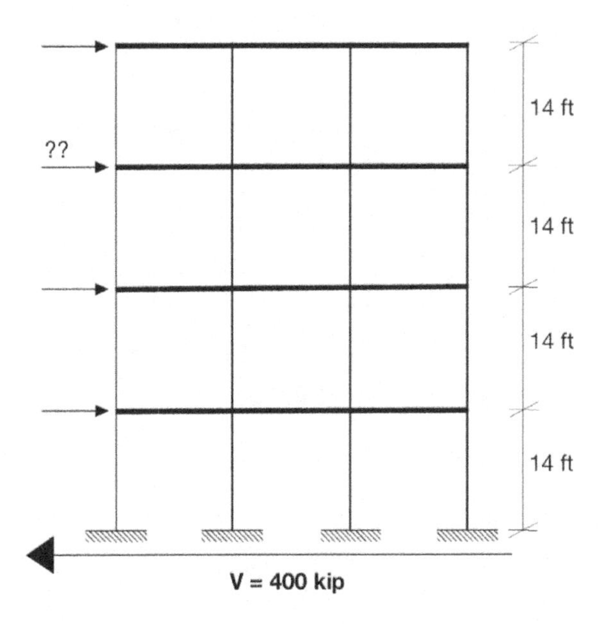

The lateral seismic force in *kip* experienced on the third floor is most nearly:

(A) 2

(B) 8.5

(C) 11

(D) 45

* Use the following spectral parameters for Miami

$S_S = 0.048$

$S_1 = 0.021$

$T_L = 8$ Sec

PROBLEM 3.3 ***Seismic Design Requirements***

A retail building located in Los Angeles, West Hollywood (*), soil type B, with masonry walls 8 *in* thick that are not part of the seismic-force-resisting system.

The horizontal reinforcements used in this wall should at least be:

(A) Two longitudinal wires of $W1.7$ spaced no more than 16 *in*

(B) No.4 bars spaced no more than 48 *in*

(C) 0.0015 of the gross cross section with maximum spacing of 24 *in*

(D) No.6 bars spaced no more than 48 *in*

* Use the following spectral parameters for Los Angeles, West Hollywood:

$S_S = 2.38$

$S_1 = 0.86$

$T_L = 8$ Sec

PROBLEM 3.4 ***Seismic Load Directions***

The direction of lateral seismic forces for a Seismic Design Category B building should be applied is as follows:

(A) Seismic forces are applied in each direction independent of the other, the forces' orthogonal interactional effects are not considered.

(B) Seismic forces should be applied in each direction independent of the other, the forces' orthogonal interactional effects should be considered.

(C) Seismic forces are applied 100% in one direction and 30% in the other perpendicular direction.

(D) Seismic forces are applied 100% in one direction and 50% in the other perpendicular direction.

PROBLEM 3.5 ***Vertical Seismic Load Distribution***

A retail building assigned to a Seismic Design Category B with equal story heights of 14 *ft* each. The building is founded on soil type C and is built using reinforced concrete ordinary moment resisting frames.

Base shear has been determined previously as 400 *kip*. The portion of effective weight per story is 1,250 *kip* except for the roof which is 1,000 *kip*.

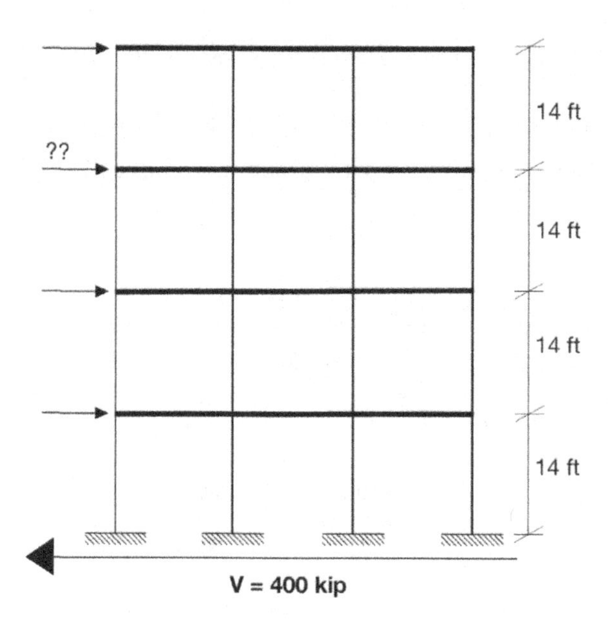

Using the Equivalent Lateral Force Analysis method, the lateral seismic force in *kip* experienced on the third floor is most nearly:

(A) 100

(B) 47.5

(C) 120

(D) 130

PROBLEM 3.6 *Horizontal Seismic Load Distribution*

A four-story building assigned to Seismic Design Category B consists of six reinforced concrete ordinary moment resisting frames in the short direction and four in the long direction spaced as shown. The interior frames have double the stiffness of the exterior frames, and center of mass coincides with center of rigidity.

The lateral seismic force experienced at the third floor is 200 *kip*.

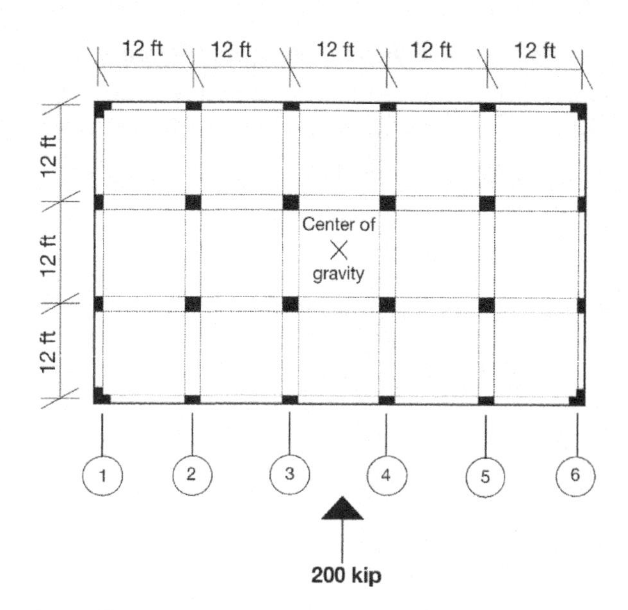

The seismic force applied on frame 4 of the third floor generated from accidental torsion in *kip* is most nearly:

(A) 5.6

(B) 2.2

(C) 6.7

(D) 3.3

PROBLEM 3.7 *Rain Load Calculation*

A roof of a building slightly sloped at 1 ft to ¼ in with clear dimensions of 100 $ft \times$ 150 ft has a primary drain and its secondary drain is situated at 3.5 in on top of the primary drain. Rain intensity in the area has been determined as 4 in/hr which generates a hydraulic head of nearly 4 in.

The total rain load on the roof in *kip* is most nearly:

(A) 21

(B) 39

(C) 312

(D) 585

PART II Structural Depth

Section 3 Load Analysis

PART II Structural Depth

Section 3 Load Analysis

PROBLEM 3.8 ***Flood Elevation Design***

The equivalent surcharge depth (*) to be added to the Design Flood Elevation (DFE) for an elliptical shaped pier with 2:1 length to width ratio located in the middle of the below canal is most nearly:

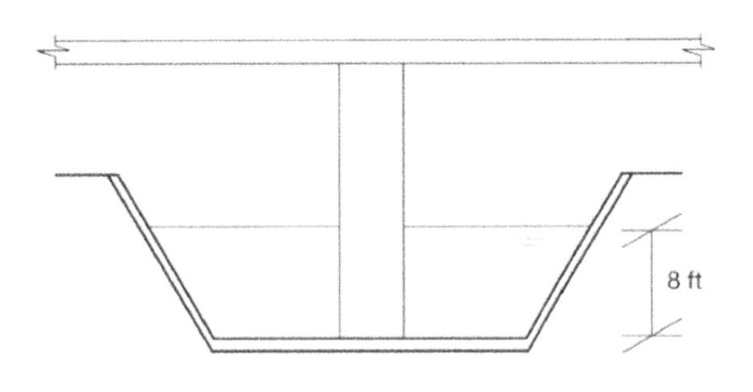

(A) More information required.

(B) $7\ ft$

(C) $2\ ft$

(D) $20\ ft$

* Use the following data:

DFE = $8\ ft$

Velocity = $10\ ft/sec$

Consider drag force coefficients as reported by NASA: *Drag of Cylinders of Simple Shapes* as follows:

Shape	Drag Coefficient
Circular pier	1.20
Elongated pier with semi-circular ends	1.33
Elliptical piers with 2:1 length to width	0.60
Square piers	2.00
Triangular nose with 30-degree angle	1.00
Triangular nose with 120-degree angle	1.72

PROBLEM 3.9 ***Retaining Wall Applicable Loads***

The figure below is for a plain concrete gravity retaining wall with an $8\ ft$ Clayey Silt backfill and base.

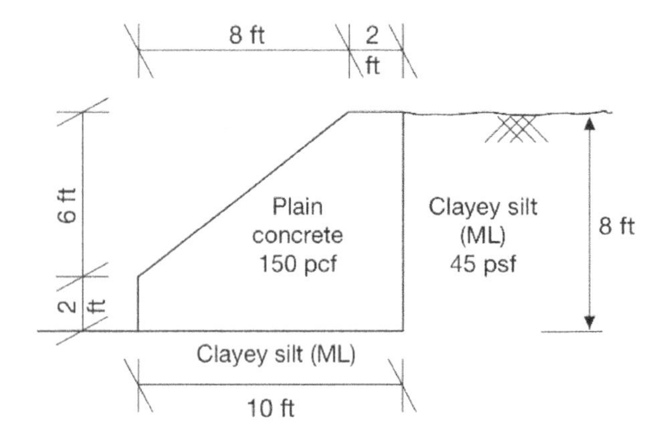

Safety Factors against sliding and overturning for this wall are as follows:

(A) 0.9 for sliding and 13.2 for overturning.

(B) 1.1 for sliding and 1.3 for overturning.

(C) 2.9 for sliding and 13.2 for overturning.

(D) More information required.

PROBLEM 3.10 ***Distribution of Pressure under Footing***

The below is a plan view for a single concrete footing with a $12\ in \times 24\ in$ column that sits directly at a property line carrying the following loads:

Load type	Service load (kip)	Ultimate load (kip)
Live load	16	19.2
Dead load	11	13.2
Total load	**27**	**32.4**

The substrate below this footing is an improved sandy gravel with a load-bearing allowable pressure of 3,000 psf.

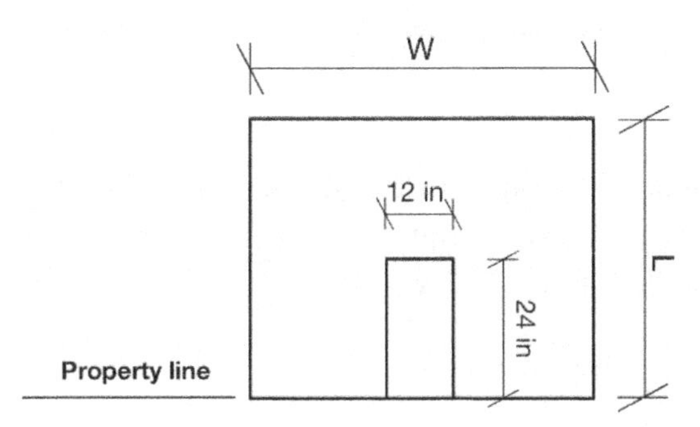

The best dimensions $W\ X\ L$ for this footing that generate a safe with no negative pressure zones is:

(A) 8.0 ft × 3.0 ft

(B) 6.0 ft × 3.0 ft

(C) 4.3 ft × 4.3 ft

(D) 3.0 ft × 8.0 ft

PROBLEM 3.11 ***Traffic Dynamic Load***

The culvert below carries an H truck loading along with a design lane load of 6.4 klf as shown:

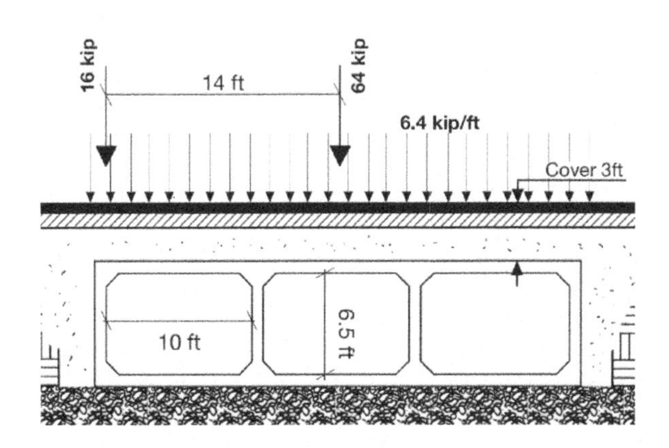

Considering the dynamic effect traffic has on structures, the culvert design shall consider the following magnified loads from left to right as shown on the figure:

(A) Truck load of 22.3- & 85.1-kip, lane load 6.4 klf.

(B) Truck load of 19.4- & 77.5- kip, lane load 6.4 klf.

(C) Truck load of 19.4- & 77.5- kip, lane load 7.7 klf.

(D) Truck load of 22.3- & 85.1- kip, lane load 8.5 klf.

PROBLEM 3.12 ***Lateral Distribution of Loads***

The below building's 100 ft × 75 ft, 7.5 ft heigh floor has a lateral force applied to its short direction as shown. The building's lateral support system is that of shear walls and it has four shear walls, with thickness of 10 in each, distributed in the below manner.

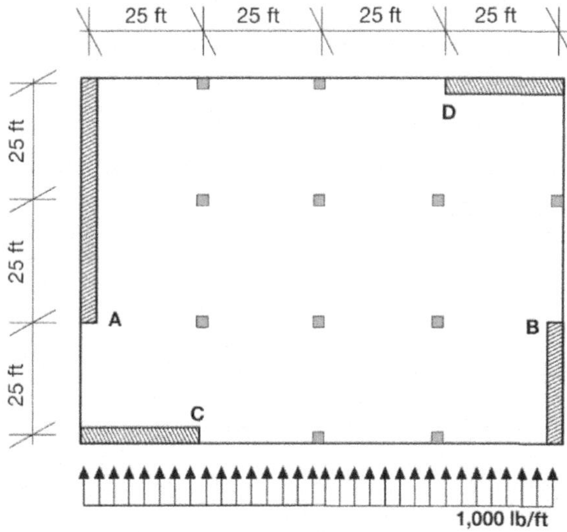

Ignoring the stiffness of columns or any other framing system, force applied on shear wall D in the long direction due to lateral force applied to the floor's short direction is most nearly:

(A) 9.3 kip ←

(B) 6.6 kip →

(C) 9.3 kip →

(D) 6.6 kip ←

PART II Structural Depth

Section 3 Load Analysis

PROBLEM 3.13 ***Load Combination***

The below concrete footing is being assessed for the safety of its strata's presumptive load bearing capacity specified in the IBC code, and has the following vertical downward loads applied on it:

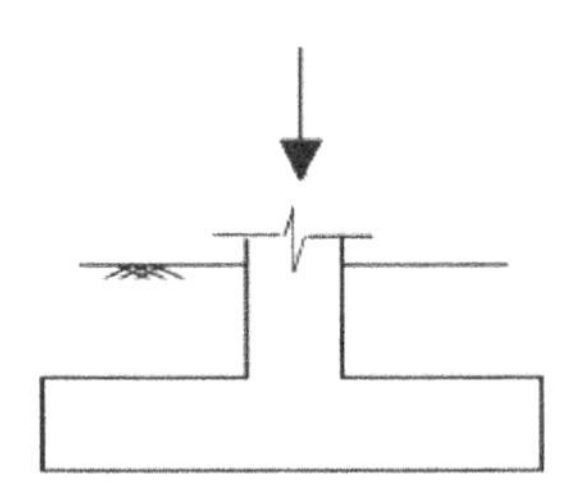

Load type	Service load (Kip)
Dead load	7.0
Live load	9.5
Roof Load (*)	5.0
Wind Load	2.5
Snow Load	1.0

The total, combined, load that shall be used for this purpose is:

(A) 21.5 kip

(B) 25.0 kip

(C) 18.6 kip

(D) 32.1 kip

* Roof Live load corresponds to a roof load of 30 psf in this question.

PROBLEM 3.14 ***Snow Load Calculation***

The below gable roofed building has a mean roof height of 25 ft located in a suburban area, fully exposed, and is used to store furniture. The roof is unheated but ventilated with $R > 25\ F \times h \times ft^2/BTU$. Ground snow load p_g of the building zone is 30 psf.

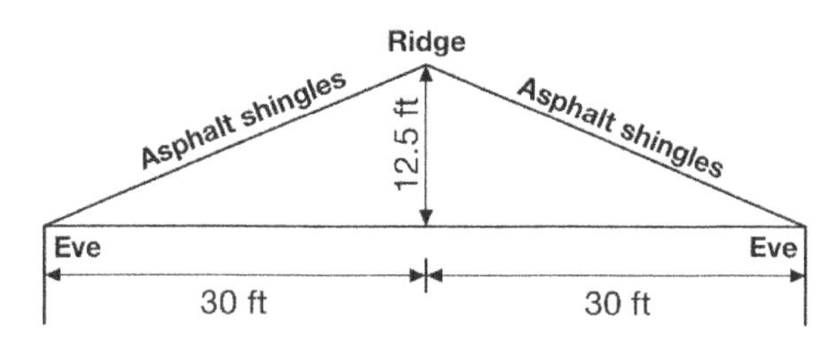

Based on this information, the below horizontal load projection mostly represents snow loads for this roof:

(A) Option A

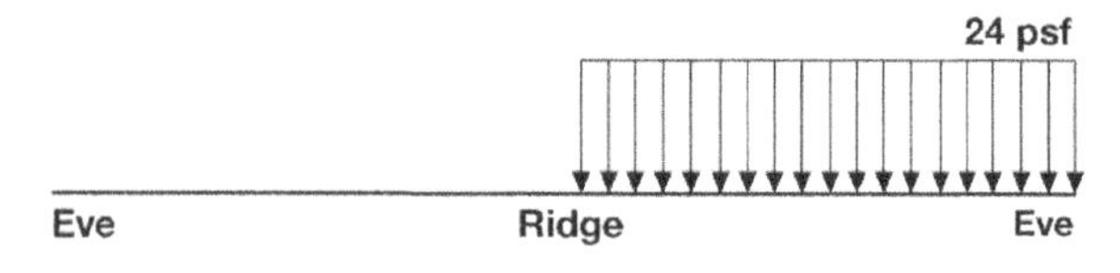

(B) Option B

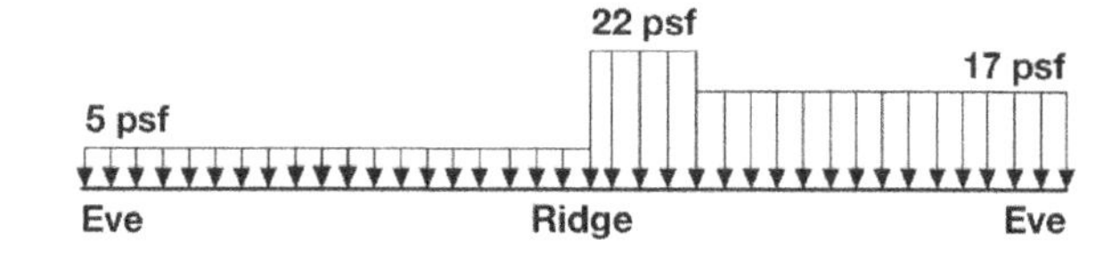

(C) Option C

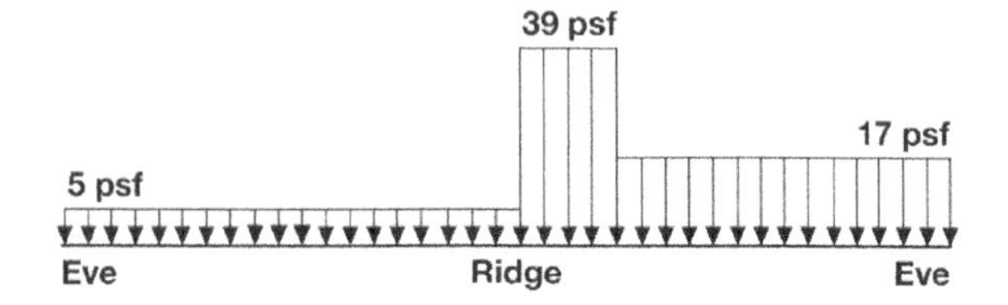

(D) Option D

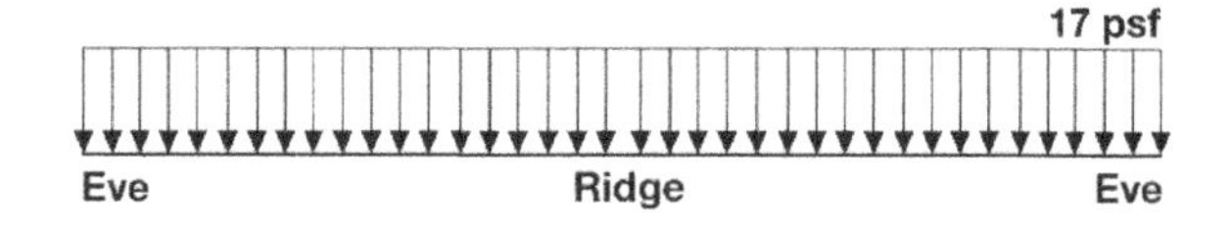

PROBLEM 3.15 ***Live Load Reduction***

The below is the first-floor layout of a one-story building. The Live Load considered on this floor is 75 psf.

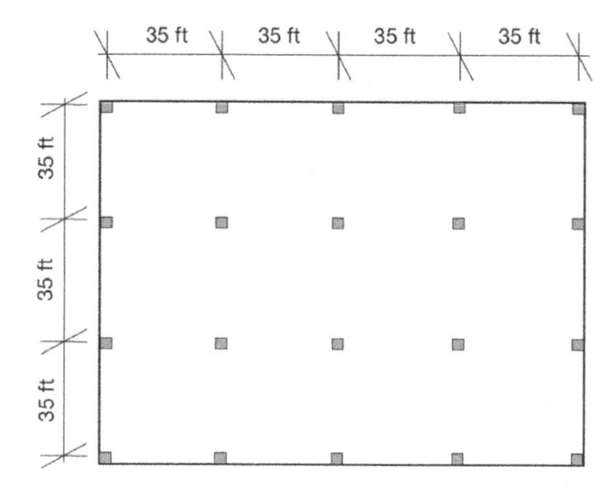

The reduced design live load for interior columns would nearly be:

(A) 35.0 psf

(B) 37.5 psf

(C) 75.0 psf

(D) 58.5 psf

PROBLEM 3.16 ***Live Load Maximization***

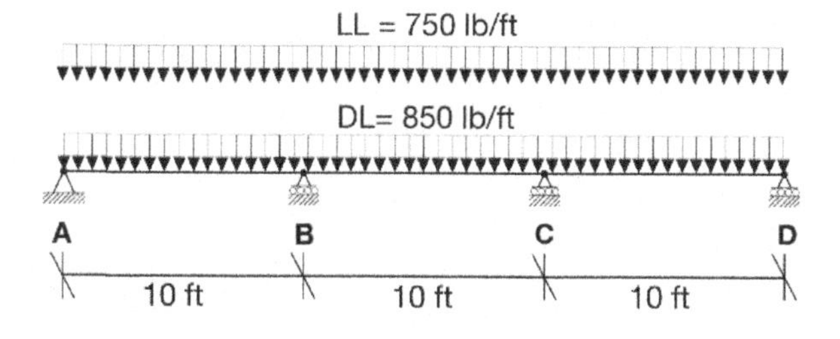

Using the Strength Design load combinations defined in the IBC or the ASCE 7 code, the maximum negative moment that can occur at support B is:

(A) 26.0 $kip.ft$

(B) 24.2 $kip.ft$

(C) 28.0 $kip.ft$

(D) 25.3 $kip.ft$

PROBLEM 3.17 ***Vehicular Collision Force***

The following shall be considered when designing a pier or an abutment to resist collision:

(A) Design the structure with an extra equivalent static force of 600 kip.

(B) Design the structure with an extra equivalent static force of 1,000 kip.

(C) Add a 42 in barrier located more than 10 ft from the structure to be protected.

(D) Provide an embankment.

PROBLEM 3.18 ***Load Combination***

A concrete footing to be designed using the strength design method. The footing is assigned to Seismic Design Category E with redundancy factor ρ of '1.3' and S_{DS} of '0.75', along with the following loads:

Load Type	*Kip*
Dead Load	5.0
Live Load	6.0
Roof Load	1.5
Effect of horizontal seismic force Q_E	3.0
Wind Load	2.0

The combined load that shall be used for this purpose is:

(A) 16.7 kip

(B) 15.0 kip

(C) 16.4 kip

(D) 17.7 kip

PROBLEM 3.19 ***Wind Pressure on Roof***
The below building is located in an urban, mostly flat, area > 5,000 $ft \times 5{,}000\ ft$ with basic wind speed of 130 mph.

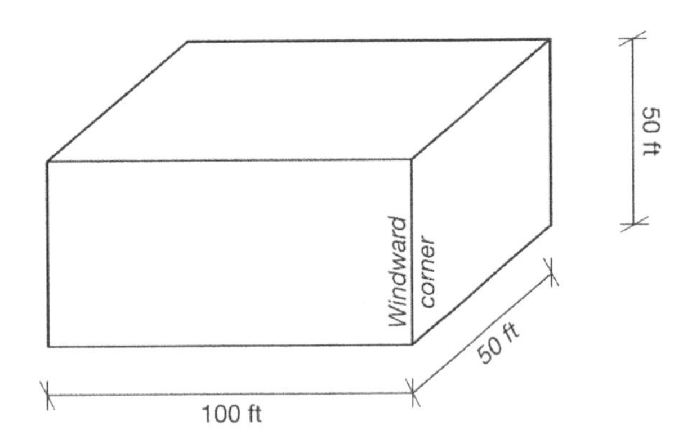

Using the simplified method, the wind pressure at the roof adjacent to the windward corner is most nearly:

(A) $-32.2\ psf$

(B) $-37.4\ psf$

(C) $-50.2\ psf$

(D) $-16.0\ psf$

PROBLEM 3.20 ***Wind Force on a Signboard***
A 30 ft long signboard situated on an escarpment in a zone with exposure C and wind speed of 160 mph.

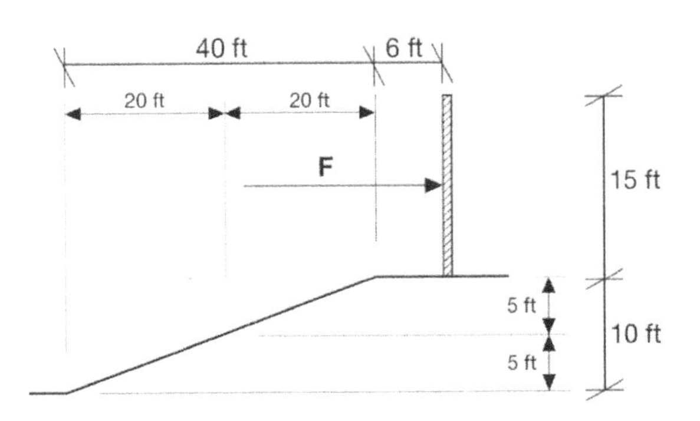

Assuming wind is perpendicular to the signboard, F is most nearly:

(A) 29 kip

(B) 21 kip

(C) 26 kip

(D) 54 kip

SOLUTION 3.1

Using the ASCE 7 code, base shear is obtained from the following equation:

$$V = C_s W$$

W is the Effective Seismic load and is defined in 12.7.2. C_s is the Seismic Response Coefficient and is calculated as follows – applicable when using the Equivalent Lateral Force Analysis method only:

$$C_s = \frac{S_{DS}}{\left(\frac{R}{I_e}\right)}$$

S_{DS} is the design spectral response acceleration parameter in the short period range = $2/3\, S_{MS}$

$$S_{MS} = F_a S_S$$

S_s is the mapped MCER Spectral response acceleration parameter at short periods. MCER is the maximum considered earthquake = 0.43 (given)

F_a is the site coefficient and can be looked up from Table 11.6 of the ASCE 7 or the IBC code.

For site class C and $S_s = 0.43 \rightarrow F_a = 1.2$

$\rightarrow S_{DS} = (2/3) \times 1.2 \times 0.43 = 0.344$

The Seismic Design Category can now be determined from Table 11.6 of chapter 11 of ASCE 7. With $0.33 < S_{DS} < 0.5$ and a risk category II (retail), the Seismic Design Category is mapped as C.

Based on the above category and number of stories, Table 12.6-1 of the ASCE 7 permits the use of the Equivalent Lateral Force Analysis method of Section 12.8.

R is the response modification factor and can be looked up from Table 12.2-1 of ASCE 7. For Ordinary reinforced concrete moment frames = 3.0.

I_e is the importance factor and can be determined from Section 11.5.1 and Table 1.5-1 from ASCE 7, in which case $I_e = 1.0$

$$C_s = \frac{S_{DS}}{\left(\frac{R}{I_e}\right)} = \frac{0.344}{\left(\frac{3}{1.0}\right)} = 0.11$$

Minimum acceptable value for C_s is as follows:

$$C_s = 0.044 S_{DS}\, I_e \geq 0.01$$

$$= 0.044 \times 0.344 \times 1.0$$

$$= 0.015 < 0.11 \rightarrow ok$$

Maximum value for C_s should be checked per 12.8.1.1 using the following steps:

Step 1 - determine the fundamental period:

$$T = T_a = C_t h_n^x$$

$$= 0.016 \times (6 \times 9)^{0.9}$$

$$= 0.58$$

T can also be calculated using (0.1×Number of floors) for less than 12 floors = $0.1 \times 6 = 0.6\ sec$

Step 2 – determine if $T < T_L$, in which case:

$$C_{s,max} = \frac{S_{D1}}{T\left(\frac{R}{I_e}\right)}$$

T_L is the long-period transition periods as determined from Section 11.4.5 and Table 22.12-22.16 and has been given in the question = *6 seconds* (given)

S_{D1} is determined from S_1 (given) and F_v Table 11.6 taken as '1.68':

$$S_{D1} = (2/3)\, S_{MS} = (2/3)\, F_v\, S_1$$

$$= (2/3) \times 1.68 \times 0.12 = 0.134$$

PART II Structural Depth

Section 3 Load Analysis

$$C_{s,max} = \frac{S_{D1}}{T\left(\frac{R}{I_e}\right)}$$

$$= \frac{0.134}{0.6\times\left(\frac{3}{1.0}\right)} = 0.075 < 0.11$$

→ $use\ C_s = 0.075$

Base shear calculation:

$V = C_s W$

$= 0.075 \times (6 \times 700 \times 1.25)$

$= 393.75\ kip$

Correct Answer is (A)

SOLUTION 3.2

ASCE 7 explains in Section 11.4.1 that when $S_1 \leq 0.04$ and $S_s \leq 0.15$, it is permitted to use Seismic Design Category A and compliance with Section 11.7 of the code shall suffice.

Section 11.7 of the code states that all buildings assigned to Seismic Design Category A shall comply with the requirements of Section 1.4 of the code.

Based on this, Section 1.4.3 specifies that lateral force F_x should be calculated as follows:

$F_x = 0.01\ W_x$

$\boldsymbol{W_x}$ being the portion of total dead load assigned per floor = 850 kip

$F_x = 0.01 \times 850\ kip = 8.5\ kip$

Correct Answer is (B)

SOLUTION 3.3

Based on the spectral response acceleration parameter of one second for this location, S_1 (0.86 sec) > 0.75, and, risk category II (retail), the Seismic Design Category for this building is E.

TMS 402-13 code, Section 7.4.5.1, specifies for Seismic Design Category E, with nonparticipating masonry not laid in running bond, should have an area of horizontal reinforcement of at least '0.0015' multiplied by the gross area of masonry with a maximum spacing of 24 in.

Correct Answer is (C)

SOLUTION 3.4

ASCE 7 Section 12.5.1 *Direction of Loading Criteria,* and 12.5.2 states that for structures assigned to Seismic Design Category B, the design seismic forces are permitted to be applied independently in each of the two orthogonal directions and orthogonal interactions effect are permitted to be ignored.

Correct Answer is (A)

SOLUTION 3.5

Table 12.6-1 of ASCE 7 permits the use of the Equivalent Lateral Force Analysis method for all structures in Seismic Design Category B (given in the question).

The lateral seismic force $\boldsymbol{F_x}$ is calculated based on Section 12.8.3 of the code as follow:

$F_x = C_{vx} V$

$$C_{vx} = \frac{w_x h_x^k}{\sum_{i=1}^{n} w_i h_i^k}$$

$\boldsymbol{C_{vx}}$ is the vertical distribution factor.

$\boldsymbol{w_{x\ or\ i}}$ is the portion of the effective seismic weight at level 'x' or 'i'.

$\boldsymbol{h}$ is the height from base to either level $'x'$ or $'i'$.

$k = 1 \; for \; T \leq 0.5 \sec \quad and$

$k = 2 \; for \; T \geq 2$

$T = 0.1 \times 4 \; floors$

$= 0.4 \sec \rightarrow use \; k = 1$

Or

$T = T_a = C_t h_n^x$

$= 0.016 \times (4 \times 14)^{0.9} = 0.6$

→ we can also use $k = 1.1$

The differences between using a $\boldsymbol{k}$ of 1 or 1.1 or 2 are not too significant. This question will continue using $k = 1.0$.

$$C_{v,3} = \frac{w_3 h_3^k}{\sum_{i=1}^{n} w_i h_i^k}$$

$$= \frac{1{,}250 \times (14 \times 3)^1}{\left[\begin{array}{c} 1250 \times (14^1 + (14 \times 2)^1 + (14 \times 3)^1) \\ + \\ (1000 \times (14 \times 4)^1) \end{array}\right]}$$

$= 0.33$

$F_3 = 0.33 \times 400 = 130 \; kip$

Correct Answer is (D)

SOLUTION 3.6

ASCE 7 Section 12.8.4.2 sets accidental torsion at 5% from the direction of the perpendicular dimension that the seismic force is applied to multiplied by the seismic force.

Based on this, accidental torsion is calculated as follows:

$M_T = 0.05 \times (5 \times 12 \; ft) \times 200 \; kip$

$= 600 \; kip.ft$

The forces distribution due to this moment is proportional to the rotational stiffness of each frame – see first figure in the extras section after the solution below.

Given that the center of gravity coincides with the center of stiffness, forces in frames 1, 2 and 3 will be opposite in direction and equal in magnitude to forces in frames 4, 5 and 6.

The share of each frame from the floor is calculated as follows:

$$F_{x,T} = \frac{M_T}{\sum k_i r_i^2} k_x r_x$$

$\boldsymbol{k}$ is the stiffness per frame and are as follows which is given in the question:

$k_1 = k_6 = k$

$k_2 = k_3 = k_4 = k_5 = 2k$

$\boldsymbol{r}$ is the distance from the center of each frame to the center of gravity (given):

$r_1 = r_6 = 30 \; ft$

$r_2 = r_5 = 18 \; ft$

$r_3 = r_4 = 6 \; ft$

$$F_{x,T} = \frac{600}{\left(\begin{array}{c} k_1 r_1^2 + k_2 r_2^2 + k_3 r_3^2 \\ + \\ k_4 r_4^2 + k_5 r_5^2 + k_6 r_6^2 \end{array}\right)} \times k_x r_x$$

$$= \frac{600}{\left(\begin{array}{c} 900k + 324(2k) + 36(2k) \\ + \\ 36(2k) + 324(2k) + 900k \end{array}\right)} \times k_x r_x$$

$$= \frac{k_x r_x}{5.4 \times k}$$

$$F_{4,T} = \frac{k_4 r_4}{5.4k} = \frac{(2k) \times 6}{5.4k} = 2.22 \; kip \;\; (= F_{3,T})$$

Correct Answer is (B)

Question extras:
The rest of the lateral forces' distribution are calculated using the same method and are shown in the first figure down below:

$$F_{1,T} = \frac{k_1 r_1}{5.4 \times k} = \frac{k \times 30}{5.4 \times k} = 5.6\ kip\ (= F_{6,T})$$

$$F_{2,T} = \frac{k_2 r_2}{5.4 \times k} = \frac{(2k) \times 18}{5.4 \times k} = 6.7\ kip\ (= F_{5,T})$$

Frames resistance due to the lateral/ horizontal forces alone and without the torsional effect is proportional to each frame's stiffness as all the frames share the same drift. See below:

$$F_{x,h} = \frac{F}{\sum k_i} k_x$$

$$\sum k_i = k + 2k + 2k + 2k + 2k + k$$

$$= 10 \times k$$

$$F_{1,h} = F_{6,h} = \frac{200}{10 \times k} \times k = 20\ kip$$

$$F_{2,3,4,5,h} = \frac{200}{10 \times k} \times 2k = 40\ kip$$

The following figure represents forces attributed to accidental torsion:

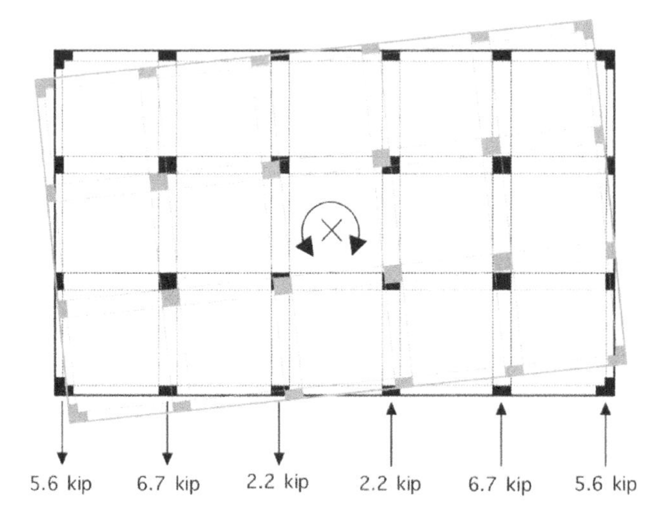

The following figure represents the drift direction attributed to lateral forces alone:

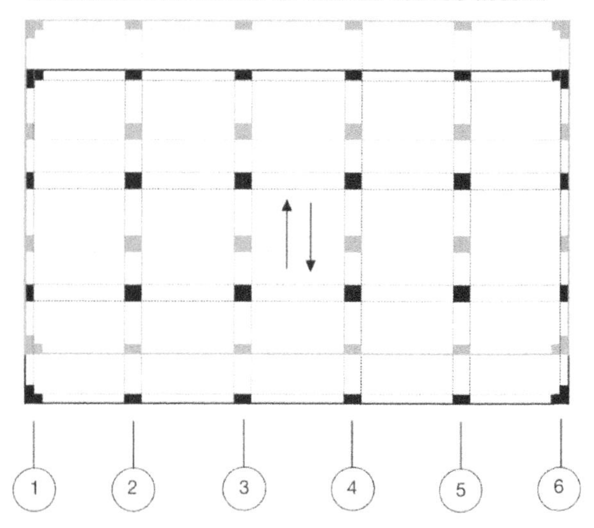

The following figure represents all forces combined:

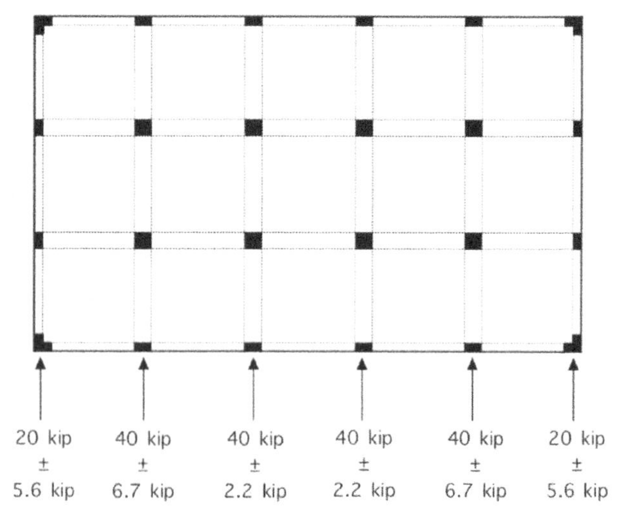

SOLUTION 3.7
Per ASCE 7-10 Chapter 10 rain load can be calculated as follows:

$$R = 5.2 \times (d_s + d_h)$$

$\boldsymbol{d_s}$ is the depth of water on the undeflected roof up to the secondary drain and ponding can be ignored for roofs sloped at ¼ $in/1\ ft$ and less.

$\boldsymbol{d_h}$ is the hydraulic head above the secondary drain which is given in this question. Newer versions of the code provide means of how to calculate $\boldsymbol{d_h}$ based on rain intensity and other factors.

$R = 5.2 \times (3.5 + 4) = 39\ psf$

$Total\ load = 39\ psf \times 100\ ft \times 150\ ft$

$= 585{,}000\ lb\ (585\ kip)$

Correct Answer is (D)

SOLUTION 3.8

Per ASCE 7 Section 5.4.3, dynamic effects of moving water shall be considered unless velocities are under 10 ft/sec, in which case, water can be converted into an equivalent hydrostatic load with an equivalent surcharge depth $\boldsymbol{d_h}$

Coefficient of drag (a) used shall not be less than $'1.25'$. It is noted however, and as given in the question, drag coefficient for the described elliptic pier is $'0.6'$. In order to comply with the code, a value of $a = 1.25$ will be used instead:

$$d_h = \frac{aV^2}{2g}$$

$$= \frac{1.25 \times \left(10\,\frac{ft}{sec}\right)^2}{2 \times 32.2\,\frac{ft}{sec^2}}$$

$$= 1.94\ ft \sim 2.0\ ft$$

Correct Answer is (C)

SOLUTION 3.9

Overturning:

Safety Factor against overturning is determined by calculating the Overturning Moment $\sum O.M.$ and the Resisting Moment $\sum R.M.$ around point 'O' as follows:

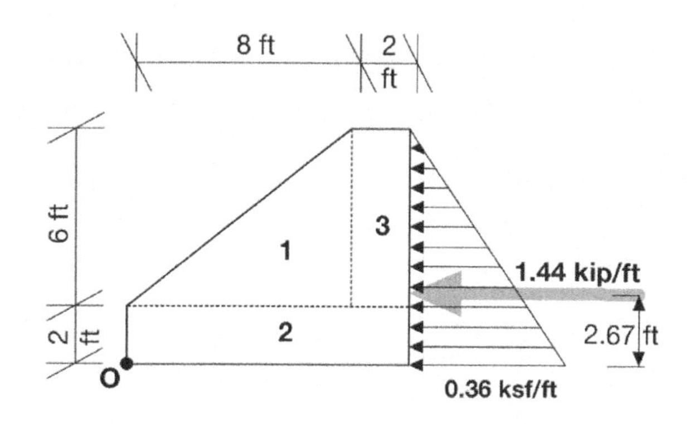

Resisting Moments $\sum R.M.$ as exhibited by the concrete wall:

Section	Volume ft^3/ft	Wt. kip/ft	Lever arm ft	R.M. $Kip.ft/ft$
1	24	3.6	5.33	19.2
2	20	3.0	5.0	15.0
3	12	1.8	9.0	16.2
Totals	**56**	**8.4**		**50.4**

$\sum R.M. = 50.4\ kip.ft/ft$

$\sum O.M. = [0.5 \times (45\ psf/ft \times 8\ ft) \times 8\ ft] \times 2.67\ ft$

$= 3{,}845\ lb.ft/ft\ (3.8\ kip.ft/ft)$

$$SF_{OT} = \frac{\sum R.M.}{\sum O.M.} = \frac{50.4}{3.8} = 13.26$$

Sliding:

Safety factor against sliding is determined by calculating the sliding force F_S and resisting force F_R as follows:

$$F_S = 0.5 \times (45\ psf/ft \times 8ft) \times 8ft \times \frac{1\ kip}{1{,}000\ lb}$$

$$= 1.44\ kip/ft$$

Resisting force is determined from IBC code Section 1806.3, Table 1806.2 with due consideration to the maximum limit in Section 1806.3.2.

Considering the above, the cohesion exhibited by clayey silts is found to be 130 psf and is used/multiplied by the bottom area of the wall. This value should not exceed half the deadload of the wall:

$$F_R = 130\,psf \times 1\,ft \times 10\,ft \times \frac{1\,kip}{1{,}000\,lb}$$

$$= 1.3\,kip/ft < \frac{8.4\,kip/ft}{2}$$

$$SF_{sliding} = \frac{F_R}{F_S} = \frac{1.3}{1.44} = 0.9$$

Correct Answer is (A)

SOLUTION 3.10

Service loads are used when checking for bearing pressures and area sizing. Ultimate loads are only used to design the concrete cross section.

This is a case of eccentricity ***e***, and the following equation can be used to ascertain that the lowest pressure in the eccentricity direction is zero, and the largest pressure does not exceed 3,000 psf.

$$Q = \frac{P}{A}\left(1 \mp \frac{6e}{L}\right)$$

<u>First condition pressure ≥ zero:</u>

$$\frac{P}{A}\left(1 - \frac{6e}{L}\right) \geq zero$$

$$1 - \frac{6e}{L} \geq zero$$

$$1 \geq \frac{6e}{L} \rightarrow e \leq L/6$$

From the following figures:

$$e = \frac{L}{2} - \frac{C}{2}$$

$$\frac{L}{6} \geq \frac{L}{2} - \frac{C}{2}$$

$$\rightarrow L \geq \frac{3C}{2} = 36\,in\,(3ft)$$

$$\rightarrow e = 6\,in\,(0.5\,ft)$$

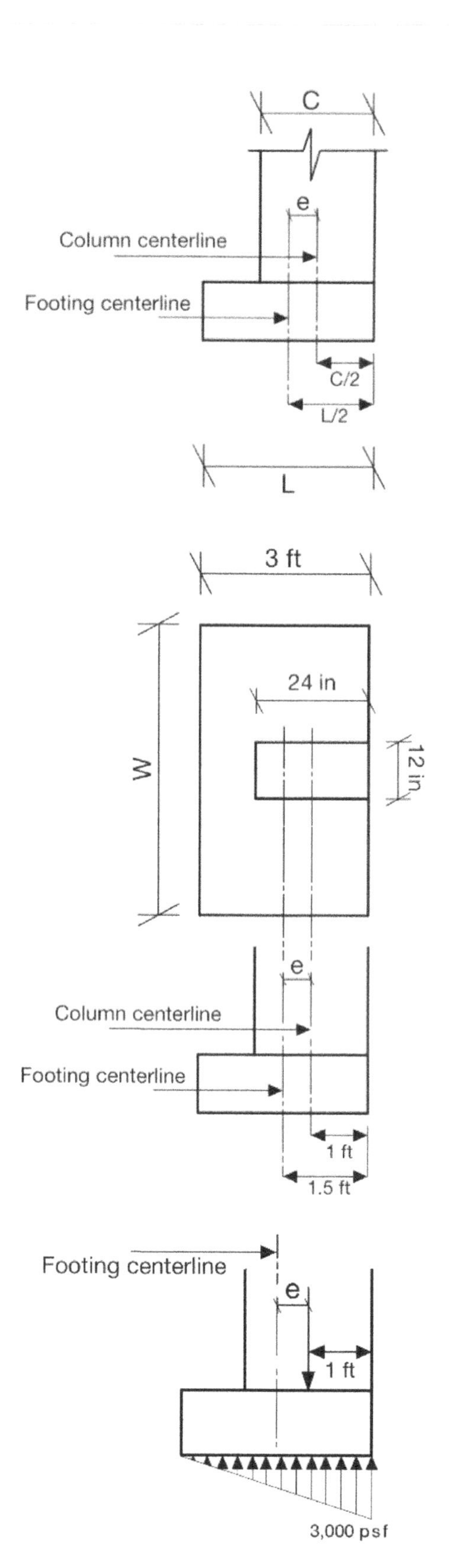

Second condition pressure ≤ 3,000 psf:

$$\frac{P}{A}\left(1+\frac{6e}{L}\right) \leq 3{,}000\ psf$$

$$\frac{27\ kip}{3\ ft \times W}\left(1+\frac{6 \times 0.5\ ft}{3\ ft}\right) \leq 3\ ksf$$

$$\rightarrow W \geq 6\ ft$$

Correct Answer is (B)

SOLUTION 3.11

AASTHO LRFD Bridge Design Specification, Section 3.6.2.2 *Buried Components* specifies the Dynamic Load Allowance IM as follows:

$$IM = 33 \times (1.0 - 0.125 D_E)$$
$$= 33 \times (1.0 - 0.125 \times 3.0\ ft)$$
$$= 21\%$$

Per 3.6.2.1, and 3.6.1.2.3, dynamic factors are applied to static and tandem loads as follows:

$$Tandem\ 1 = 1.21 \times 16 = 19.4\ kip$$

$$Tandem\ 2 = 1.21 \times 64 = 77.5\ kip$$

$$Lane\ Load = 6.4\ klf$$

Correct Answer is (B)

SOLUTION 3.12

Stiffness offered by shear walls is based on their height-to-length ratio as follows:

- $h/l \leq 0.3$
  Flexural and shear stiffness can be neglected, and lateral loads' distribution relies on walls' cross-sectional areas.
- $h/l \geq 3.0$
  Flexural stiffness should be considered, but shear deformations ignored. Lateral loads distribution is based on walls' moment of inertias.
- $0.3 > h/l < 3.0$
  Both shear and flexural stiffness should be considered. In this case, and for simplicity, an effective moment of inertia I_{eff} can be used for distribution purposes.

  I_{eff} can be found in Figure 3.10.26 of the *PCI Design Handbook 7th* edition and other sources as well.

$h/l = 7.5/25 = 0.3$ → cross sectional area of walls will be used to determine center of rigidity.

Short direction (y-axis):

Given that 'wall A' and 'wall B' are both resisting walls in the short direction, only those two walls' cross-sectional areas will be considered in determining the center of gravity.

Taking datum to the left-most of the floor:

$$= \frac{50 \times \frac{10}{12} \times \left(\frac{10/12}{2}\right) + 25 \times \frac{10}{12} \times \left(100 + \frac{10/12}{2}\right)}{50 \times \frac{10}{12} + 25 \times \frac{10}{12}}$$

$$= 33.75\ ft$$

Distance from center of gravity to center of rigidity is as follows:

$$= \left(50 + \frac{10/12}{2}\right) ft - 33.75\ ft = 16.7\ ft$$

Long direction (x-axis):

Both 'wall D' and 'wall C' have the same longitudinal cross-sectional area and are placed in a symmetrical position. Center of gravity coincides with the center of rigidity then.

Taking center of gravity in the middle of the rectangular shaped floor, the below figure depicts the eccentricity the question seeks.

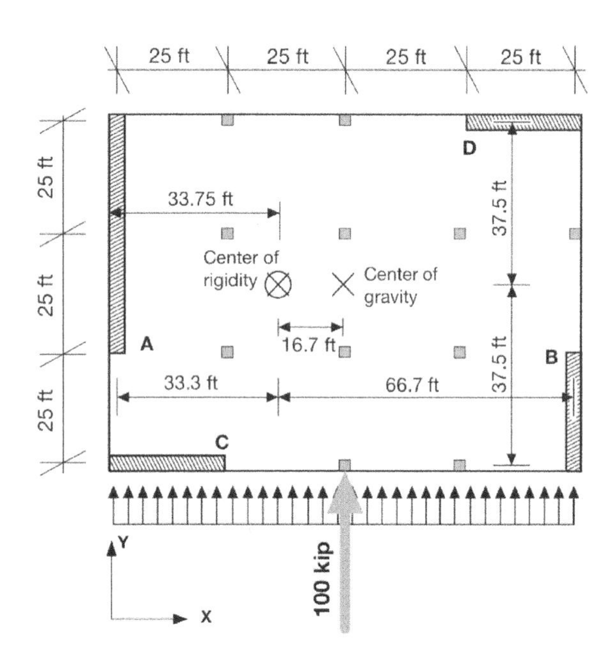

Torsional moment M_T:

$$M_T = 16.7 \times 100\ kip = 1{,}670\ kip.ft$$

ASCE 7 Section 12.14.8.3.1 deals with flexible diaphragm lateral forces distributions and relies on tributary area for distribution. 12.14.8.3.2 specifies that rigid diaphragms shall consider relative stiffness of vertical components for lateral distribution, which will be the case in this question.

PCI Design Handbook Section 3.5.7 can be referred to for equations that cover such distribution as well.

Taking x-axis for the long direction and y-axis for the short:

$$F_x = \mp \frac{M_T\, r_y\, k_x}{\Sigma(K_y r_x{}^2 + K_x r_y{}^2)}$$

$\boldsymbol{M_T}$ is the torsional moment, $\boldsymbol{k}$ is rigidity (in this case cross-sectional area) in the specified direction, and $\boldsymbol{r_x}$ and $\boldsymbol{r_y}$ are distances to the required wall from center of stiffness.

$$F_D = \frac{-1{,}670\ kip.ft \times 37.5\ ft \times \left(25 \times \frac{10}{12}\right) ft^2}{\left(\begin{array}{c}\left(50\times\frac{10}{12}\right)\times 33.3^2 + \left(25\times\frac{10}{12}\right)\times 66.7^2 \\ + \\ \left(25\times\frac{10}{12}\right)\times 37.5^2 + \left(25\times\frac{10}{12}\right)\times 37.5^2\end{array}\right) ft^4}$$

$$= \frac{-1{,}304{,}687.5\ kip.ft^4}{197{,}482.7\ ft^4}$$

$$= -6.6\ kip \leftarrow$$

Correct Answer is (D)

Question extras:

The rest of the lateral forces' distribution are calculated using the same method for the long direction as follows:

Forces in the long direction (x-axis):

$$F_C = \frac{1{,}670\ kip.ft \times 37.5\ ft \times \left(25 \times \frac{10}{12}\right) ft^2}{\left(\begin{array}{c}\left(50\times\frac{10}{12}\right)\times 33.3^2 + \left(25\times\frac{10}{12}\right)\times 66.7^2 \\ + \\ \left(25\times\frac{10}{12}\right)\times 37.5^2 + \left(25\times\frac{10}{12}\right)\times 37.5^2\end{array}\right) ft^4}$$

$$= +6.6\ kip \rightarrow$$

Forces in the short direction (y-axis):

$$F_y = \mp \frac{V_y\, K_y}{\Sigma K_y} \mp \frac{M_T\, r_x\, k_y}{\Sigma(K_y r_x{}^2 + K_x r_y{}^2)}$$

V_y is the resultant force applied to the center of gravity.

$$F_A = \left[\begin{array}{c}(+)\dfrac{100\ kip \times \left(50\times\frac{10}{12}\right) ft^2}{50\times\frac{10}{12} ft^2 + 25\times\frac{10}{12} ft^2} \\ + \\ (-)\dfrac{1{,}670\ kip.ft \times 33.3\ ft \times \left(50\times\frac{10}{12}\right) ft^2}{197{,}482.7\ ft^4}\end{array}\right]$$

$$= +\frac{4{,}166.67\ kip.ft^2}{62.5\ ft^2} - \frac{2{,}317{,}125\ kip.ft^4}{197{,}482.7\ ft^4}$$

$$= +66.7 - 11.7$$

$$= +55\ kip \uparrow$$

$$F_B = \left[(+)\frac{100\ kip \times \left(25 \times \frac{10}{12}\right) ft^2}{50 \times \frac{10}{12} ft^2 + 25 \times \frac{10}{12} ft^2} + (+)\frac{1{,}670\ kip.ft \times 66.7\ ft \times \left(25 \times \frac{10}{12}\right) ft^2}{197{,}482.7\ ft^4} \right]$$

$$= +\frac{2{,}083.3\ kip.ft^2}{62.5\ ft^2} + \frac{2{,}320{,}604.2\ kip.ft^4}{197{,}482.7\ ft^4}$$

$$= +33.3 + 11.7$$

$$= +45\ kip \uparrow$$

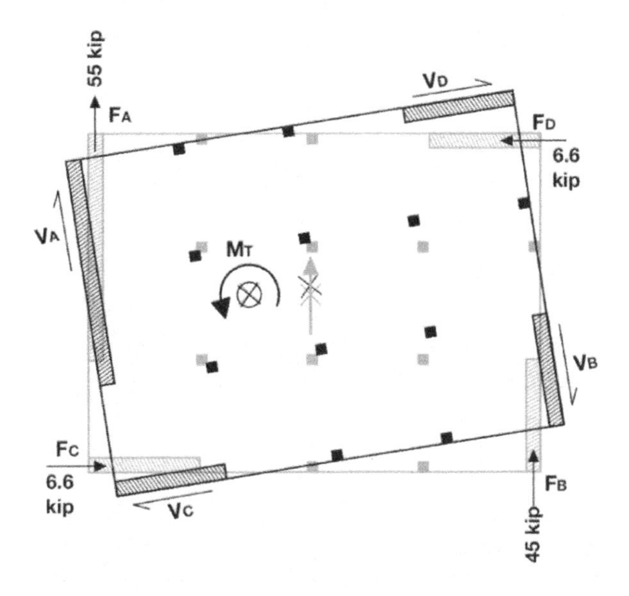

SOLUTION 3.13

IBC code Section 1806.1 specifies that the load combination that shall be used with presumptive load bearing values is the allowable stress design combinations of Section 1605.3, and those are as follows(*):

	Combination based on allowable stress	Result (kip)
1	D + F 7 + 0	7.0
2	D + H + F + L 7 + 0 + 0 + 14.5	21.5
3	D + H + F + (L_r or S or R) 7 + 0 + 0 + (0 or 1 or 0)	8.0
4	D + H + F + 0.75 L + 0.75 (L_r or S or R) 7 + 0 + 0 + 0.75×14.5 + 0.75 (0 or 1 or 0)	18.6
5	D + H + F + (0.6 W or 0.7 E) 7 + 0 + 0 + (0.6×2.5 or 0)	8.5
6	D + H + F + 0.75 (0.6 W) + 0.75 L + 0.75 (L_r or S or R) 7 + 0 + 0 + 0.75 (0.6×2.5) + 0.75×14.5 + 0.75 (0 or 1 or 0)	19.8
7	D + H + F + 0.75 (0.7 E) + 0.75 L + 0.75 S 7 + 0 + 0 + 0.75 (0) + 0.75×14.5 + 0.75×1	18.6
8	0.6 D + 0.6 W + H 0.6×7 + 0.6×2.5 + 0	5.7
9	0.6 (D + F) + 0.7 E + H 0.6 (7 + 0) + 0.7 (0) + 0	4.2

L_r as defined in the IBC is the Roof Live Load for 30 psf roofs or less. In which case, L_r is taken as zero as the roof supported carries 30 psf.

Because of that, the Roof Load identified in this question is added to the Live Load to make a new a total Live Load of 14.5 psf.

Total load used to check for bearing capacity in this case is generated from Load Combination 2 which is 21.5 psf.

Correct Answer is (A)

* ACI 318-14.3.2.2 specifies that footing areas are determined using unfactored loads.

PART II Structural Depth

Section 3 Load Analysis

SOLUTION 3.14

Using Chapter 7 of ASCE 7, *Snow Loads,* the following sections should be referred to in order:

Section 7.3 Flat Roof Snow Load p_f:

Flat roof snow load is calculated using the following equation:

$p_f = 0.7C_eC_tI_sp_g$

Exposure factor C_e, based on the information given in the question, and in reference to Section 26.7 of the code, is '0.9'.

Thermal factor C_t with ventilated roofs and $R > 1.1$ is '1.1'.

Importance factor I_s for the risk category defined is '0.8' as defined per Section 1.5.2.

Based on all the above, Flat Roof Snow Load p_f is calculated as follow:

$p_f = 0.7 \times 0.9 \times 1.1 \times 0.8 \times 30 = 17\ psf$

Section 7.4 Sloped Roof Snow Loads p_s:

Roof slope factor C_s for cold roofs with $C_t = 1.1$, un-slippery surface (i.e., asphalt shingles) and a slope of 12.5 on 30 (i.e., 5 on 12), and in reference to Figure 7.2b = 1.0.

Based on this information, Sloped Snow Load p_s is calculated as follows:

$p_s = C_sp_f = 1.0 \times 17.0 = 17.0\ psf$

Section 7.6.1 Unbalanced Snow Loads for Hips & Gable Roofs:

The slope of this roof is 5 on 12 and hence qualifies for unbalanced loading (i.e., the snow load to the left side of the ridge is different than that to its right or vice versa). 'Option D' as a solution to this question in this case would be wrong.

Distance from eave to ridge W is $30\ ft > 20\ ft$, which makes 'Option A' that unloads one side and loads the other with (I_sp_g) a wrong choice as well – see Figure 7.5 of ASCE 7 code for more clarity on this matter.

Because $W > 30\ ft$, the leeward shall be loaded with $0.3p_s = 5\ psf$, and the Winward shall be loaded with $p_s = 17.0\ psf$ along with a rectangular surcharge that has the following value:

$\frac{h_d\,\gamma}{\sqrt{S}} = \frac{1.9 \times 17.9}{\sqrt{2.4}} = 22\ psf$

Drift height h_d of $1.9\ ft$ was interpolated from Figure 7.9. γ is snow density and is calculated from Equation 7.7-1 as:

$0.13p_g + 14 = 17.9\ psf$

S is the roof slope run for a rise of one = $2.4\ ft$.

In a similar fashion, the horizontal run of the surcharge from the ridge is calculated as follows:

$\frac{8\sqrt{S}\,h_d}{3} = \frac{8 \times \sqrt{2.4} \times 1.9}{3} \approx 6\ ft$

Using Figure 7.5 from the ASCE 7 code, the above surcharge is added to the horizonal projection of the load as follows:

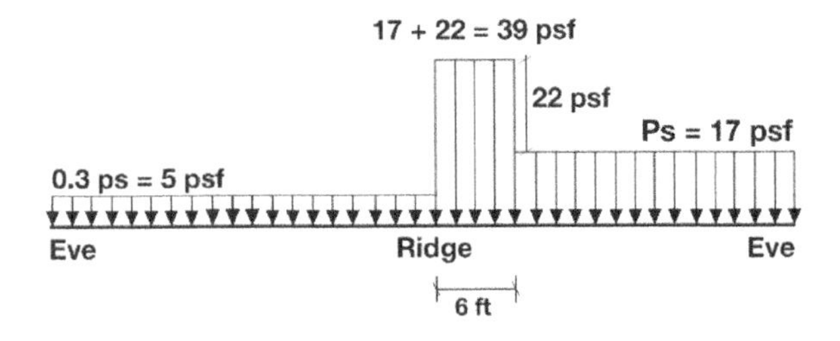

Correct Answer is (C)

SOLUTION 3.15

Both the IBC and the ASCE 7 codes can be used to solve this question.

Using the ASCE 7 code in this case, given the uniform Live Load applied is less than 100 psf, Section 4.7.2 can be used.

Tributary area $\boldsymbol{A_T}$ for internal columns is 1,225 ft^2 as shown below, $\boldsymbol{K_{LL}}$ for interior columns is '4' per Table 4.2.

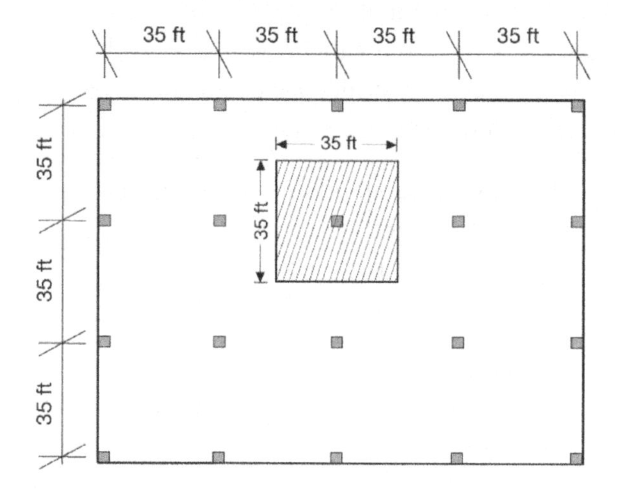

The value of $K_{LL} \times A_T$ is $1{,}225 \times 4 = 4{,}900 > 400\ ft^2$ which means that is it permitted to reduce uniform Live Loads as follows:

$$L = L_o\left(0.25 + \frac{15}{\sqrt{K_{LL}\,A_T}}\right)$$

$$= 75\ psf \times \left(0.25 + \frac{15}{\sqrt{4{,}900}}\right) = 34.8\ psf$$

However, Section 4.7.2 specifies that for members supporting one floor, the reduced load cannot be less than:

$$0.5L_o = 0.5 \times 75$$

$$= 37.5\ psf > 34.8\ psf \rightarrow Not\ ok$$

Hence use $L = 37.5\ psf$

Correct Answer is (B)

SOLUTION 3.16

Using the *Load and Resistance Factor Design* (LRFD) method, otherwise known as the *Strength Design* method, the Ultimate Load for a DL and a LL combination is:

$$1.2\ \text{DL} + 1.6\ \text{LL}$$

IBC Section 1607.11 specifies that in order to achieve the maximum stress at locations under consideration, the minimum applied loads shall be that of the Dead Load applied to all spans in combination with Live Loads.

Live Loads in this case should be applied to selected spans such that the greatest effect in that location is generated.

Reviewing the diagrams provided in Chapter 4 of the *NCEES handbook* for continuous beams with three spans, the below load combination generates the maximum negative moment at 'B':

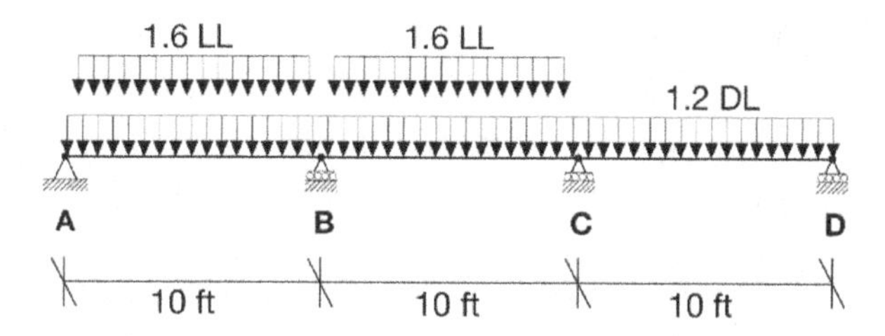

Applying load factors for this combination provides the following load diagram:

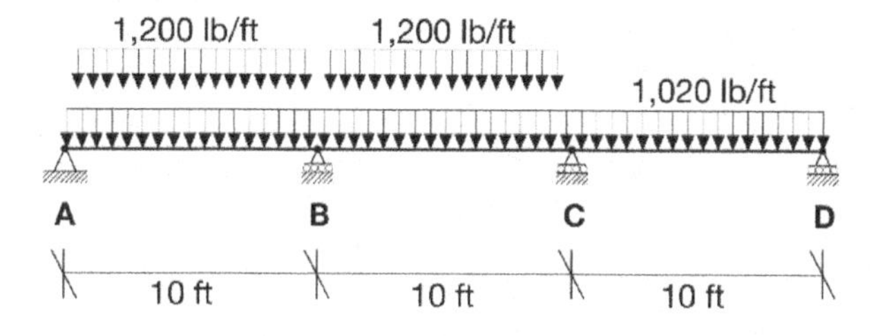

$$M_{B,max} = M_{B,DL,max} + M_{B,LL,max}$$

$$= -0.10\ wl^2 + -0.117\ wl^2$$

$$= -0.1 \times 1{,}020 \times 10^2 + -0.117 \times 1{,}200 \times 10^2$$

$$= -24{,}240\ lb.ft\ (-24.2\ kip.ft)$$

Correct Answer is (B)

SOLUTION 3.17

In reference to the *AASHTO LRFD Bridge Design Specification* Section 3.6.5.1. When abutments and piers are located within 30 ft to the edge of a roadway, and the owner wishes to provide protection, a static load of 600 kip can be assumed to act on the structure in a direction of $zero$ to 15° with the edge of the pavement – previous versions of the code stipulate a load of 400 kip applied in any direction.

The other methods mentioned in options C and D are true in case the recommendation was to redirect or absorb the collision load, not to resist it.

Correct Answer is (A)

SOLUTION 3.18

ASCE 7 code Section 12.4.2.3 specifies the following load combination replacement for combination 5 and 6 for the basic combination of Section 2.3.2.

This is also applicable when flood and ice loads are not used. Which is the case here.

	Combination based on allowable stress	Result (kip)
1	1.4 D 1.4×5	7.0
2	1.2 D + 1.6 L + 0.5 (L_r or S or R) 1.2×5 + 1.6×6 + 0.5 (1.5 or 0 or 0)	16.4
3	1.2 D + 1.6 (L_r or S or R) + (L or 0.5W) 1.2×5+ 1.6(1.5 or 0 or 0) + (6 or 0.5×2)	14.4
4	1.2 D + 1.0 W + L + 0.5 (L_r or S or R) 1.2×5 + 1 × 2 + 6 + 0.5 (1.5 or 0 or 0)	14.8
5	(1.2 + 0.2 S_{DS}) D + ρ Q_E + L + 0.2S (1.2 + 0.2×0.75)×5 + 1.3 × 3 + 6 + 0	16.7
6	(0.9 – 0.2S_{DS}) D + ρ Q_E + 1.6 H (0.9 – 0.2×0.75)×5 + 1.3 × 3 + 0	7.7
7	0.9 D + 1.0 E 0.9×5 + 1 × 3	7.5

The design shall proceed using the maximum load generated which belongs to load combination 5 with 16.7 kip.

Correct Answer is (A)

SOLUTION 3.19

Wind load calculation is all laid out in Chapters 26 to 31 of the ASCE 7 as follows – this information is also repeated and further detailed in the last section of this book:

Chapter 26 contains most the basic parameters and <u>shall frequently be referred to</u> as a starting point for any question on wind loads.

Chapter 27 explains the directional procedure, a slightly lengthier procedure that applies to all building shapes and heights.

Chapter 28 refers to the envelope procedure for low rise $\leq 60\,ft$ buildings, and its methods are summarized in two parts:

- Part I is a slightly more detailed procedure.
- Part II is a simplified procedure with conditions.

Chapter 29 tackles other structures and buildings' appurtenances.

Chapter 30 for components and cladding.

Chapter 31 provides the specs for the wind tunnel procedure.

This question requests the use of the simplified procedure which is detailed in Chapter 28 Part II. The basic parameters needed in this question are all gathered from Chapter 26 as follows:

Section 26.7 assigns this building to an exposure category B as it is located in an urban area (Section 26.7.2), and the surface roughness prevails in the upwind direction for over 2,600 ft when the building is > 30 ft high (Section 26.7.3).

The topographic factor K_{zt} is determined from Section 26.8 as '1.0' given the structure is not on a ridge and does not meet all the conditions specified in Section 26.8.1 – i.e., the wind will not speed up due to a ridge, escarpment, or a hill.

In reference to Figure 28.6-1 of the code, the simplified design wind pressure for exposure category B along with a height of 30 ft for zone E as shown in the figure below is as follows:

$$p_{S30} = -32.2\ psf$$

p_{s30} is adjusted using the adjustment factor λ to convert the pressure at 30 ft and exposure B to any other heights and/or exposures.

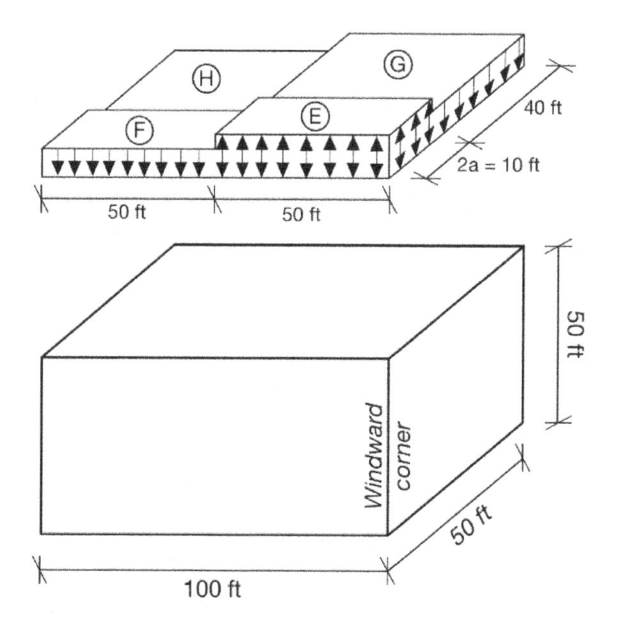

The adjustment factor λ is taken from the bottom of Figure 28.6-1 as follows:

$$\lambda = 1.16$$

$$p_S = \lambda K_{zt} p_{S30}$$

$$= 1.16 \times 1.0 \times (-32.2) = -37.4\ psf$$

Correct Answer is (B)

SOLUTION 3.20

In reference to the summary of wind load Chapters presented in the solution of the previous question, this question shall use Chapter 29 (other structures) of the ASCE 7 with due reference to Chapter 26 to gather the basic parameters needed in Chapter 29 equations.

Section 29.4.1 specifies the following force for free standing walls and signboards:

$$F = q_h G C_f A_s \quad (in\ lb)$$

q_h is the velocity pressure and is calculated from equation 29.3.2 of the same chapter.

$$q_h = 0.00256\, K_z\, K_{zt}\, K_d V^2 \quad (in\ lb/ft^2)$$

K_z is the pressure exposure coefficient and is taken from Table 29.3-1 as '0.85' for areas with exposure category C and a structure height of 15 ft.

K_{zt} is the topographic factor defined and found in Section 26.8.2, Figure 26.8-1 by applying the following equation:

$$K_{zt} = (1 + K_1 K_2 K_3)^2$$

Topographic multipliers K_1, K_2 and K_3 are gathered from the first table of Figure 26.8-1 for escarpments located in exposure category C, taking $H = 10\,ft, L_h = 20\,ft, x = 6\,ft, z = 10 + 15 = 25\,ft$ as follows:

- For $H/L_h = 0.50, K_1 = 0.43$
- For $x/L_h = 0.30$, $K_2 = 0.93$ (using interpolation)
- For $z/L_h = 1.25$, $K_3 = 0.05$ (using interpolation)

$K_{zt} = (1 + 0.43 \times 0.93 \times 0.05)^2 = 1.04$

$\boldsymbol{K_d}$ is the directionality factor and is collected from Section 26.6, Table 26.1.46-1 as '0.85' for free standing signs.

<u>Back to velocity pressure calculation:</u>

$$q_h = 0.00256 \times 0.85 \times 1.04 \times 0.85 \times 160^2$$
$$= 49.2\ lb/ft^2$$

<u>Back to the wind force equation:</u>
$\boldsymbol{G}$ is the gust factor and can be assumed as '0.85' per Section 26.9.1.

$\boldsymbol{C_f}$ is the net force coefficient collected from Figure 29.4-1 for case A, and for:

$$s\ (15\ ft) = h\ (15\ ft)$$

and

$$\frac{B}{s} = \frac{30}{15} = 2$$

$$\rightarrow C_f = 1.4$$

In which case the resultant force $\boldsymbol{F}$ occurs at a distance above the centroid of the sign equals to $0.05h$ $(i.e., 0.75\ ft)$ as follows:

$$F = q_h G C_f A_s$$
$$= 49.2\ lb/ft^2 \times 0.85 \times 1.4 \times (15 \times 30) ft^2$$
$$= 26{,}347\ lb\ (26\ kip)$$

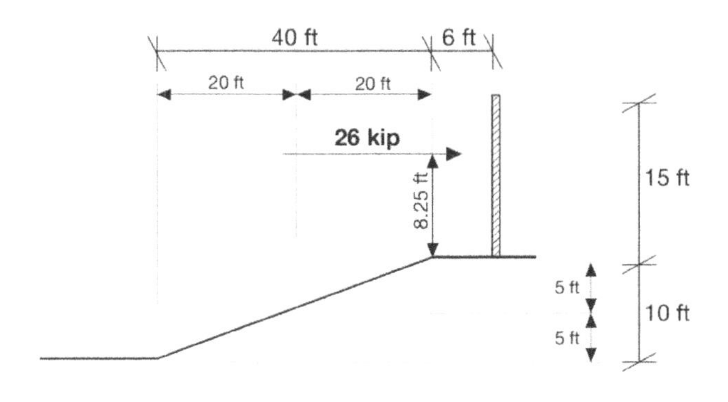

Correct Answer is (C)

SECTION 4

Structural Design

Problems & Solutions

PROBLEM 4.1 ***Critical Bucking Load Nonsway Frames***

The architect decided to remove two beams from the second floor of this normal weight concrete building to create a double volume as shown. The beams removed (D3 to C3 and D4 to C4) are clouded.

Dimensions for all beams and columns are $12\ in \times 12\ in$ and concrete compressive strength is $f_c' = 3{,}500\ psi$.

Clear distance between floors:

- First floor 14 ft
- Second floor 12 ft

End moments for exterior columns are almost equivalent, and those exterior columns are bent in single curvature.

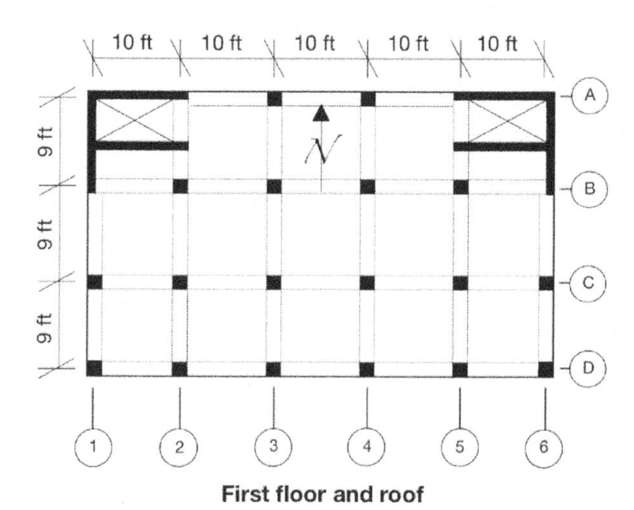

First floor and roof

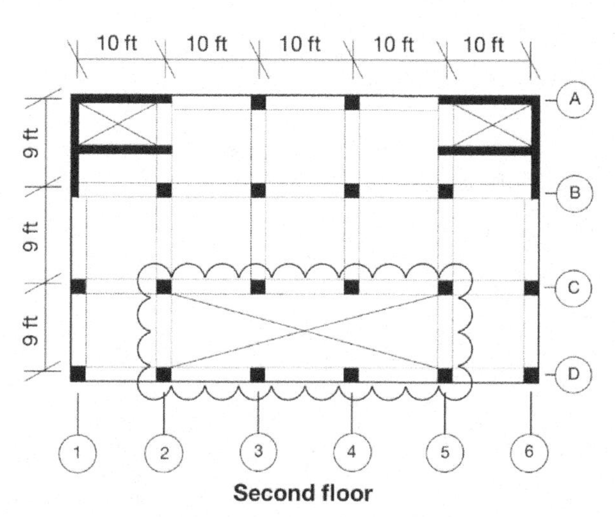

Second floor

Reduction in the critical buckling load as applied from the roof due to this change in kip for column D4 is most nearly:

(A) 470

(B) 130

(C) 320

(D) 600

PROBLEM 4.2 ***Critical Bucking Load Sway Frames***

The architect decided to remove two beams from the second floor of this normal weight concrete building to create double volume as shown. The beams removed (D3 to C3 and D4 to C4) are clouded.

Dimensions for all beams and columns are $12\ in \times 12\ in$ and concrete compressive strength is $f_c' = 3{,}500\ psi$.

Clear distance between floors:

- First floor 14 ft
- Second floor 12 ft

There are short stubs below the first floor 4 ft length to transfer loads to the respective footings.

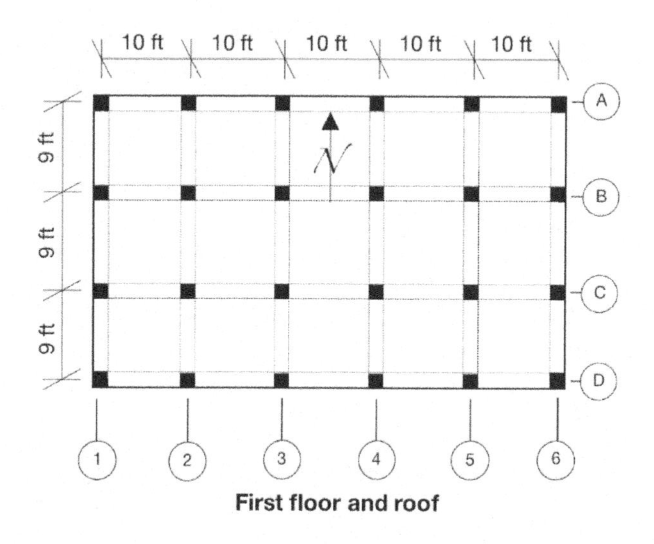

First floor and roof

Second floor

Reduction in the critical buckling load as applied from the roof due to this change for column D4 (*) in kip is most nearly:

(A) 70

(B) 124

(C) 180

(D) 55

* Use 1st floor column as base for calculating the *before* load capacity.

PROBLEM 4.3 ***One Way Slab Thickness***

The below is a one-way slab supported by $10\ in$ wide and $20\ in$ deep beams. The following reinforced concrete properties were used for those slabs:

- $f'_c = 3{,}500\ psi$
- $f_y = 50\ ksi$
- $\gamma_{concrete} = 95\ pcf$

All spans should have same thicknesses for MEP purposes.

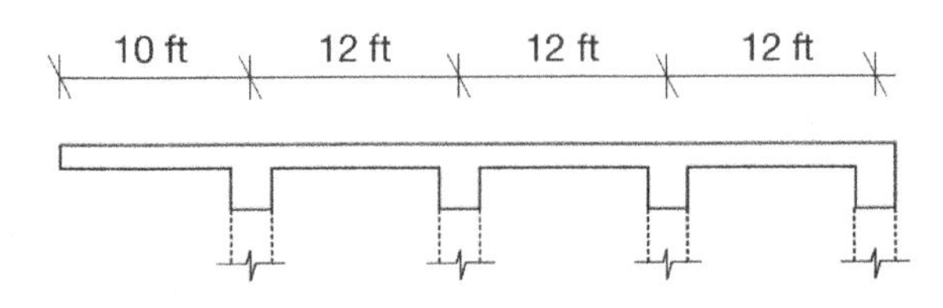

Based on the above information, the minimum thickness for the above one-way slabs should be:

(A) 11.5 in

(B) 12.5 in

(C) 10.0 in

(D) 16.0 in

PROBLEM 4.4 ***Two Way Slab Thickness***

The below slab – section at top and plan to the bottom – is designed to sit on a grid of $12\ in$ wide by $14\ in$ deep beams with $f_y = 60\ ksi$ & $f'_c = 4\ ksi$, and normal weight concrete has been used all the way.

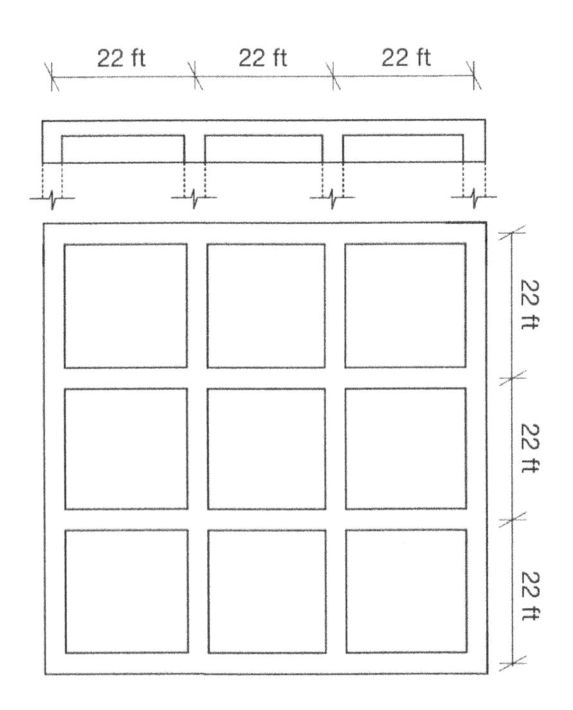

Based on this information, minimum slab thickness should be:

(A) 5.0 in

(B) 7.0 in

(C) 8.5 in

(D) 9.5 in

PROBLEM 4.5 *Single Footing Thickness*

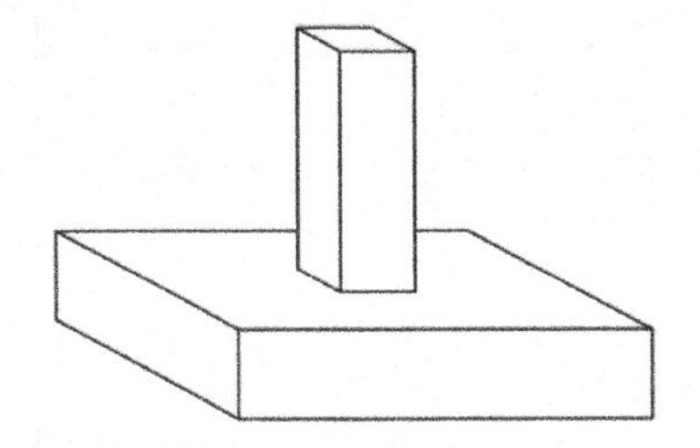

The minimum thickness for a single, $6\,ft \times 6\,ft$ plain concrete, footing is:

(A) $4.0\,in$

(B) $6.0\,in$

(C) $7.0\,in$

(D) $8.0\,in$

PROBLEM 4.6 *Concrete Slab Moment Calculation*

The below reinforced concrete slab is $8.0\,in$ thick and supported by beams (*) all around as shown. The top figure is a cross-section, and the bottom figure is a plan with:

- Dead Load = self-weight + $40\,pcf$
- Live Load = $50\,pcf$
- $\gamma_{concrete} = 150\,pcf$
- Supporting beams' width = $12\,in$

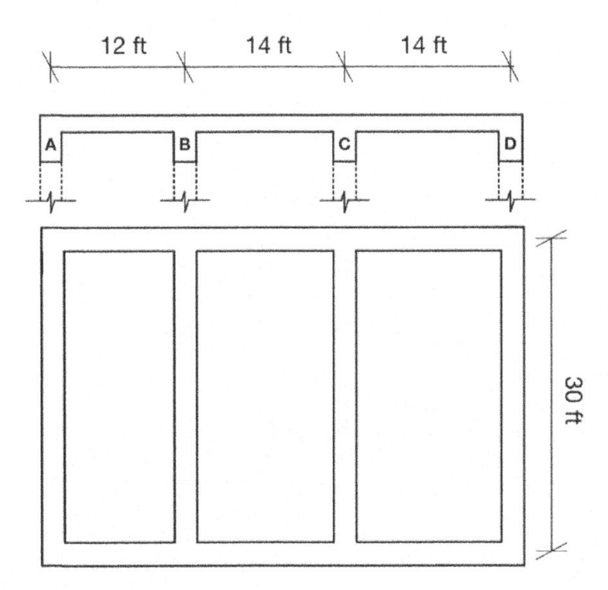

Based on this information, the negative ultimate moment in the short direction of panels at beam B in $kip.ft/ft$ is most nearly:

(A) 3.2

(B) 2.7

(C) 3.8

(D) 4.4

* Dimensions shown are center to center for all beams.

PROBLEM 4.7 *Flat Slab Moment*

The below reinforced concrete flat slab is $8\,in$ thick and is supported by columns (*) with no beams. Considering the following design attributes:

- Dead Load = self-weight + $40\,psf$
- Live Load = $50\,psf$
- $\gamma_{concrete} = 150\,pcf$
- Columns are all $12\,in \times 12\,in$

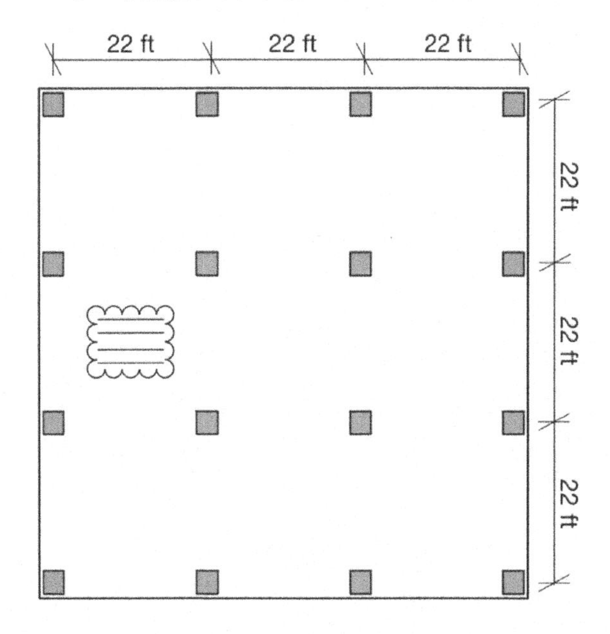

Positive moment generated in the clouded area in $kip.ft/ft$ and in the direction shown as represented by reinforcing rebars is most nearly:

(A) 5.6

(B) 11.3

(C) 7.6

(D) 3.8

* Dimensions shown are center to center for all columns.

PROBLEM 4.8 ***Flat Slab Shear Analysis***
Given the following information for a flat slab supported with square columns only:

- Effective depth $d = 8.5\ in$
- $f_y = 60\ ksi$
- $f_c' = 4{,}000\ psi$
- $18\ in \times 18\ in$ columns

The nominal two-way shear strength provided by concrete for an internal column in psi is most nearly:

(A) 329

(B) 253

(C) 246

(D) 191

PROBLEM 4.9 ***Punching Shear caused by Circular Columns***
The below large, reinforced concrete, flat slab is 9.5 in thick with an effective depth $d = 8\ in$, and is supported by circular columns (*) and has the following design attributes:

- Dead Load = self-weight + 40 psf
- Live Load = 50 psf
- $\gamma_{concrete} = 150\ pcf$
- Supporting columns are 13.5 in dia.

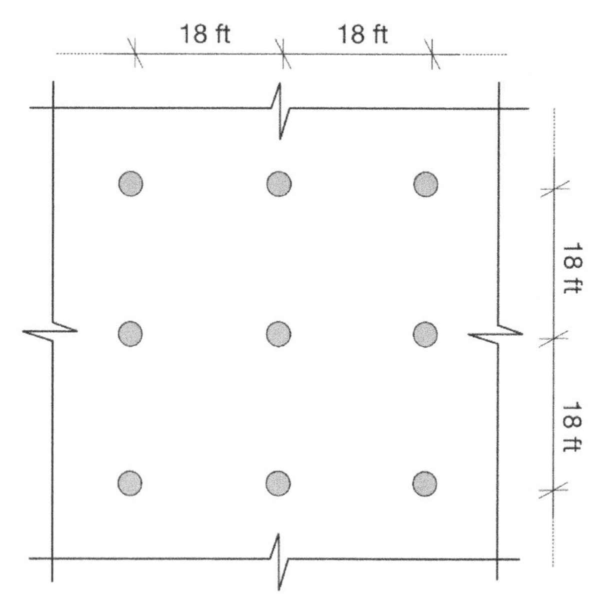

Assuming no moment transferred to columns, the ultimate punching shear stress caused by an internal column in psi is:

(A) 63

(B) 126

(C) 202

(D) 136

* Dimensions shown are center to center for all columns.

PROBLEM 4.10 ***Concrete Section Design***
The ultimate shear and moment load applied to the below concrete beam are $V_u = 150\ kip$ and $M_u = 730\ kip.ft$ respectively.

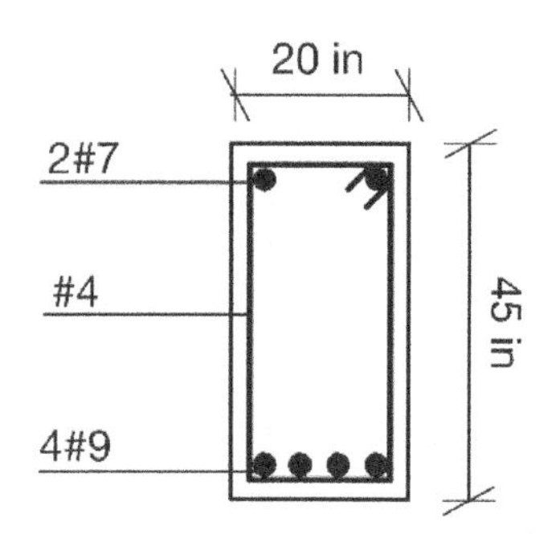

Using #4 stirrups with $f_y = 40\ ksi$, $f_c' = 4\ ksi$, and a specified concrete cover of 1 ½ in, the spacing that should be provided for those stirrups is most nearly:

(A) 4 in

(B) 7.5 in

(C) 17.5 in

(D) 10 in

PROBLEM 4.11 ***Concrete Section Design***

The ultimate shear and moment applied to the below concrete beam are $V_u = 30\ kip$ and a positive $M_u = 100\ kip.ft$.

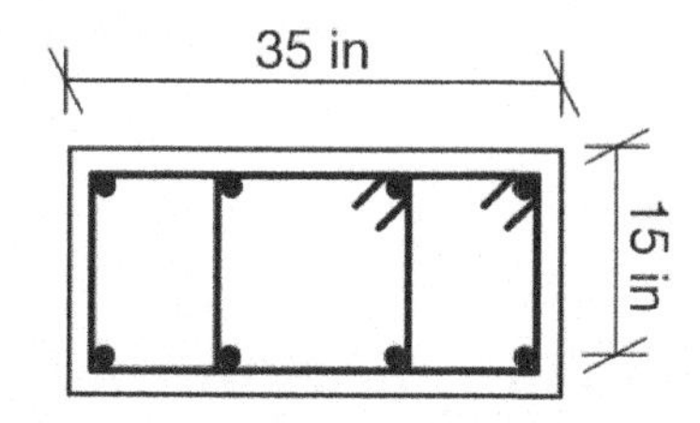

Using $f_y = 60\ ksi$ for main positive reinforcements, concrete strength of $f_c' = 4\ ksi$, the required positive/bottom area of steel for this section in in^2 is most nearly:

(A) 1.75

(B) 1.66

(C) 1.50

(D) 2.50

PROBLEM 4.12 ***Masonry Wall Bearing Analysis***

The below is a masonry wall that is not laid in a running bond with fully grouted 10 in thick blocks.

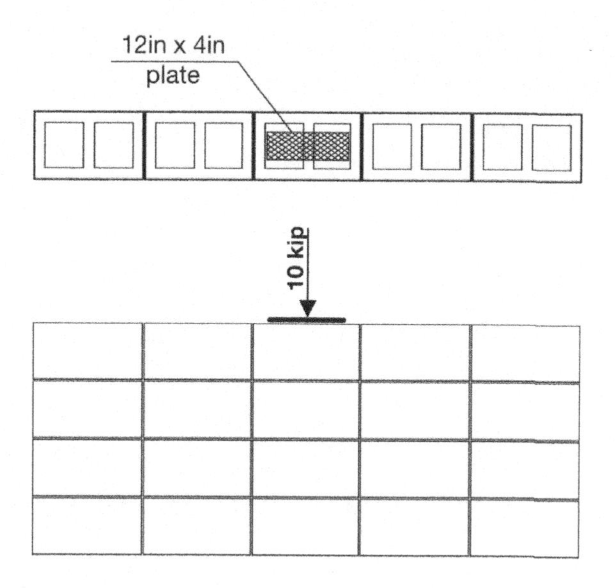

The bearing stress under this load in psi is most nearly:

(A) 127.5

(B) 107.5

(C) 208.3

(D) 78.1

PROBLEM 4.13 ***Masonry Wall Bearing Design***

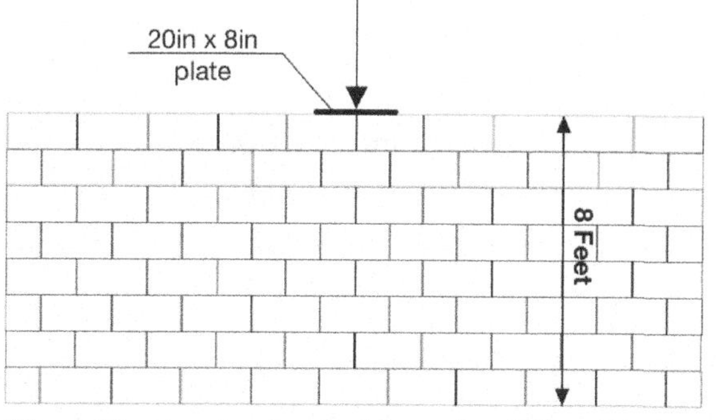

The effective horizontal length of distribution for the above shown concentrated load is most nearly:

(A) 68 in

(B) 116 in

(C) 192 in

(D) 48 in

PROBLEM 4.14 ***Masonry Beam Section Design***

In the below concrete masonry section, the following concrete and steel properties apply:

- $f_y = 60\ ksi$
- $f'_m = 2{,}800\ psi$
- $E_s = 29{,}000\ ksi$

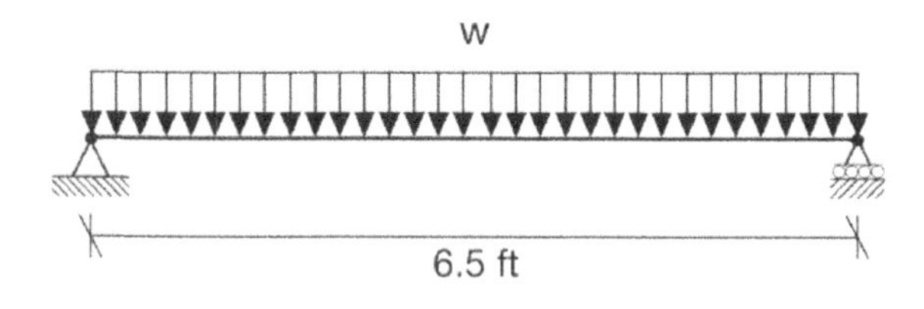

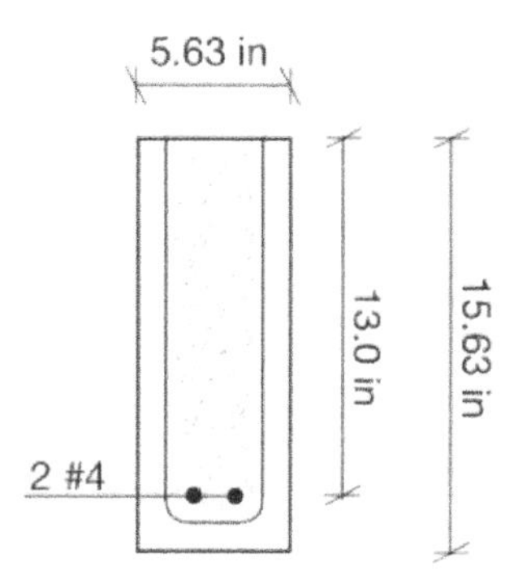

Based on the above information, and using the ASD method, load w (inclusive of self-weight) in lb/ft is most nearly:

(A) 2,553

(B) 4,716

(C) 2,363

(D) 4,764

PROBLEM 4.15 ***Masonry Wall Design***

The below is an elevation view for a 6 in thick basement wall, hollow bricks used, not grouted, unreinforced concrete masonry, and spanning horizontally as shown.

The wall cross section is shown below it and is modelled as a simply supported vertical member.

The properties of masonry and the soil behind the wall are as follows:

- $f'_m = 1{,}800\ psi$
- $E_m = 1{,}620{,}000\ psi$
- $\gamma_{soil} = 130\ pcf$
- $\emptyset'_{soil} = 35^o$

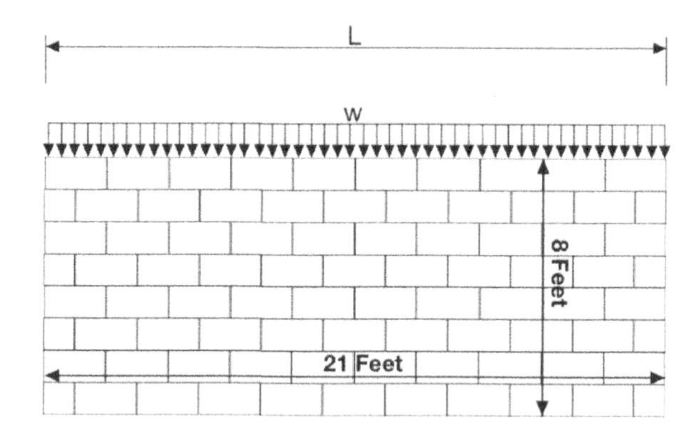

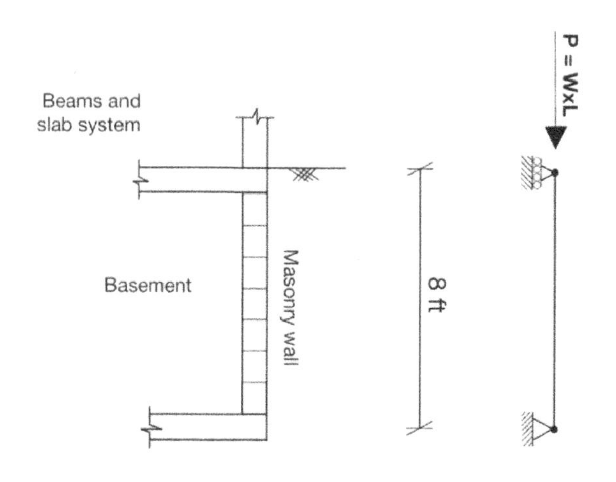

Ignoring the wall self-weight, and using the Allowable Stress Design method ASD, the maximum distributed vertical load w the wall can withstand in kip/ft is most nearly:

(A) 8

(B) 55

(C) 10

(D) 20

PROBLEM 4.16 ***Steel Section Column Design***

The owner of the below steel building would like to remove portion of the second floor between axis 'C' and axis 'D' to create more vertical space. The steel building consists of two floors. The 1st floor height is 20 ft, and the 2nd floor height is 10 ft.

Perimeter columns are all $HP10 \times 42$, $F_y = 50\ ksi$. Columns on axis 'D' are all braced in the weak direction at 10 ft intervals. All end connections (at roof top and foundation level) can be assumed as pinned. Columns are continuous from top to bottom – i.e., they do not terminate at any floor.

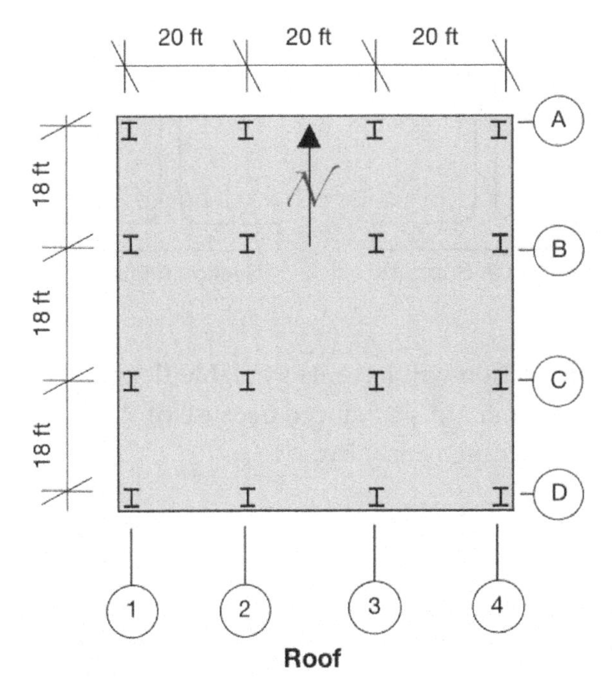

Roof

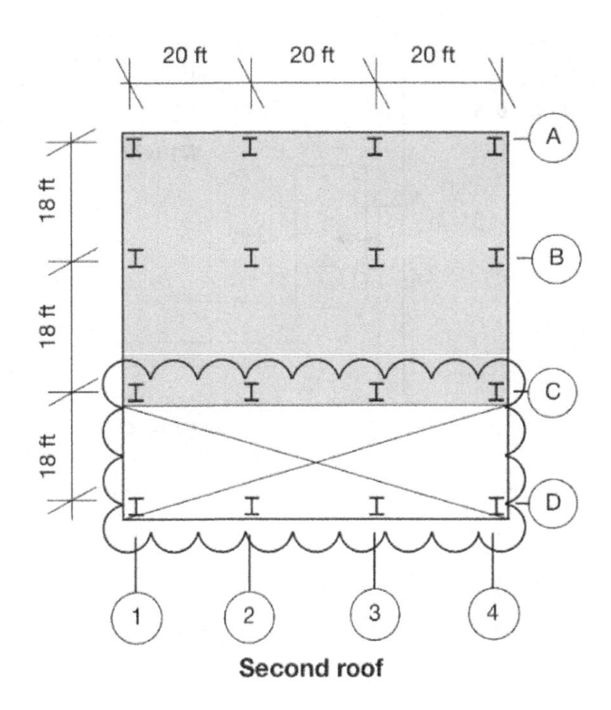

Second roof

The reduction in the available compressive strength for column D2 in kip due to this change using LRFD method is:

(A) 9.5

(B) 115

(C) 320

(D) Change is unadvisable

PROBLEM 4.17 ***Steel Shear Connection Analysis***

The below beam/column shear connection has an $F_y = 50\ ksi\ \&\ F_u = 65\ ksi$. The connecting L angle's $F_y = 36\ ksi\ \&\ F_u = 58\ ksi$. All bolts are group 'A', slip critical standard holes.

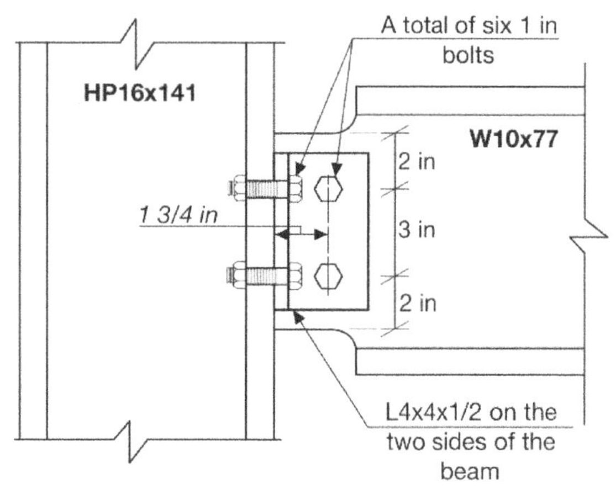

The shear capacity for the above connection is most nearly:

(A) 43 kip (ASD)
64 kip (LRFD)

(B) 46 kip (ASD)
69 kip (LRFD)

(C) 85 kip (ASD)
128 kip (LRFD)

(D) 89 kip (ASD)
133 kip (LRFD)

PROBLEM 4.18 ***Steel Corbel Analysis***

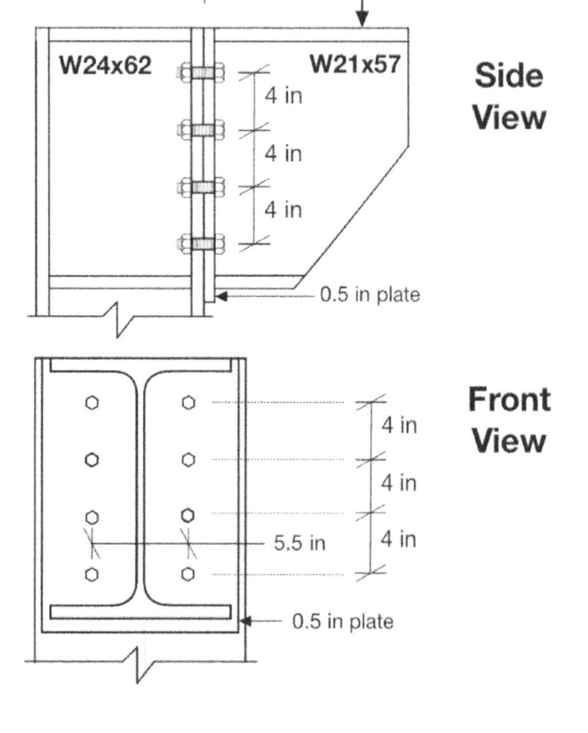

The above corbel has eight ¾ in $A307$ bolts. Considering the available bearing for those bolts only, the maximum load P this connection can withstand is most nearly:

(A) 55 kip (ASD)
80 kip (LRFD)

(B) 45 kip (ASD)
70 kip (LRFD)

(C) 220 kip (ASD)
345 kip (LRFD)

(D) 170 kip (ASD)
230 kip (LRFD)

PROBLEM 4.19 ***Steel Section with Web Opening***

The below is a simply supported steel section $W24 \times 250$, $F_y = 50\,ksi$, with a $10\,in \times 25\,in$ opening midspan situated to have pipes pass through.

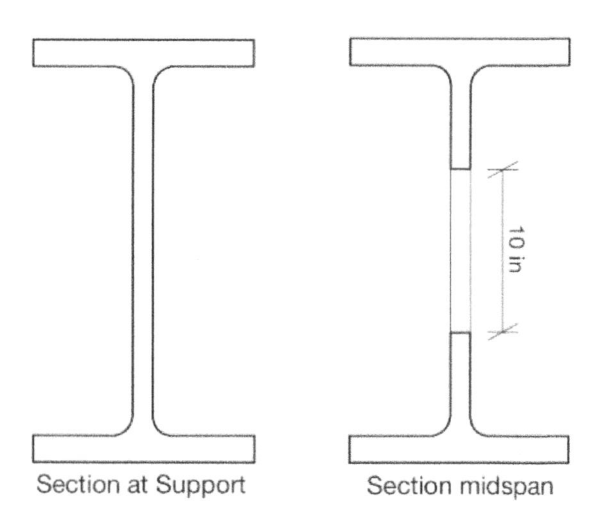

This section will have its available flexural strength in $kip.ft$ reduced because of the web opening nearly by:

(A) 50 (LRFD)
30 (ASD)

(B) 75 (LRFD)
50 (ASD)

(C) 100 (LRFD)
65 (ASD)

(D) 125 (LRFD)
80 (ASD)

PROBLEM 4.20 ***Steel Section with Reinforced Opening***

The below simply supported $W24 \times 250$ steel section has a $14\ in \times 34\ in$ opening midspan situated to have pipes pass through. The opening is reinforced with four $2\ in \times$ ½ in plates as shown with the same yield strength of the W section ($50\ ksi$).

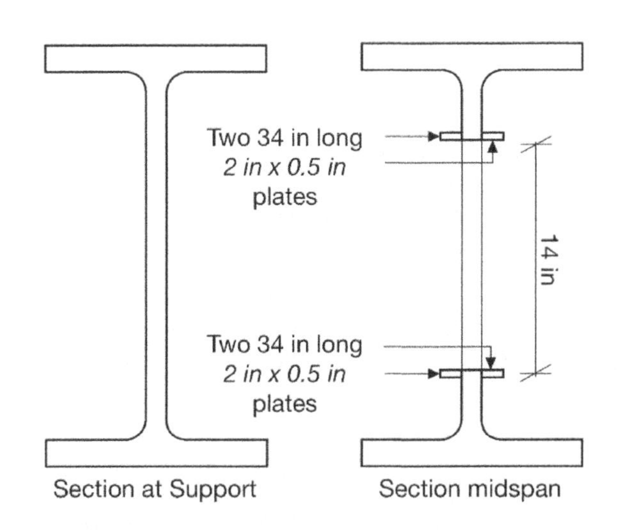

The overall available flexural strength of the resulting section at mid span in $kip.ft$ is nearly:

(A) 2,700 (LRFD)
1,800 (ASD)

(B) 2,675 (LRFD)
1,880 (ASD)

(C) 2,765 (LRFD)
1,840 (ASD)

(D) 2,800 (LRFD)
1,860 (ASD)

PROBLEM 4.21 ***Steel Corbel Analysis***

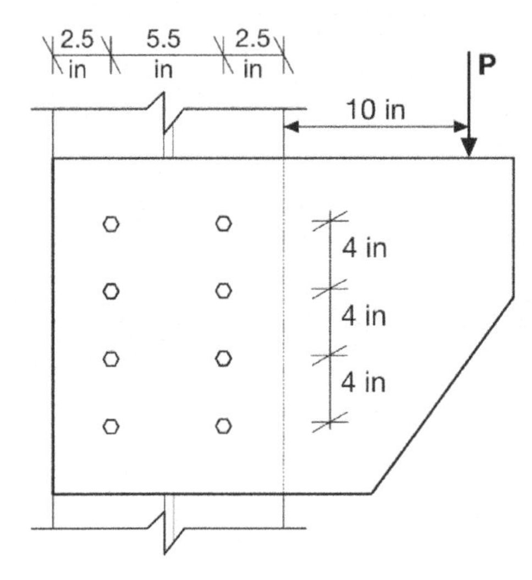

The maximum load P for the above eight ¾ in $A307$ bolt group is most nearly:

(A) 47.8 kip (ASD)
71.8 kip (LRFD)

(B) 19.4 kip (ASD)
29.2 kip (LRFD)

(C) 11.9 kip (ASD)
17.9 kip (LRFD)

(D) 14.3 kip (ASD)
21.5 kip (LRFD)

PROBLEM 4.22 ***Truss Connection Weld Design***

The below is a typical connection for a large steel truss project. Weld used all over the project is fillet size $5/16\ in$, electrode $E60$.

Tension load on the two diagonals is $90\ kip$ (ultimate) or $60\ kip$ (allowable).

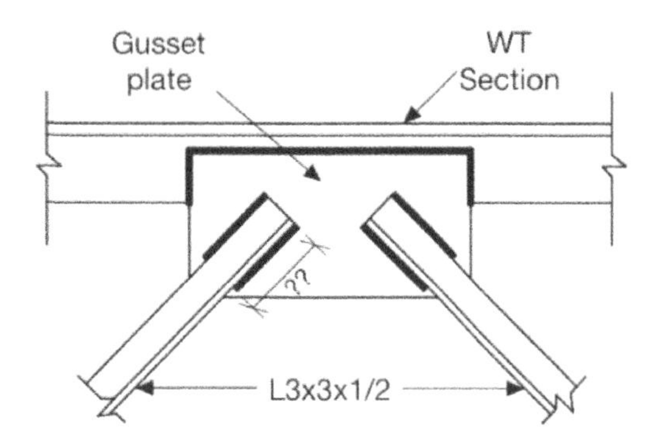

The minimum length of the fillet weld shown is most nearly:

(A) 2 in

(B) 8 in

(C) 4 in

(D) 16 in

PROBLEM 4.23 ***Gantry Crane WT Section Design***

The below is a gantry crane installed between two columns in an industrial building. The main beam is an inverted WT section ($F_y = 50\ ksi$) and the two ends are simply supported.

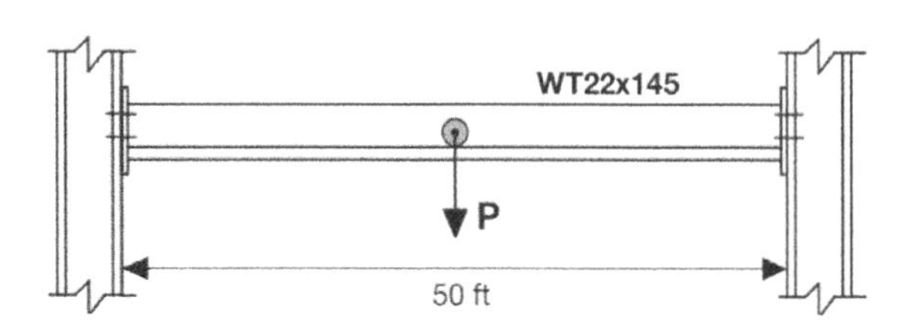

The maximum allowable load P this beam can carry in flexure is most nearly:

(A) 18 kip

(B) 26 kip

(C) 40 kip

(D) 6 kip

PROBLEM 4.24 ***Steel WT Section Design***

The below is a WT beam with a concentrated load applied at its midspan. The beam is simply support with $F_y = 50\ ksi$.

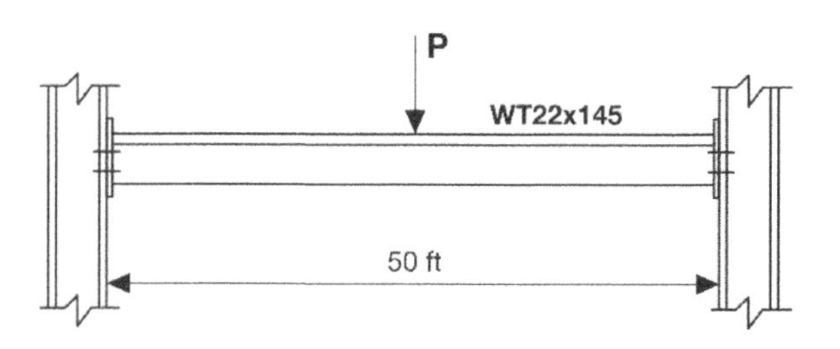

The maximum load P this beam can carry is most nearly:

(A) 6 kip (ASD)
10 kip (LRFD)

(B) 26 kip (ASD)
40 kip (LRFD)

(C) 30 kip (ASD)
44 kip (LRFD)

(D) 32 kip (ASD)
47 kip (LRFD)

PROBLEM 4.25 ***Steel W Section Design***

Select three economical sections from the below list for a column with effective length of 20 ft subjected to the following service loads: $M_x = 250\ kip.ft$ and $P = 300\ kip$ inclusive of self-weight:

(A) $W36 \times 330$

(B) $W30 \times 116$

(C) $W24 \times 207$

(D) $W27 \times 129$

(E) $W18 \times 119$

(F) $W14 \times 120$

(G) $W24 \times 76$

PROBLEM 4.26 *Steel Plate in Tension*

The below is ½ *in* steel plate with a yield strength $F_y = 50\ ksi$ and ultimate strength of $F_u = 65\ ksi$ connected to a larger member in tension.

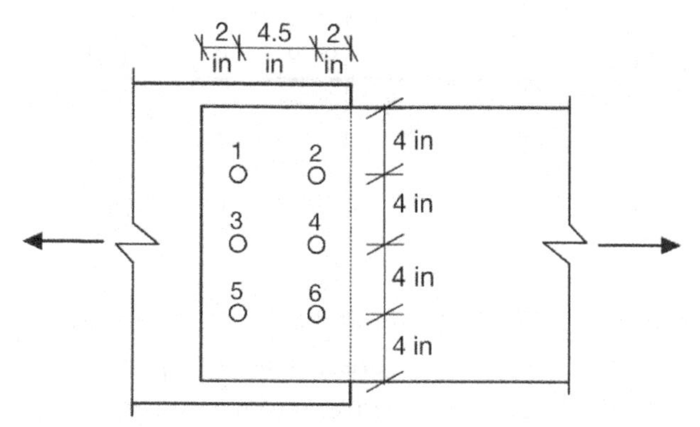

Considering the diameter of each of the connecting six bolts is ½ *in*, the maximum ultimate tension in *kip* the plate alone can withstand is most nearly:

(A) 311

(B) 345

(C) 360

(D) 460

PROBLEM 4.27 *Glulam Wooden Beam Design*

The below glulam beam is made from layers of southern pine timber grade N2M12, placed in a building with 19% moisture, and net finished dimensions as shown below:

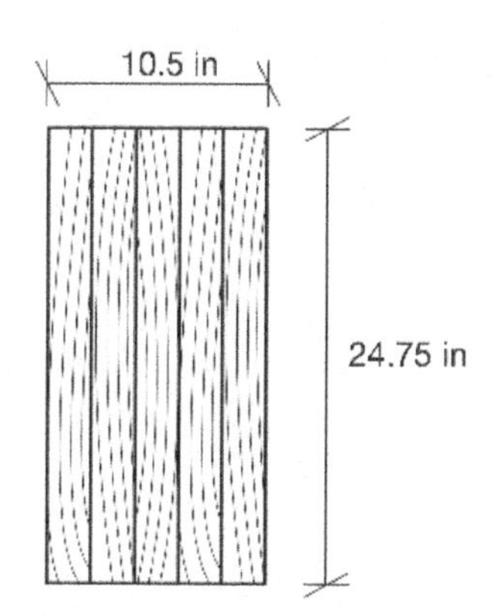

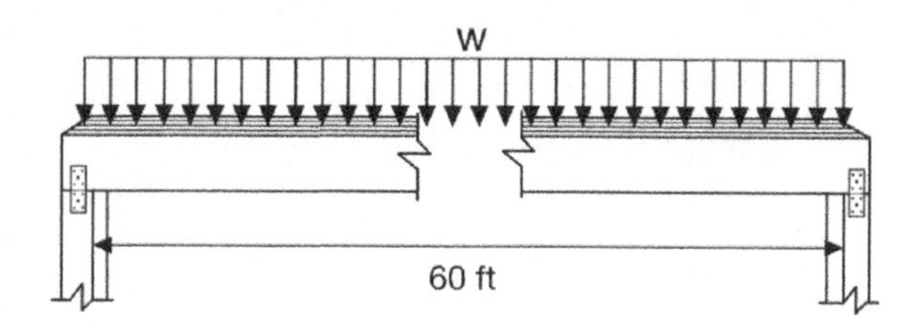

The distributed service permanent dead load w in lb/ft this beam can withstand is most nearly:

(A) 100

(B) 240

(C) 120

(D) 300

PROBLEM 4.28 *Sawn Lumber Column Design*

The available service load during construction for a $10\ ft$ long $2\ in \times 4\ in$ Sawn Douglas Fir-Larch No 3 Lumber (*) column supported with two hinges top and bottom is most nearly:

(A) 590 *lb*

(B) 1,080 *lb*

(C) 710 *lb*

(D) 1,880 *lb*

* Assume all adjustment factors equal to unity apart from the Column Stability Factor.

PROBLEM 4.29 *Lumber Loss of Strength due to Fire*

A $14\ in \times 24\ in$ solid sawn wooden lumber beam with a nominal char rate $\beta_n = 1.5\ in/hr$ has its strong axis moment of inertia reduced by ____ in^4 when fully exposed to fire for $1.5\ hours$.

(A) 4,485

(B) 10,115

(C) 10,985

(D) 9,455

PROBLEM 4.30 ***Adjustment Factors for Glulam***

The below Section is a Western Species structural glued laminated timber "Glulam". The section represents a simply support beam 25 ft long with uniform loading atop.

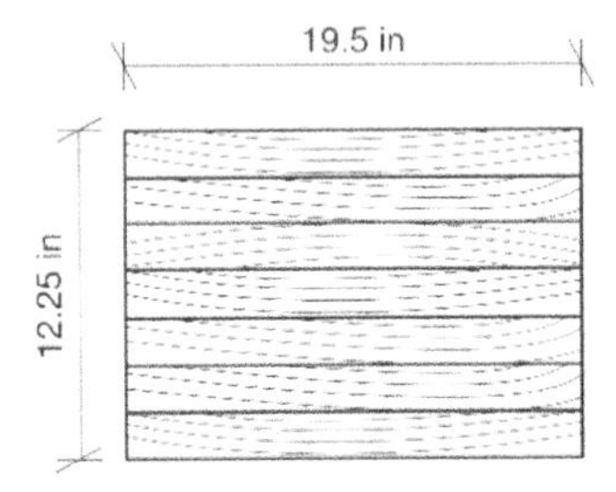

The volume adjustment factor for this beam is most nearly:

(A) 0.67

(B) 0.57

(C) 0.86

(D) 0.93

PROBLEM 4.31 ***Wooden Truss Joint***

Loading for the bottom members is parallel to the 3.5 in × 8 in sawn lumber's, Douglas Fir-Larch, grains. Live load is for normal occupancy, normal temperature and 16% moisture. Also use a modulus of elasticity for Douglas Fir-Larch of $E = 1{,}400\ ksi$.

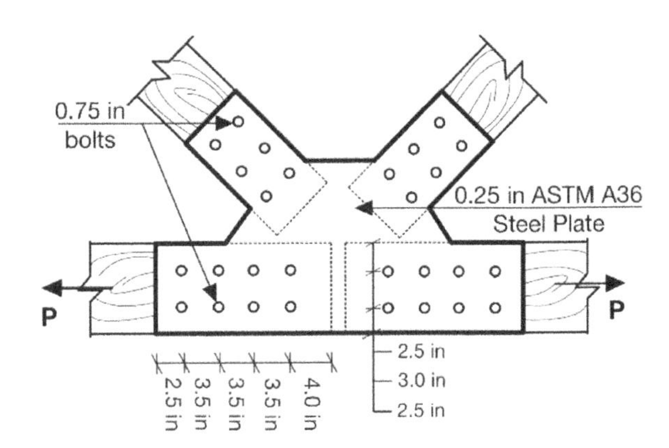

Taking the lateral capacity of bolts at the bottom cord alone, considering the plate is placed from one side only, maximum service load P in kip is:

(A) 13.4

(B) 13.0

(C) 9.8

(D) 7.4

PROBLEM 4.32 ***Effective Char Depth***

The effective char depth for Cross Laminated Timer CLT, 2.5 $hours$ exposure, 1¼ in lamination thickness, 5 layers of laminations with nominal char rate $\beta_n = 1.5\ in/hr$, is most nearly:

(A) 4.8 in

(B) 3.9 in

(C) 4.6 in

(D) 4.0 in

SOLUTION 4.1

Critical buckling load P_c is calculated per ACI 318-14 code Section 6.6.4.4.2 as follows:

$$P_c = \frac{\pi^2 (EI)_{eff}}{(k\ell_u)^2}$$

E_c is calculated per ACI 318 Chapter 19 Section 19.2.2.1.b, and I_g is calculated per Section 6.6.3.1 as follows:

$$E_c = 57{,}000\sqrt{f'_c}$$

$$= 57{,}000\sqrt{3{,}500}$$

$$= 3.4 \times 10^6\ psi$$

$$I_g = \frac{12 \times 12^3}{12} = 1{,}728\ in^4$$

$$(EI)_{eff} = \frac{0.4E_cI_g}{1+\beta_{dns}}$$

β_{dns} is the ratio of maximum factored sustained axial load to the maximum factored axial load associated with the same load combination. For simplicity, the code allows it to be taken as '0.6' per Section R6.6.4.4.4 of the code:

$$= \frac{0.4 \times (3.4 \times 10^6\ psi) \times 1{,}728\ in^4}{1+0.6}$$

$$= 1.47 \times 10^9\ lb.in^2$$

Calculate critical buckling load before removing the beams:

This building has shear walls integrated with it which prevents it from swaying – i.e., braced against sidesway. Hence, k shall be calculated using Section R6.2.5 Figure (a) for *Nonsway Frames*. However, Section 6.6.4.4.3 allows k to be taken as '1.0' for nonsway members.

Because the north direction of the building was the one affected by removing the two beams, the critical buckling load in that direction will be the one affected with this change.

Given that the first floor is the highest floor, it shall be taken into consideration during the *before* calculations.

Check if this column is slender to determine if the critical buckling load applies to it using Section 6.2.5 of the code:

$$\frac{k\ell_u}{r} \le 34 + 12\left(\frac{M1}{M2}\right) \qquad \text{(ACI 318-14 6.2.5b)}$$

$$\frac{k\ell_u}{r} \le 40 \qquad \text{(ACI 318-14 6.2.5c)}$$

ℓ_u is the unsupported length for the column taking center to center of supports, which is/are as follows:

$= 15\ ft$ (before removing beams)

$= 28\ ft$ (after removing the beams)

r radius of gyration and is permitted to be taken as '0.3' times the dimension in the direction of stability that is being considered for rectangular columns per Section 6.2.5.1:

$$= 0.3 \times 12 = 3.6\ in$$

$\left(\frac{M1}{M2}\right)$ is the ratio of end moments and is negative in case of single curvature. Given those two moments are almost equal as noted in the question, this value will be taken as '– 1.0'.

$$\frac{k\ell_u}{r} = \frac{1.0 \times 15ft \times 12\frac{in}{ft}}{3.6\ in} = 50$$

$$50 > 34 + 12\left(\frac{M1}{M2}\right) \quad (= 34 - 12)$$

$$> 22$$

Per ACI 318-14 6.2.5b & 6.2.5c slenderness effect cannot be neglected. Rechecking for slenderness as the column becomes longer after removing the two beams is not required given that slenderness is implied in both cases.

$$P_c = \frac{\pi^2 (EI)_{eff}}{(k\ell_u)^2}$$

$$= \frac{\pi^2 \times (1.47\times10^9\ lb.in^2)}{\left(1.0\times\left(15ft\times 12\frac{in}{ft}\right)\right)^2} \times \frac{1\ kip}{1{,}000\ lb}$$

$$= 447.8\ kip$$

Calculate critical buckling load after removing the beams:

$$P_c = \frac{\pi^2 (EI)_{eff}}{(k\ell_u)^2}$$

$$= \frac{\pi^2 \times (1.47\times10^9\ lb.in^2)}{\left(1.0\times\left(28ft\times 12\frac{in}{ft}\right)\right)^2} \times \frac{1\ kip}{1{,}000\ lb}$$

$$= 128.5\ kip$$

Difference between the two buckling loads is as follows:

$447.8 - 128.5 = 319.3\ kip \quad reduction$

Correct Answer is (C)

SOLUTION 4.2

Critical buckling load P_c is calculated per ACI 318-14 code Section 6.6.4.4.2 as follows:

$$P_c = \frac{\pi^2 (EI)_{eff}}{(k\ell_u)^2}$$

E_c is calculated per ACI 318 Chapter 19 Section 19.2.2.1.b, and I_g is calculated per Section 6.6.3.1 as follows:

$$E_c = 57{,}000\sqrt{f_c'}$$

$$= 57{,}000\sqrt{3{,}500}$$

$$= 3.4\times10^6\ psi$$

$$I_g = \frac{12\times12^3}{12} = 1{,}728\ in^4$$

$$(EI)_{eff} = \frac{0.4E_cI_g}{1+\beta_{dns}}$$

β_{dns} is the ratio of maximum factored sustained axial load to the maximum factored axial load associated with the same load combination. For simplicity, the code allows it to be taken as '0.6' per R6.6.4.4.4

$$= \frac{0.4\times(3.4\times10^6\ psi)\times1{,}728\ in^4}{1+0.6}$$

$$= 1.47\times10^9\ lb.in^2$$

Calculate critical buckling load before removing the beams (at first floor):

This building is not braced against sidesway, k shall be calculated using R6.2.5 Figure (b) for *Sway Frames* and should be > 1.0.

Because the north direction of the building was the one affected by removing the two beams, the critical buckling load in that direction will be the one affected with this change.

Given that the first floor is the highest floor, it shall be taken into consideration during the *before* calculations (given in the question).

*Determine **k** and the effective length for columns:*

Calculate Ψ at the 1st floor (where there is a stub below), also, at the 2nd floor intersection with the 1st floor.

$$\Psi = \frac{\left(\frac{\Sigma EI}{\ell_c}\right) of\ columns}{\left(\frac{\Sigma EI}{\ell}\right) of beams}$$

I for this particular purpose is calculated based on 6.6.3.1.1(a) – unlike the calculation of $(EI)_{eff}$:

$I_{Columns} = 0.7 \times I_g$

$= 0.7 \times 1{,}728\ in^4$

$= 1{,}210\ in^4$

$I_{Beams} = 0.35 \times I_g$

$= 0.35 \times 1{,}728\ in^4$

$= 605\ in^4$

ℓ_c is the length of compression members measured center to center of the joints, which is/are as follows:

$= 12\ ft + 6\ in + 6\ in$
$= 13\ ft$ (Second floor)

$= 14\ ft + 6\ in + 6\ in$
$= 15\ ft$ (First floor)

$= 4\ ft + 6\ in + 6\ in$
$= 5\ ft$ (below first/stub)

ℓ is the length of beams measured center to center of the joints $= 9\ ft$ for the roof beam 'C4' to 'D4'.

$$\Psi_{2nd\ floor} = \frac{\Sigma\left(\frac{EI}{\ell_c}\right)}{\Sigma\left(\frac{EI}{\ell}\right)}$$

$$= \frac{\Sigma\left(\frac{I}{\ell_c}\right)}{\Sigma\left(\frac{I}{\ell}\right)}$$

$$= \frac{\left(\frac{1{,}210\ in^4}{13\ ft} + \frac{1{,}210\ in^4}{15\ ft}\right)}{\left(\frac{605\ in^4}{9\ ft}\right)}$$

$= 2.6$

$$\Psi_{1st\ floor} = \frac{\Sigma\left(\frac{EI}{\ell_c}\right)}{\Sigma\left(\frac{EI}{\ell}\right)} = \frac{\Sigma\left(\frac{I}{\ell_c}\right)}{\Sigma\left(\frac{I}{\ell}\right)}$$

$$= \frac{\left(\frac{1{,}210\ in^4}{15\ ft} + \frac{1{,}210\ in^4}{5\ ft}\right)}{\left(\frac{605\ in^4}{9\ ft}\right)}$$

$= 4.8$

Using Figure R6.2.5 (b):

$k = 1.9 > 1.0 \rightarrow OK$

Checking if this column is slender to determine if the critical buckling load applies to it using Section 6.2.5 of the code:

$$\frac{k\ell_u}{r} \le 22 \qquad \text{(ACI 318-14 6.2.5a)}$$

ℓ_u is the unsupported length for the column taking center to center of supports, which is/are as follows:

$= 13\ ft$(before removing beams 2nd floor)
$= 15\ ft$ (before removing beams 1st floor)
$= 28\ ft$ (after removing beams)

r radius of gyration and is permitted to be taken as '0.3' times the dimension in the direction of stability that is being considered for rectangular columns per Section 6.2.5.1:

$= 0.3 \times 12 = 3.6\ in$

$$\frac{k\ell_u}{r} = \frac{1.9 \times 15\ ft \times 12\frac{in}{ft}}{3.6\ in}$$

$= 95 > 22 \rightarrow$ Slender column

Per ACI 318-14 6.2.5b & 6.2.5c slenderness effect cannot be neglected. Rechecking for slenderness when the column becomes longer after removing the two beams (i.e., $28\ ft$) is not required given that slenderness is implied in both cases.

Critical load at the first floor before removing the beams is therefore calculated as follows:

$$P_c = \frac{\pi^2 (EI)_{eff}}{(k\ell_u)^2}$$

$$= \frac{\pi^2 \times (1.47 \times 10^9\ lb.in^2)}{\left(1.9 \times \left(15\ ft \times 12\frac{in}{ft}\right)\right)^2} \times \frac{1\ kip}{1{,}000\ lb}$$

$= 124.0\ kip$

PART II Structural Depth

Section 4 Structural Design

PART II Structural Depth

Section 4 Structural Design

Calculate critical buckling load after removing the beams:

*Determine **k** and the effective length for columns:*

Calculate Ψ at the roof end and at the 1st floor end floor (where there is a stub below).

$$\Psi_{roof} = \frac{\Sigma\left(\frac{EI}{\ell_c}\right)}{\Sigma\left(\frac{EI}{\ell}\right)}$$

$$= \frac{\Sigma\left(\frac{I}{\ell_c}\right)}{\Sigma\left(\frac{I}{\ell}\right)}$$

$$= \frac{\left(\frac{1,210\ in^4}{28ft}\right)}{\left(\frac{605\ in^4}{9ft}\right)}$$

$$= 0.64$$

$$\Psi_{1st\ floor} = \frac{\Sigma\left(\frac{EI}{\ell_c}\right)}{\Sigma\left(\frac{EI}{\ell}\right)}$$

$$= \frac{\Sigma\left(\frac{I}{\ell_c}\right)}{\Sigma\left(\frac{I}{\ell}\right)}$$

$$= \frac{\left(\frac{1,210\ in^4}{28\ ft} + \frac{1,210\ in^4}{5\ ft}\right)}{\left(\frac{605\ in^4}{9\ ft}\right)}$$

$$= 4.24$$

Based on Fig R6.2.5 (b):

$$k = 1.55 > 1.0 \rightarrow ok$$

Critical load after removing the beams is therefore calculated as follows:

$$P_c = \frac{\pi^2 (EI)_{eff}}{(k\ell_u)^2}$$

$$= \frac{\pi^2 \times (1.47\times10^9\ lb.in^2)}{\left(1.55\times\left(28\ ft \times 12\frac{in}{ft}\right)\right)^2} \times \frac{1\ kip}{1,000\ lb}$$

$$= 53.5\ kip$$

Difference between the two buckling loads is as follows:

$$124.0 - 53.5 = 70.5\ kip\ \ reduction$$

Correct Answer is (A)

SOLUTION 4.3

Using the ACI 318-14 code, the minimum thickness for one-way slabs shall be determined using Section 7.3.1.1, Table 7.3.1.1, Sections 7.3.1.1.1 and 7.3.1.1.2. Thicknesses lesser than what is specified in those sections shall be supported with deflection calculations.

For a cantilever span – which is the critical section of all per Table 7.3.1.1:

$$h_{min} = \frac{\ell}{10}$$

$$= \frac{10ft - \frac{10}{2}in \times \frac{1\ ft}{12\ in}}{10}$$

$$= 0.96\ ft\ (11.5\ in)$$

Where $\boldsymbol{\ell}$ is the clear span in this case.

Given that a different grade of steel to the one that Table 7.3.1.1 specifies is used, $\boldsymbol{h_{min}}$ calculated per this table shall be multiplied by:

$$0.4 + \frac{f_y}{100,00} = 0.9$$

Also, given that lightweight concrete is used, the resultant shall be multiplied by the greatest of the following:

$$1.65 - 0.005\ w_c = 1.65 - 0.005 \times 95$$

$$= 1.175$$

Or

1.09

→ $h'_{min} = 11.5\,in \times 0.9 \times 1.175$

$= 12.2\,in$

Correct Answer is (B)

SOLUTION 4.4

Using the ACI 318-14 code, the minimum thickness for two-way slabs with beams on all sides of panels shall be determined using Section 8.3.1.2. Thicknesses lesser than what is specified in those sections shall be supported with deflection calculations.

$$\alpha_f = \frac{(EI)_{beam}}{(EI)_{slab}} \quad \text{(ACI 318-14 8.10.2.7b)}$$

$$= \frac{I_{beam}}{I_{slab}}$$

$\boldsymbol{\alpha_f}$ is the ratio of flexural stiffness of beam section to flexural stiffness of a width of slab bounded laterally by centerlines of adjacent panels, if any, on each side of the beam.

In which case, moment of inertia for beam and slab should be for the uncracked section, i.e., $\boldsymbol{I_{gross}}$ to be used.

The width of the slab above a beam supporting a center panel is as follows:

$= 22\,ft - 12\,in\;(beam\;width)$

$= 252\,in$

The width of the slab above an edge beam is as follows:

$= 11\,ft - 6\,in\;(half\;\;beam\;width)$

$= 126\,in$

Assume a slab thickness of 6 in to calculate an initial value of $\boldsymbol{\alpha_f}$ as follows:

	For a central beam	For an edge beam
α_f	$\frac{\frac{12\times14^3}{12}}{\frac{252\times6^3}{12}} = 0.6$	$\frac{\frac{12\times14^3}{12}}{\frac{126\times6^3}{12}} = 1.2$

Based on the above, $\boldsymbol{\alpha_{fm}}$, which is the average value for all beams on edges of a panel, shall be calculated and used in Table 8.3.1.2 to determine the minimum thickness.

α_{fm}	For a central panel - all beams around $\alpha_f = 0.6$	$\frac{0.6\times4}{4} = 0.6$
	For an edge panel from one side	$\frac{\begin{bmatrix}0.6\times3\\+\\1.2\times1\end{bmatrix}}{4} = 0.75$
	For an edge panel from two sides	$\frac{\begin{bmatrix}0.6\times2\\+\\1.2\times2\end{bmatrix}}{4} = 0.9$

Using Table 8.3.1.2 with $0.2 < \alpha_{fm} \leq 2.0$:

$$h_{min} = \frac{\ell_n\left(0.8 + \frac{f_y}{200{,}000}\right)}{36 + 5\beta\left(\alpha_{fm} - 0.2\right)}$$

$$or$$

$$5.0\,in$$

Where $\boldsymbol{\ell_n}$ is the clear span in the long direction, which is 21.0 ft (252 in), and $\boldsymbol{\beta}$ is the ratio of long to short spans which is '1.0' in this case. Also use $\alpha_{fm} = 0.6$ for a maximum value.

$$= \frac{252\,in \times \left(0.8 + \frac{60{,}000}{200{,}000}\right)}{36 + 5\times1\times(0.6 - 0.2)}$$

$$= 7.3\,in$$

$\boldsymbol{\alpha_f}$ is recalculated using a slab thickness of 7.3 in instead of the initial assumption of 6 in for confirmation:

	For a central beam	For an edge beam
α_f	$\frac{\frac{12\times14^3}{12}}{\frac{252\times7.3^3}{12}} = 0.34$	$\frac{\frac{12\times14^3}{12}}{\frac{126\times7.3^3}{12}} = 0.7$

α_{fm}	For a central panel - all beams around $\alpha_f = 0.34$	$\frac{0.34\times4}{4} = 0.34$
	For an edge panel from one side	$\frac{\begin{bmatrix}0.34\times3\\+\\0.7\times1\end{bmatrix}}{4} = 0.4$
	For an edge panel from two sides	$\frac{\begin{bmatrix}0.34\times2\\+\\0.7\times2\end{bmatrix}}{4} = 0.5$

Similar to the above, the minimum value of $\alpha_{fm} = 0.34$ for a maximum value is used in calculating h_{min}:

$$h_{min} = \frac{252\ in \times \left(0.8 + \frac{60{,}000}{200{,}000}\right)}{36 + 5 \times 1 \times (0.34 - 0.2)}$$

$$= 7.55\ in$$

Per 8.3.1.2.1 and with an edge beam $\alpha_f < 0.8$, the minimum slab thickness in that panel shall be increased by 10%.

$$h_{min} = 7.55 \times 1.1 = 8.3\ in \sim 8.5\ in$$

Correct Answer is (C)

SOLUTION 4.5

The minimum footing thickness shall at least be 8 *in* per ACI 318-14 Section 14.3.2.1.

Correct Answer is (D)

SOLUTION 4.6

$$DL_{Slab} = 8\ in \times \frac{1\ ft}{12\ in} \times 150\ pcf$$

$$= 100\ psf$$

$$DL_{Total} = 100 + 40 = 140\ psf$$

$$\frac{Slab\ long\ dimension}{Slab\ short\ dimension} = \frac{30}{14} > 2$$

Based on the above:

- All members are prismatic.
- Loads are uniformly distributed.
- LL < 3 DL
- There are three spans.
- The longer of two adjacent spans does not exceed the shorter by more than 20%
- Slabs aspect ratio > 2

This slab can be designed as a one-way slab and ACI 318-14 Table 6.5.2 can be used.

$$w_u = 1.2\ DL + 1.6\ LL = 248\ psf$$

$$= 248 \frac{lb}{ft}/ft$$

$$M_u = \frac{w_u \ell_n^2}{11}$$

Where $\boldsymbol{\ell_n}$ is the average of adjacent two clear spans, in this case:

$$\ell_n = \frac{(12\ ft - 1\ ft) + (14\ ft - 1\ ft)}{2} = 12\ ft$$

$$M_u = \frac{248 \frac{lb}{ft}/ft \times (12\ ft)^2}{11}$$

$$= 3{,}247\ lb.ft/ft\ \ (3.2\ kip.ft/ft)$$

Correct Answer is (A)

SOLUTION 4.7

The aspect ratio for all panels in this floor < 2, hence, this flat slab shall be designed as a two-way slab.

$$DL_{Slab} = 8\ in \times \frac{1\ ft}{12\ in} \times 150\ pcf$$

$$= 100\ psf$$

$DL_{Total} = 100 + 40 = 140\ psf$

$\frac{Slab\ (long)\ dimension}{Slab\ (short)\ dimension} = \frac{22}{22} = 1$

Considering the following:

- There are at least three continuous spans in each direction.
- Panels are rectangular with aspect ratio < 2.
- All loads are due to gravity.
- LL < 2 DL

The Direct Design Method of Section 8.10 from ACI 318-14 can be used.

$q_u = 1.2\ DL + 1.6\ LL = 248\ psf$

The Direct Design Method starts with determining the width of the column strip and the middle strip. The width of the column strip is the smallest of $0.5\ell_1$ or $0.5\ell_2$ with ℓ_1 and ℓ_2 being the slab dimensions where moment is being determined, and the dimension perpendicular to it, respectively (see 8.4.1.5 or more details).

Due to symmetry, both columns and middle strips are 11 ft wide.

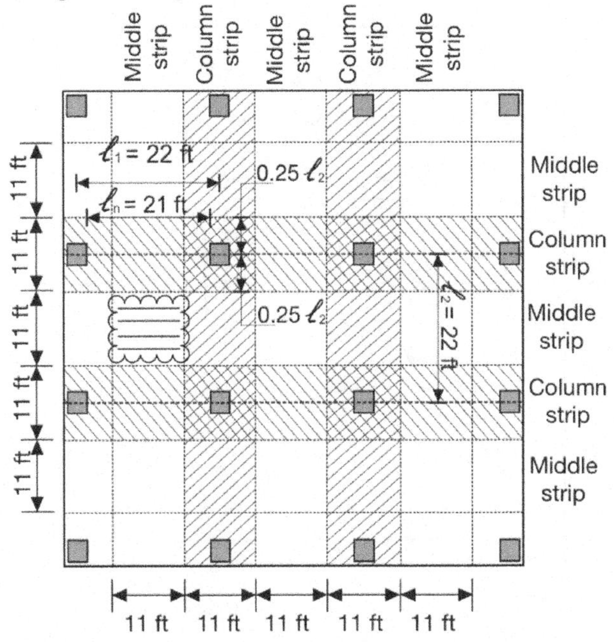

The absolute sum of positive and average negative moment:

$$M_o = \frac{q_u \ell_2 \ell_n^{\ 2}}{8}$$

$$= \frac{248\ psf \times 22\ ft \times (21\ ft)^2}{8} \times \frac{1\ kip}{1{,}000\ lb}$$

$$= 301\ kip.ft$$

The positive moment portion of this moment assigned to the edge/end span for a column strip is determined using coefficients from Table 8.10.4.2.

For a slab with no beams and no edge beams, the end span positive moment coefficient is '0.52':

$$M_{+ve,end\ span} = 0.52 \times 301\ kip.ft$$

$$= 156\ kip.ft$$

The portion of this positive moment in an end span column strip is determined from coefficients from Table 8.10.5.5 where $[\alpha_f = (EI)_{beam}/(EI)_{slab} = 0]$ because there are no beams.

The coefficient for $\alpha_{f1}\ell_1/\ell_2 = 0$, and $\ell_1/\ell_2 = 1.0$ as gathered from Table 8.10.5.5 is '0.6':

$$M_{+ve,end\ span,Col.strip} = 0.6 \times 156\ kip.ft$$

$$= 94\ kip.ft$$

The remainder of this moment is resisted by the two halves of the middle strip adjacent to this column strip (8.10.6.1) – one on its right side and the other on its left side. The entire middle strip resists two halves in a symmetrical situation as it takes the other half from the column strip on its other side (8.10.6.2).

$$M_{+ve,end\ span,mid.strip} = 156 - 94$$
$$= 62\ kip.ft$$

The clouded location represented in the question is located at a middle strip in the requested direction, hence:

$$M_{+ve,end\ span,mid.strip} = \frac{62\ kip.ft}{11\ ft}$$
$$= 5.6\ kip.ft/ft$$

Correct Answer is (A)

SOLUTION 4.8

Referring to the ACI 318-14 code Section 22.6.5, nominal concrete two-way shear strength is the minimum of the following:

$$v_c = 4\lambda\sqrt{f_c'}$$
$$= 4 \times 1 \times \sqrt{4{,}000}$$
$$= 253\ psi$$

Or

$$v_c = \left(2 + \frac{4}{\beta}\right)\lambda\sqrt{f_c'}$$
$$= \left(2 + \frac{4}{1}\right) \times 1 \times \sqrt{4{,}000}$$
$$= 379.5\ psi$$

Or

$$v_c = \left(2 + \frac{\alpha_s\,d}{b_o}\right)\lambda\sqrt{f_c'}$$
$$= \left(2 + \frac{40 \times 8.5}{4 \times (18 + 8.5)}\right) \times 1 \times \sqrt{4{,}000}$$
$$= 329\ psi$$

Where $\boldsymbol{\beta}$ is the aspect ratio of the column dimensions, $\boldsymbol{\lambda}$ is '1.0' for normal weight concrete, and $\boldsymbol{b_o}$ is the perimeter at the critical section which is $\boldsymbol{d/2}$ from the face of the column.

The nominal value is therefore $v_c = 253\ psi$

Correct Answer is (B)

SOLUTION 4.9

Punching/two-way shear critical section is at $\boldsymbol{d/2}$ from the column face (ACI 318-14 22.6.4.1). Circular columns are treated as square supports with the same cross-sectional area (ACI 318-14 R8.10.1.3), see below:

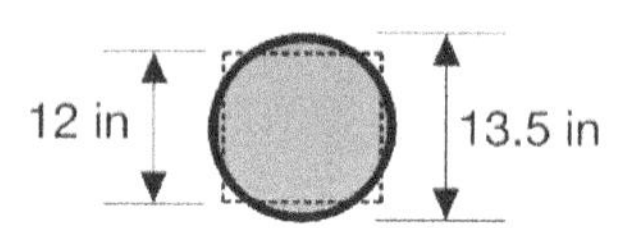

Hence, the critical section equals to:

$$= 12\ in + 2 \times \left(\frac{d}{2}\right)$$
$$= 12\ in + 2 \times \left(\frac{8\ in}{2}\right)$$
$$= 20\ in$$

The perimeter where punching stress is measured against:

$b_o = 20\ in \times 4 = 80\ in$

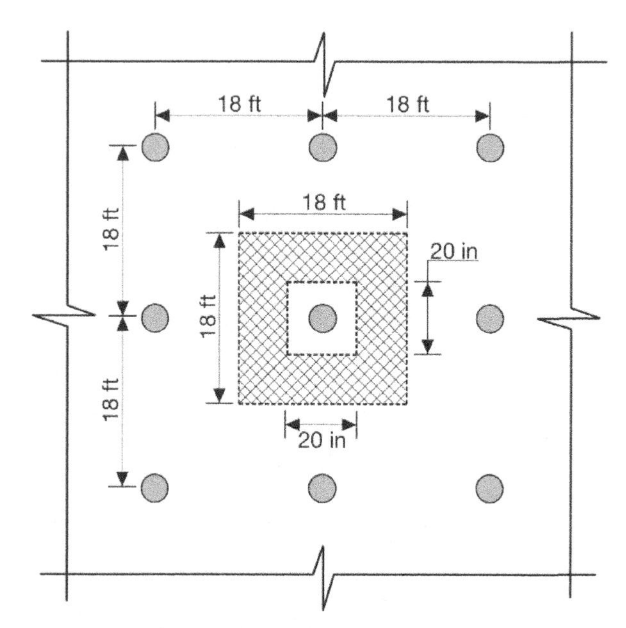

The above shaded area is the tributary and effective area which will most likely cause punching for internal columns when loaded:

$$A = 18\,ft \times 18\,ft - \frac{20\,in \times 20\,in}{144\,\frac{in^2}{ft^2}}$$

$$= 321.2\,ft^2$$

$$DL_{Slab} = 9.5\,in \times \frac{1\,ft}{12\,in} \times 150\,pcf$$

$$= 118.75\,psf$$

$$DL_{Total} = 118.75 + 40$$

$$= 158.75\,psf$$

$$q_u = 1.2\,DL + 1.6\,LL$$

$$= 270.5\,psf$$

$$V_u = 270.5\,psf \times 321.2\,ft^2$$

$$= 86{,}884.6\,lb$$

$$v_u = \frac{V_u}{b_0\,d}$$

$$= \frac{86{,}884.6\,lb}{80\,in \times 8\,in}$$

$$= 135.75\,psi$$

Correct Answer is (D)

SOLUTION 4.10

Determine the effective depth ***d***. Given the specified concrete cover of 1.5 *in* and a dia of reinforcements of 1.13 *in* for # 9 and 0.5 *in* dia of a #4 rebar stirrup:

$$d = 45 - \left(1\tfrac{1}{2} + \tfrac{1}{2} + \frac{1.13}{2}\right) \approx 42.5\,in$$

Using Section 22.5 of the ACI 318-14 code, the following equation should be satisfied:

$$V_n = V_c + V_s$$

$\boldsymbol{V_c}$ is the concrete shear resistance and is the lowest of the following using Table 22.5.5.1 from the code:

$$\left(1.9\lambda\sqrt{f_c'} + 2{,}500\rho_w \frac{V_u d}{M_u}\right) b_w d =$$

$$\left[\begin{pmatrix} 1.9 \times 1 \times \sqrt{4{,}000} \\ + \\ 2{,}500 \times \frac{4 \times 1\,in^2}{20 \times 42.5\,in^2} \\ \times \frac{150\,kip \times 42.5\,in}{730\,kip.ft \times 12\frac{in}{ft}} \end{pmatrix} \\ \times \\ 20\,in \times 42.5\,in\right] \times \frac{1\,kip}{1{,}000\,lb}$$

$$= 110\,kip$$

In this equation, M_u should occur simultaneously with V_u, since this was not given in the question, will move to the next equation.

Or

$$(1.9\lambda\sqrt{f_c'} + 2{,}500\rho_w)b_w d$$

$$= \left[\begin{pmatrix} 1.9 \times 1 \times \sqrt{4{,}000} \\ + \\ 2{,}500 \times \frac{4 \times 1\,in^2}{20 \times 42.5\,in^2} \end{pmatrix} \\ \times \\ 20\,in \times 42.5\,in\right] \times \frac{1\,kip}{1{,}000\,lb}$$

$$= 112\,kip$$

Or

$$3.5\lambda\sqrt{f_c'}\,b_w d$$

$$= \frac{3.5 \times 1 \times \sqrt{4{,}000} \times 20 \times 42.5 \times 1\,kip}{1{,}000\,lb}$$

$$= 188\,kip$$

PART II Structural Depth

Section 4 Structural Design

PART II Structural Depth

Where $\boldsymbol{\lambda}$ is '1.0' for normal weight concrete, and $\boldsymbol{\rho_w}$ is the steel ratio = $As/b_w d$.

$\boldsymbol{V_c}$ in this case is the lowest of the above, disregarding the first option due to the comment provided underneath it:

$V_C = 112\ kip$

$\boldsymbol{V_s}$ is the reinforcement shear resistance and is calculated as follows using Equations 22.5.10.1 & 22.5.10.2:

$$V_s \geq \frac{V_u}{\emptyset} - V_c = \frac{150}{0.75} - 112 = 88\ kip$$

Spacing $\boldsymbol{s}$ is therefore derived as follows:

$$V_s = \frac{A_v f_{yt} d}{s}$$

$$s = \frac{A_v f_{yt} d}{V_s}$$

$$= \frac{(2\times0.2)\times 40\ ksi \times 42.5\ in}{88\ kip}$$

$$= 7.7\ in$$

In which case $\frac{A_v}{s} = \frac{2\times0.2}{7.7} = 0.05$

Section 4 Structural Design

Check for minimum shear reinforcements for the section using Section 9.6.3.3 of the code:

$\frac{A_{v,min}}{s}$ should be the greatest of:

$$0.75\sqrt{f_c'}\ \frac{b_w}{f_{yt}}$$

$$= 0.75 \times \sqrt{4{,}000} \times \frac{20\ in}{40{,}000\ psi}$$

$$= 0.02$$

Or

$$50\ \frac{b_w}{f_{yt}} = 50 \times \frac{20\ in}{40{,}000\ psi} = 0.025$$

$$\rightarrow \frac{A_{v,min}}{s} = 0.025 < 0.05$$

In which case spacing of $s = 7.7\ in \cong 7.5\ in$ can be used.

Correct Answer is (B)

SOLUTION 4.11

Because the question is using ultimate loads, the ultimate design method is used which is partly described in ACI 318-14 Section 22.2 and the *NCEES handbook* as well.

$$M_n = \frac{M_u}{\emptyset} = 0.85\ f_c'\ ab(d - \frac{a}{2})$$

$$\frac{100\ kip.ft \times 12\ \frac{in}{ft}}{0.9} =$$

$$0.85 \times 4\ ksi \times a \times 35\ in \times \left(15\ in - \frac{a}{2}\right)$$

This can be reduced into the following quadratic equation:

$$a^2 - 30a + 22.4 = 0$$

The above quadratic equation is solved as follows:

$$root\ a = \frac{30 \pm \sqrt{30^2 - 4\times1\times22.4}}{2\times1} = 0.77\ in$$

Use the following equation to extract $A_{s,required}$:

$$a = \frac{A_s\ f_y}{0.85\ f_c'\ b}$$

$$A_s = \frac{0.85\ f_c'\ ab}{f_y}$$

$$= \frac{0.85 \times 4{,}000\ psi \times 0.77\ in \times 35\ in}{60{,}000\ psi}$$

$$= 1.5\ in^2$$

Substitute A_s in the following equation to confirm M_n:

$M_n = \frac{M_u}{\emptyset} = A_s\, f_y\, (d - \frac{a}{2})$

$M_u = \emptyset\, A_s\, f_y\, \left(d - \frac{a}{2}\right) =$

$0.9 \times 1.5\, in^2 \times 60\, ksi \times \left(15\, in - \frac{0.77}{2} in\right) \times \frac{1\, ft}{12\, in}$

$= 98.7\, kip.ft$

Close enough to 100 $kip.ft$ → OK

Check minimum reinforcements per ACI 318-14 Section 9.6.1.2:

$A_{s,min}$ should be the greatest of:

$$\frac{3\sqrt{f_c'}}{f_y} b_w d = \frac{3\sqrt{4{,}000}}{60{,}000} \times 35\, in \times 15\, in$$

$$= 1.66\, in^2$$

Or

$$\frac{200}{f_y} b_w d = \frac{200}{60{,}000} \times 35\, in \times 15\, in$$

$$= 1.75\, in^2$$

Provided that both values are larger than the calculated A_s, the required A_s in this case should be $A_{s,min}$:

$A_{s,required} = 1.75\, in^2$

Correct Answer is (A)

SOLUTION 4.12

TMS 402 *Building Code Requirements and Specifications for Masonry Structures* specifies in Section 8.1.5 that bearing stresses σ_{br} shall not exceed $0.33 f_m'$ and the bearing area A_{br} is defined in Section 4.3.4.

Section 4.3.4 specifies that A_{br} shall not exceed $A_1\sqrt{A_2/A_1}$ or $2A_1$ given that A_2 is the area of the lower base of the largest cone of a right pyramid with the loaded area A_1 acting as its upper base. It also specifies that walls not laid in running bonds, which is the case in this question, A_2 shall terminate at head joints.

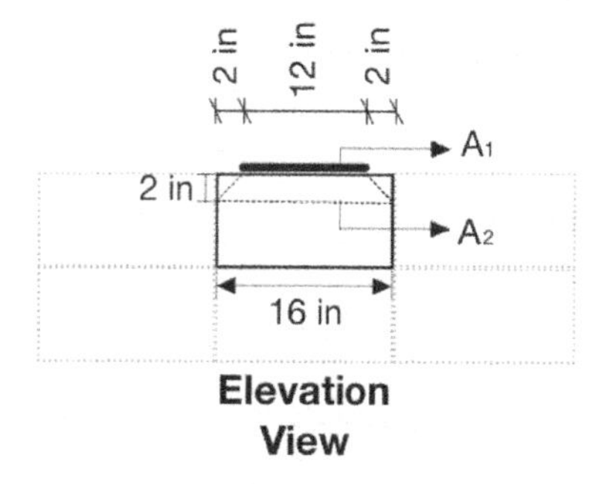

Elevation View

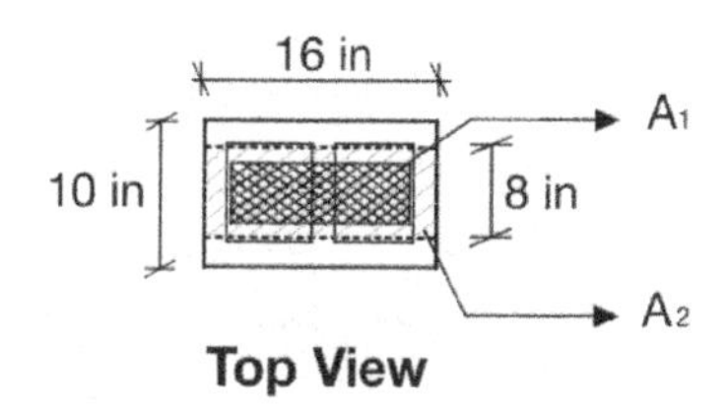

Top View

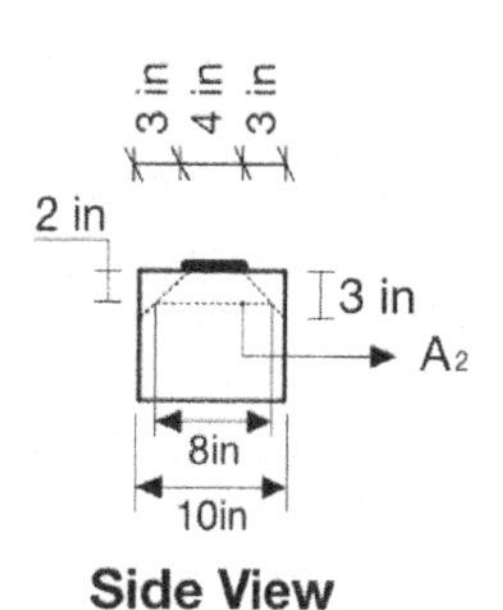

Side View

Given in the NCM TEK *National Concrete Masonry Association Specification*, the dimension of a 10 *in* block is 16 *in* long and 8 *in* height as shown in the above figures.

Based on this, and in reference to the above cross section and the hypothetical cone shape for A_2, the following can be derived:

$$A_1 = 12 \times 4 = 48\ in^2$$

$$A_2 = 16 \times 8 = 128\ in^2$$

$$A_{br} = \min\left(2A_1\ or\ A_1\sqrt{A_2/A_1}\right)$$

$$= \min(96\ or\ 78.4) = 78.4\ in^2$$

$$\sigma_{br} = \frac{10{,}000\ lb}{78.4\ in^2} = 127.6\ psi$$

Correct Answer is (A)

SOLUTION 4.13

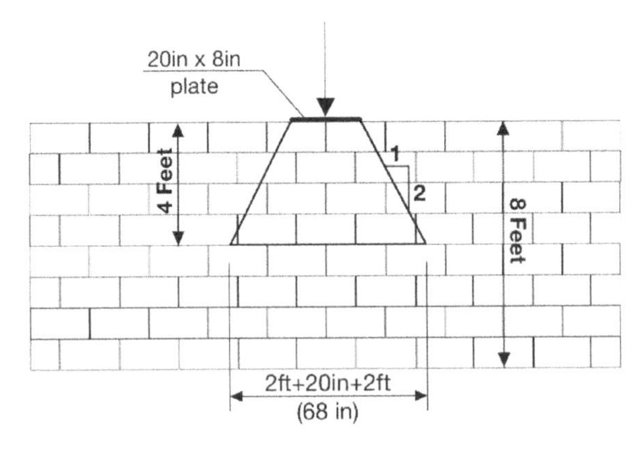

The effective length of distribution is determined using TMS 402 *Building Code Requirements and Specifications for Masonry Structures* Section 5.1.3 as depicted in the above figure as follows:

$$L_{eff} = 2 \times \left[2\ ft \times 12 \frac{in}{ft}\right] + 20\ in$$

$$= 68\ in$$

Correct Answer is (A)

SOLUTION 4.14

A check on the section's moment and shear capacity is performed based upon which load w can be calculated as the smallest of the following:

$$M_s = \frac{w\,l^2}{8} \rightarrow w = \frac{8M_s}{l^2}$$

Or

$$V_s = \frac{w\,l}{2} \rightarrow w = \frac{2\,V_s}{l}$$

Calculate allowable moment:

Per Section 4.2.2 of TMS 402 code:

$$E_s = 29{,}000{,}000\ psi$$

$$E_m = 900 \times f'_m = 2{,}520{,}000\ psi$$

Per Section 8.3.3.1 and 8.3.4.2.2 of TMS 402 code:

$f_s = 32{,}000\ psi$ for grade 60 steel

$$f_c = 0.45 \times f'_m = 1{,}260\ psi$$

The *NCEES handbook* Section 4.3.2.3 can be used as well as follows:

$$n = \frac{E_s}{E_m} = \frac{29{,}000{,}000}{2{,}520{,}000} = 11.5$$

$$\rho = \frac{A_s}{bd} = \frac{2 \times 0.2\ in^2}{5.63\ in \times 13.0\ in} = 0.0055$$

$$k = \sqrt{(\rho n)^2 + 2\rho n} - \rho n$$

$$= \left(\sqrt{(0.0055 \times 11.5)^2 + 2 \times 0.0055 \times 11.5} - 0.0055 \times 11.5\right)$$

$$= 0.3$$

$$j = 1 - \frac{k}{3} = 1 - \frac{0.3}{3} = 0.9$$

Allowable moment in this case is the lowest of:

$M_s = f_s\, A_s\, j\, d$

$= 32{,}000\ psi \times (2 \times 0.2)\ in^2 \times 0.9 \times 13\ in \times \frac{1\ ft}{12\ in}$

$= 12{,}480\ lb.ft$

Or

$M_s = \frac{f_c\ (or\ f'_m)\ k\ b\ j\ d^2}{2} =$

$\frac{1{,}260\ psi \times 0.3 \times 5.63\ in \times 0.9 \times (13\ in)^2}{2} \times \frac{1\ ft}{12\ in}$

$= 13{,}487\ lb.ft$

Allowable moment that is used in this case is the least of the above:

$M_s = 12{,}480\ lb.ft$

Allowable load due to allowable moment:

$w = \frac{8\ M_s}{l^2}$

$= \frac{8 \times 12{,}480\ lb.ft}{(6.5\ ft)^2}$

$= 2{,}363\ lb/ft$

Calculate allowable shear:
Section 8.3.5.1.2 of the TMS 402 code is used to calculate the allowable shear.

With reference to the commentary of the same section, value of M/Vd_v is permitted to be taken as '1.0' which leads to the use of Equation 8-27 as follows:

$F_v = (2\sqrt{f'_m})\ \gamma_g$

Where $\gamma_g = 1$ for fully grouted sections which is the case in this question, and $F_{vs} = 0$ given that no stirrups were provided.

$= (2\sqrt{2{,}800}) \times 1 = 105.8\ psi$

Using shear Equation 8-24 from Section 8.3.5.1.1:

$V_s = F_v\ A_{nv}$

$= 105.8\ psi \times 5.63\ in \times 13\ in$

$= 7{,}743.5\ lb$

Allowable load due to allowable shear:

$w = \frac{2\ V_s}{l}$

$= \frac{2 \times 7{,}743.5\ lb}{6.5\ ft}$

$= 2{,}382\ lb/ft$

The lowest value of w from all the above is used in this case:

$w = 2{,}363\ lb/ft$

Correct Answer is (C)

SOLUTION 4.15

Using the *Allowable Stress Design* method ASD presented in Chapter 8 of the TMS 402 code Section 8.2.4, the following equations shall be satisfied:

$$\frac{f_a}{F_a} + \frac{f_b}{F_b} \le 1$$

$$P \le (1/4)P_e$$

Where f_a/F_a is the ratio of the actual compressive stress to the allowable compressive stress of the wall and f_b/F_b is the same but for the flexural/bending stress of the wall. P_e is the wall's Euler buckling load.

Flexural/wall bending:

Maximum moment the wall experiences is calculated using lateral earth pressure and Rankine's at rest coefficient:

$K_o = 1 - \sin \emptyset'$

$= 1 - \sin(35^o) = 0.43$

Pressure at the bottom of the wall is therefore:

$= 130\, pcf \times 8\, ft \times 0.43$

$= 447.2\, psf/ft$

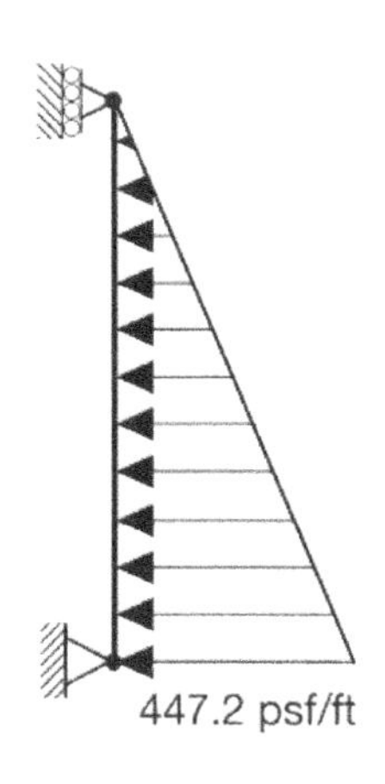

Using the moment diagrams from *NCEES handbook* Chapter 4, the maximum moment for the above load distribution is calculated as follows and occurs at $0.577h$ measured from the top of the wall:

$M_{max} = 0.128wh$

$= 0.128 \times 447.2\, \frac{psf}{ft} \times 8\, ft$

$= 458\, lb.ft/ft$

$= 5{,}495\, lb.in/ft$

$f_b = \frac{M_{max}}{S_n}$ Equation 8-18

Where S_n is taken from Table 3 of the NCMA TEK Section Properties of Concrete Masonry as $46.3\, in^3/ft$.

$= \frac{5{,}495\, in.lb/ft}{46.3\, in^3/ft} = 118.7\, psi$

$F_b = \left(\frac{1}{3}\right) f'_m$

$= \frac{1{,}800\, psi}{3} = 600\, psi$

Axial load and allowable axial load:

$f_a = \frac{P}{A_n} = \frac{P}{24\, in^2/ft \times 21\, ft} = \frac{P}{504}\, psi$

Where A_n is taken from Table 3 of the NCMA TEK Section Properties of Concrete Masonry Walls as $24\, in^2/ft$. And the radius of gyration $\boldsymbol{r}$ is taken from the same table and equals to $2.08\, in$.

Calculate h/r ratio to determine allowable axial stress as follows:

$\frac{h}{r} = \frac{8\, ft \times 12\, in/ft}{2.08\, in} = 46 < 99$

$F_a = \left(\frac{1}{4}\right) f'_m \left[1 - \left(\frac{h}{140r}\right)^2\right]$

$$= \left(\left(\frac{1}{4}\right) \times 1{,}800\, psi \times \left[1 - \left(\frac{8\, ft \times 12\, in/ft}{140 \times 2.08\, in}\right)^2\right] \right)$$

$= 401\, psi$

Flexure and axial combined:

$\frac{f_a}{F_a} + \frac{f_b}{F_b} \le 1$

$$\frac{P/504}{401\,psi} + \frac{118.7\,psi}{600\,psi} \leq 1$$

$P = 162{,}121\ lb$

Check the Euler load P_e:

$$P_e = \frac{\pi^2 E_m I_n}{h^2}\left(1 - 0.577\frac{e}{r}\right)^3$$

Where:

$E_m = 1{,}620{,}000\ psi$ - concrete masonry

$I_n = 130.3\ in^4/ft$ taken from the NCMA TEK Section Properties.

$e = 0$ given that no eccentricity was reported in the question.

$P_e =$

$$\frac{\pi^2 \times 1{,}620{,}000\ psi \times (130.3\times 21) in^4}{\left(8\ ft \times 12\frac{in}{ft}\right)^2} \times (1-0)^3$$

$= 4{,}747{,}183\ lb$

$P = \frac{P_e}{4} = 1{,}186{,}796\ lb$

P is taken as the lowest of the above two values:

$P = min(162{,}121\ lb\ and\ 1{,}186{,}183\ lb)$

$= 162{,}121\ lb$

$w = \frac{162{,}121\ lb}{21\ ft} = 7{,}720\ lb/ft\ (7.7 kip/ft)$

Correct Answer is (A)

SOLUTION 4.16

Shape $HP10 \times 42$ section properties as collected from the AISC *Steel Construction Manual*:

$r_x = 4.13\ in$
$r_y = 2.41\ in$
$A_g = 12.41\ in^2$
$F_y = 50\ ksi$
$E = 29{,}000\ ksi$

Before removing part of the floor:

Length of the laterally unbraced member (L):

$L_x = 20\ ft\ in\ the\ strong\ direction$
$L_y = 10\ ft\ in\ the\ weak\ direction$

Given that the effective length factor $k = 1.0$ for pinned connections, the following can be established from the AISC Specification Section E2 *Effective Length* and Section E3 *Flexural Buckling of Members*:

$$4.71\sqrt{\frac{E}{F_y}} = 4.71\sqrt{\frac{29{,}000\ ksi}{50\ ksi}} = 113$$

$$\frac{L_{c\,(x)}}{r_x} = \frac{1*20\ ft*12\ in/ft}{4.13\ in} = 58$$

$$\frac{L_{c\,(y)}}{r_y} = \frac{1*10\ ft*12\ in/ft}{2.41\ in} = 50$$

x-axis governs, based on the above:

$$\frac{L_{c\,(x)}}{r_x} < 4.71\sqrt{\frac{E}{F_y}}$$

$$F_{cr} = \left(0.658^{\frac{F_y}{F_e}}\right)F_y$$

$$F_e = \frac{\pi^2 E}{\left(\frac{L_{c,x}}{r_x}\right)^2}$$

$$= \frac{\pi^2 * 29{,}000\ ksi}{\left(\frac{1\times 20\ ft \times 12\ in/ft}{4.13\ in}\right)^2} = 84.75\ ksi$$

$$F_{cr} = \left(0.658^{\frac{50}{84.75}}\right) \times 50\ ksi$$
$$= 39\ ksi$$

$$\phi_c P_n = 0.9 \times F_{cr} \times A_g$$
$$= 0.9 \times 39\ ksi \times 12.4\ in^2$$
$$= 435.9\ kip$$

After removing part of the floor:

Length of the laterally unbraced member (L):

$L_x = 30\ ft\ in\ the\ strong\ direction$

$L_y = 10\ ft\ in\ the\ weak\ direction$

$$4.71\sqrt{\frac{E}{F_y}} = 4.71\sqrt{\frac{29{,}000\ ksi}{50\ ksi}} = 113$$

$$\frac{L_{c\,(x)}}{r_x} = \frac{1 \times 30\ ft \times 12\ in/ft}{4.13\ in} = 87.2$$

x-axis governs, based on the above:

$$\frac{L_{c\,(x)}}{r_x} < 4.71\sqrt{\frac{E}{F_y}}$$

$$F_{cr} = \left(0.658^{\frac{F_y}{F_e}}\right) F_y$$

$$F_e = \frac{\pi^2 * 29{,}000\ ksi}{\left(\frac{1*30\ ft*12\ in/ft}{4.13\ in}\right)^2} = 37.7\ ksi$$

$$F_{cr} = \left(0.658^{\frac{50}{37.7}}\right) \times 50\ ksi$$
$$= 28.7\ ksi$$

$$\phi_c P_n = 0.9 \times F_{cr} \times A_g$$
$$= 0.9 \times 28.7\ ksi \times 12.4\ in^2$$
$$= 320.2\ kip$$

The difference between the two cases:

$435.9\ kip - 320.2\ kip = 115.7\ kip$

Correct Answer is (B)

There is an alternative/shorter method using the AISC *Steel Construction Manual* Tables in Part 4 Table 4.2:

Before removing part of the floor:

The values in the tables of the AISC manual are based on effective lengths with respect to minor axis kL_y :

$kL_y = 10\ ft$

$kL_x = 20\ ft$

From AISC manual Table 4.2 for HP10×42:

$$r_x/r_y = 1.71$$

The equivalent effective length for the x-axis in this case is:

$$\frac{kL_x}{r_x/r_y} = \frac{20\ ft}{1.71} = 11.7\ ft > 10\ ft$$

Use a length of 11.7 ft to determine the available strength of the section from Table 4.2 by interpolation:

$$\phi_c P_n = 435.9\ kip$$

After removing part of the floor:

$kL_y = 10\ ft$

$kL_x = 30\ ft$

The equivalent effective length for the x-axis in this case is:

$$\frac{kL_x}{r_x/r_y} = \frac{30\ ft}{1.71} = 17.55\ ft > 10\ ft$$

Use a length of 17.55 ft to determine the available strength of the section from Table 4.2 by interpolation:

$$\phi_c P_n = 320.2\ kip$$

The difference between the two cases:

$435.9\ kip - 320.2\ kip = 115.7\ kip$

SOLUTION 4.17

Table 10-1 of the AISC *Steel Construction Manual* provides shear connections' capacities in such cases. In doing so, the table takes care of bolt shear Limit States, bolt bearing on angles, shear yielding of angles, shear rupture of angles and block shear rupture of angles.

The following attributes of the connection in the question are plugged into the manual's Table 10.1 found on page 10-45:

- $W10 \times 77$, $t_w = 0.5\ in$
- 2 rows of group A, Slip Critical bolts in STD holes.
- $L_{ev} = 2\ in$ with both flanges coped.
- $L_{eh} = 1.75\ in.$
- Angles thickness = $0.5\ in$

Based on the above, Table 10-1 of the manual provides the following readings:

- Bolts and angles available strength are as follows:
  - $46.1\ kip$ using ASD
  - $69.2\ kip$ using LRFD

- Beam web available strength
  - $85.3 \times 0.5 = 42.7\ kip$ using ASD
  - $128 \times 0.5 = 64\ kip$ using LRFD

To determine the available shear capacity for this connection, the minimum available shear strength from either the web or the bolts & angle shall be used. In this case the beam web controls. Answer is therefore as follows:

- Using ASD = $42.7\ kip$
- Using LRFD = $64\ kip$

Correct Answer is (A)

SOLUTION 4.18

The conservative method presented in Part 7 of the AISC *Steel Construction Manual* is used to solve this question. A solution using the non-conservative method is also presented at the end of the solution.

The conservative method is explained in Case II in the manual where the neutral axis of the bolt group is assumed to be located at the center of gravity of the group – i.e., in the middle in this case.

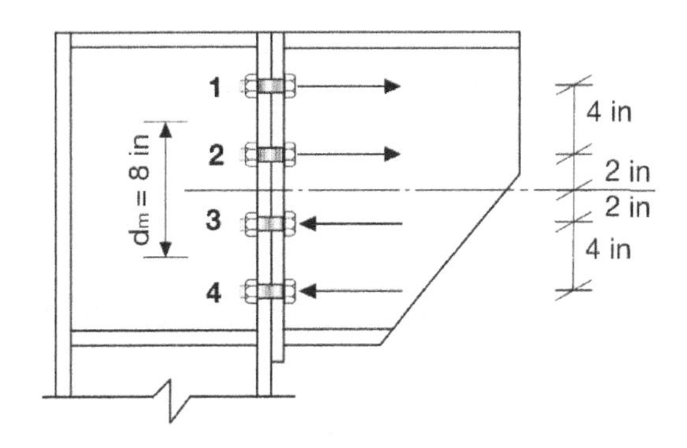

With this assumption, the four bolts above the neutral axis will theoretically be subjected to higher tension forces compared to the actual scenario where neutral axis is at a lower location creating a larger lever arm.

<u>Using LRFD method:</u>

From Table 7.1 and 7.2 of the AISC manual, a singly loaded bolt available shear capacity is:

$F_{nv} = 27\ ksi$ – nominal shear strength
$A_{bolt} = 0.442\ in^2$
$\emptyset = 0.75$
$r_{uv} = \emptyset r_n = 8.97\ kip$

And its tensile capacity is:

$F_{nt} = 45\ ksi$ – nominal tensile strength
$A_{bolt} = 0.442\ in^2$
$\emptyset = 0.75$
$r_{ut} = \emptyset r_n = 14.9\ kip$

The four bolts located above the neutral axis are subjected to tensile forces along with shear forces. In this case, the nominal tensile stress should be modified to include the effect of shear stress per Section J3.7 of the AISC *Specification for Structural Steel Buildings* as follows:

$$F'_{nt} = 1.3F_{nt} - \frac{F_{nt}}{\emptyset F_{nv}} f_{rv} \leq F_{nt}$$

In this case f_{rv} is the actual shear stress per bolt using LRFD, and as we are trying to evaluate capacity the following can be assumed → $f_{rv} = \emptyset F_{nv}$

$$F'_{nt} = 1.3F_{nt} - F_{nt}$$

$$= 1.3 \times 45\ ksi - 45\ ksi$$

$$= 13.5\ ksi$$

Recalculating r_{ut} using F'_{nt} for the four bolts above the neutral axis:

$$r'_{ut} = \emptyset F'_{nt} A_{bolt}$$

$$= 0.75 \times 13.5\ ksi \times 0.442\ in^2$$

$$= 4.48\ kip \text{ per bolt}$$

With four bolts above the neutral axis (tension) using Equation 7-14a from the manual:

$$P_{u,tension} = \frac{n' \times r'_{ut} \times d_m}{e}$$

$$= \frac{4 \times 4.48\ kip \times 8\ in}{10\ in}$$

$$= 14.3\ kip$$

For the entire set of eight bolts (only shear):

$$P_{u,shear} = n \times r_{uv} = 8 \times 8.97\ kip = 71.8\ kip$$

$$\boldsymbol{P_u = 14.3\ kip + 71.8\ kip = 86.1\ kip}$$

Using ASD method:

From Table 7.1 and 7.2 of the AISC manual, a singly loaded bolt available shear capacity is:

$F_{nv} = 27\ ksi$ – nominal shear strength
$A_{bolt} = 0.442\ in^2$
$\Omega = 2.00$
$r_{av} = r_n/\Omega = 5.97\ kip$

And its tensile capacity is:

$F_{nt} = 45\ ksi$ – nominal tensile strength
$A_{bolt} = 0.442\ in^2$
$\Omega = 2.00$
$r_{at} = r_n/\Omega = 9.94\ kip$

The four bolts located above the neutral axis are subjected to tensile forces along with shear forces. In this case, the nominal tensile stress should be modified to include the effect of shear stress per Section J3.3 of the AISC Specification for Structural Steel Buildings as follows:

$$F'_{nt} = 1.3F_{nt} - \frac{\Omega F_{nt}}{F_{nv}} f_{rv} \leq F_{nt}$$

In this case f_{rv} is the actual shear stress per bolt using ASD, and as we are trying to evaluate capacity the following can be assumed → $\Omega f_{rv} = F_{nv}$

$$F'_{nt} = 1.3F_{nt} - \frac{\Omega F_{nt}}{F_{nv}} \frac{F_{nv}}{\Omega}$$

$$= 1.3 \times 45\ ksi - 45\ ksi$$

$$= 13.5\ ksi$$

Recalculating r_{at} using F'_{nt} for the four bolts above the neutral axis:

$$r'_{at} = F'_{nt} A_{bolt}/\Omega$$

$$= 13.5\ ksi \times 0.442\ in^2/2$$

$$= 3.0\ kip \text{ per bolt}$$

With four bolts above the neutral axis (in tension), Equation 7-14a generates the following force:

$$P_{a,tension} = \frac{n' \times r'_{at} \times d_m}{e}$$

$$= \frac{4 \times 3\ kip \times 8\ in}{10\ in} = 9.6\ kip$$

For the entire set of eight bolts (only shear):

$$P_{a,shear} = n \times r_{av} = 8 \times 5.97\ kip = 47.8\ kip$$

$$\boldsymbol{P_a = 9.6\ kip + 47.8\ kip = 57.4\ kip}$$

Correct Answer is (A)

The below is an alternative solution using the non-conservative (longer) method presented in Part 7 of the AISC manual. The non conservative method is referred to as Case I in the manual.

Case I starts with assuming a trial location for a compression block. The depth of this block is readjusted using iterations with the aim of achieving equilibrium between the tension in bolts above the assumed neutral axis and this block. Upon achieving equilibrium, the combined moment of inertia of the bolt group along with the compression block is calculated, based upon which shear and tension forces per bolt is evaluated.

Assume width of compression block:

$$b_{eff} = 8 \times t_{plate} = 4\ in$$

Assume the depth of the block:

$$d = d_{section}/6$$

$$= 21\ in/6 = 3.5\ in$$

In reference to Part 7 of the AISC manual, $'y'$ is the distance from the bolt group centroid to the neutral axis located at the top of the compression block. In which case $y = 7.0\ in$.

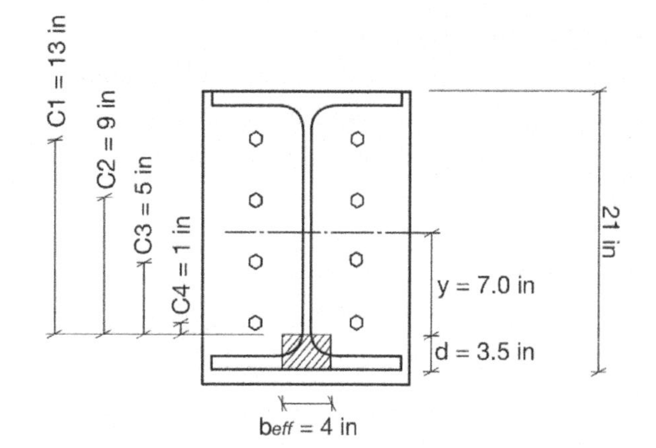

Dimensions of the compression block are verified by taking moment around the bottom leg of the block for the bolts in tension, and for the block itself which is in compression. Those two should achieve equilibrium. This process is summarized in Equation 7.11 form the manual:

$$\left(\sum A_b\right) \times y = b_{eff} \times d \times d/2$$

$$(8 \times 0.442) \times 7 = 4 \times 3.5 \times 3.5/2$$

$$24.8 \approx 24.5$$

The combined moment of inertia of the group about the neutral axis located at the top of the compression block is:

$$I_x = \frac{b_{eff} \times d^3}{3} + \Sigma A_b C^2$$

$$= \frac{b_{eff} \times d^3}{3} + A_b \begin{pmatrix} C1^2 + C2^2 \\ + \\ C3^2 + C4^2 \end{pmatrix}$$

$$= \frac{4\ in \times (3.5\ in)^3}{3} +$$

$$2 \times 0.442\ in^2 \times \begin{pmatrix} 13^2 + 9^2 \\ + \\ 5^2 + 1^2 \end{pmatrix}$$

$$= 301.2\ in^4$$

Now that the neutral axis and the moment of inertia of the group have both been determined, we can move ahead with providing a solution using the LRFD method or the ASD method.

Using LRFD method:
From Table 7.1 and 7.2 of the AISC manual, and the earlier solution of this question:

$F_{nv} = 27\ ksi$ – nominal shear strength
$F_{nt} = 45\ ksi$ – nominal tensile strength
$A_{bolt} = 0.442\ in^2$
$\emptyset = 0.75$
$r_{uv} = \emptyset r_n = 8.97\ kip$
$r_{ut} = \emptyset r_n = 14.9\ kip$

All the bolts are located above the neutral axis, and they all are subjected to tensile forces along with shear forces. In this case, the nominal tensile stress is modified to include the effect of shear stress as follows:

$$F'_{nt} = 1.3F_{nt} - \frac{F_{nt}}{\emptyset F_{nv}} f_{rv} \leq F_{nt}$$

In this case f_{rv} is the actual shear stress per bolt using LRFD, and as we are trying to evaluate capacity the following can be assumed → $f_{rv} = \emptyset F_{nv}$

$$F'_{nt} = 1.3F_{nt} - F_{nt}$$
$$= 1.3 \times 45\ ksi - 45\ ksi$$
$$= 13.5\ ksi$$

Recalculating r_{ut} using F'_{nt} for the eight bolts above the neutral axis:

$$r'_{ut} = \emptyset F'_{nt} A_{bolt}$$
$$= 0.75 \times 13.5\ ksi \times 0.442\ in^2$$
$$= 4.48\ kip \text{ per bolt}$$

Using the AISC manual Equation 7-12a to determine tensile force per bolt:

$$r_{ut} = r'_{ut} = \left(\frac{P_u ec}{I_x}\right) A_b$$

Where c is the distance from neutral axis to the furthest bolt in the group = $13\ in.$

$$P_{u,tension} = \frac{r'_{ut}\ I_x}{ecA_b}$$
$$= \frac{4.48\ kip \times 301.2\ in^4}{10\ in \times 13\ in \times 0.442\ in^2}$$
$$= 23.5\ kip$$

For the entire set of eight bolts (only shear):

$$P_{u,shear} = n \times r_{uv} = 8 \times 8.97\ kip = 71.8\ kip$$

$\boldsymbol{P_u = 23.5\ kip + 71.8\ kip = 95.3\ kip}$
Compared to 86.1 kip using the conservative method

Using ASD method:
From Table 7.1 and 7.2 of the AISC manual, and the earlier solution of this question:

$F_{nv} = 27\ ksi$ – nominal shear strength
$F_{nt} = 45\ ksi$ – nominal tensile strength
$A_{bolt} = 0.442\ in^2$
$\Omega = 2.00$
$r_{av} = r_n/\Omega = 5.97\ kip$
$r_{at} = r_n/\Omega = 9.94\ kip$

All the bolts are located above the neutral axis and they all are subjected to tensile forces along with shear forces. In this case, the nominal tensile stress is modified to include the effect of shear stress as follows:

$$F'_{nt} = 1.3F_{nt} - \frac{\Omega F_{nt}}{F_{nv}} f_{rv} \leq F_{nt}$$

In this case f_{rv} is the actual shear stress per bolt using ASD, and as we are trying to evaluate capacity the following can be assumed → $\Omega f_{rv} = F_{nv}$

$$F'_{nt} = 1.3F_{nt} - \frac{\Omega F_{nt}}{F_{nv}} \frac{F_{nv}}{\Omega}$$
$$= 1.3 \times 45\ ksi - 45\ ksi$$
$$= 13.5\ ksi$$

Recalculating r_{at} using F'_{nt} for the four bolts above the neutral axis:

$$r'_{at} = F'_{nt} A_{bolt}/\Omega$$

$= 13.5\ ksi \times 0.442\ in^2/2$

$= 3.0\ kip$ per bolt

Using the AISC Equation 7-12a to determine tensile force per bolt:

$$r_{at} = r'_{at} = \left(\frac{P_a ec}{I_x}\right) A_b$$

Where $'c'$ in this case is the distance from neutral axis to the furthest bolt in the group = 13 $in.$

$$P_{a,tension} = \frac{r'_{at} I_x}{ecA_b}$$

$$= \frac{3.0\ kip \times 301.2\ in^4}{10\ in \times 13\ in \times 0.442\ in^2}$$

$$= 15.7\ kip$$

For the entire set of eight bolts (only shear):

$$P_{a,shear} = n \times r_{av} = 8 \times 5.97\ kip = 47.8\ kip$$

$\boldsymbol{P_a = 15.7\ kip + 47.8\ kip = 63.5\ kip}$

Compared to 57.4 *kip using the conservative method*

SOLUTION 4.19

$W24 \times 250$ section properties as collected from the AISC *Steel Construction Manual*:

$Z_x = 744\ in^3$ (plastic section modulus)

$t_w = 1.04\ in$ (web thickness)

The plastic section modulus $\boldsymbol{Z_x}$ will decrease by an amount that is equivalent to the plastic section modulus of the part that has been removed from the web.

The cross-section dimensions of the part that is removed from the web:

$$h_o = 10\ in\ \&\ t_w = 1.04\ in$$

Where $\boldsymbol{h_o}$ is the height of the cut in the web.

Using LRFD method:

Plastic section modulus of the removed part is calculated as follows:

$$Z_{removed} = \frac{t_w \times h_o^2}{4} = \frac{1.04\ in \times (10\ in)^2}{4} = 26\ in^3$$

$$Z_{net} = Z_x - Z_{removed}$$

$$= 744\ in^3 - 26\ in^3 = 718\ in^3$$

Available flexural strength *before* cutting through the web is as follows:

$$M_n = \phi_b M_p$$

$$= \phi_b Z_x F_y$$

$$= 0.9 \times 744\ in^3 \times 50\ ksi \times \frac{1\ ft}{12\ in}$$

$$= 2{,}790\ kip.ft$$

Available flexural strength *after* cutting through the web is as follows:

$$M'_n = \phi_b M'_p$$

$$= \phi_b Z_{net} F_y$$

$$= 0.9 \times 718\ in^3 \times 50\ ksi \times \frac{1\ ft}{12\ in}$$

$$= 2{,}692.5\ kip.ft$$

Reduction in the available flexural strength is therefore:

$$2{,}790.0 - 2{,}692.5 = 97.5\ kip.ft$$

Using ASD method:

$$M'_n = M'_p/\Omega_b$$

$$= Z_{net} F_y/\Omega_b$$

Reduction in the available flexural strength is therefore:

$$\frac{\left(\frac{97.5\ kip.ft}{\phi_b}\right)}{\Omega_b} = \frac{\left(\frac{97.5\ kip.ft}{0.9}\right)}{1.67} = 65\ kip.ft$$

Correct Answer is (C)

The above solution is valid as long as the centroid of the opening coincides with the centroid of the section. In the absence of this, eccentricity must be accounted for in the final calculation.

Alternative solution:
The below reference provides means and methods for designing steel and composite beams with web openings:

> *Darwin, D. (1990), Steel and Composite Beams with Web Openings. American Institute of Steel Construction AISC. Steel Design Guide Series 2.*

The following equation found in the abovementioned guide calculates the modified available flexural strength when web reinforcements are not provided:

$$M_m = M_p\left[1 - \frac{\Delta A_s\left(\frac{h_o}{4} + e\right)}{Z}\right]$$

$\boldsymbol{M_m}$ maximum moment.

$\boldsymbol{\Delta A_s}$ the cross-section area that is cut from the web $= h_o \times t_w$

$\boldsymbol{h_o}$ height of opening.

$\boldsymbol{e}$ eccentricity - i.e., the distance from the center of the opening to the centroid of the overall section.

Substituting the figures derived in this example into Darwin (1990) equation – taking $\boldsymbol{e}$ as zero – generates the same results presented earlier.

$$\phi_b M_m = 2{,}790 kip.ft \times \left[1 - \frac{10\,in \times 1.04\,in \times \left(\frac{10\,in}{4} + 0\right)}{744\,in^3}\right]$$

$$= 2{,}962.5\ kip.ft$$

SOLUTION 4.20
$W24 \times 250$ section properties as collected from the AISC *Steel Construction Manual*:

$Z_x = 744\,in^3$ (plastic section modulus)
$t_w = 1.04\,in$ (web thickness)

The plastic section modulus $\boldsymbol{Z_x}$ for the section shall be modified to account for: (1) a decrease by an amount that is equivalent to the plastic section modulus of the part that has been taken out from the web, (2) an increase to account for the added plastic section modulus for the additional reinforcing plates.

To calculate the plastic section modulus, the sums of all areas above or below the centroid of the overall section are multiplied by the distance from the centroid of those individual areas to the centroid of the entire section.

In that sense, the plastic section modulus for a rectangular cross section is $bh^2/4$.

Decrease in the plastic modulus:
The cross-section dimensions of the part that has been taken out from the web:

$$h_o = 14\,in\ \&\ t_w = 1.04\,in$$

Where $\boldsymbol{h_o}$ is the height of the cut in the web.

The plastic section modulus of the removed part is calculated as follows:

$$Z_{removed} = \frac{t_w \times {h_o}^2}{4} = \frac{1.04\,in \times (14\,in)^2}{4} = 51\,in^3$$

Increase in the plastic modulus due to the addition of plates:

$A_r = 2 \times 0.5 \times 2 = 2.0\,in^2$

$\bar{y} = 7.25\,in$

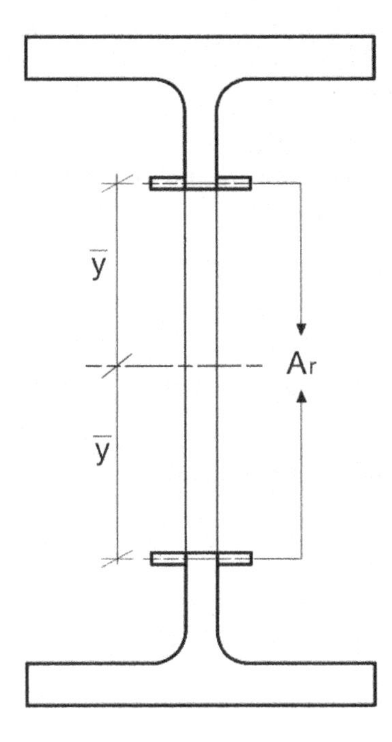

A_r is the cross-sectional area of the reinforcing plates either top or bottom
$= 2 \times A_r$

$\bar{y}$ the distance from the centroid of each plate to the centroid of the overall section.

$$Z_{addional} = 2 \times A_r \times \bar{y}$$
$$= 2 \times 2.0\ in^2 \times 7.25\ in$$
$$= 29.0\ in^3$$

$$Z_{net} = Z_x - Z_{removed} + Z_{addional}$$
$$= 744\ in^3 - 51\ in^3 + 29\ in^3 = 722\ in^3$$

$$M'_n = \phi_b M'_p$$
$$= \phi_b Z_{net} F_y$$
$$= 0.9 \times 722\ in^3 \times 50\ ksi \times \frac{1\ ft}{12\ in}$$
$$= 2{,}707.5\ kip.ft$$

Using ASD method:

$$M'_n = M'_p/\Omega_b$$
$$= Z_{net}F_y/\Omega_b$$
$$= \frac{722\ in^3 \times 50\ ksi}{1.67} \times \frac{1\ ft}{12\ in}$$
$$= 1{,}800\ kip.ft$$

Correct Answer is (A)

The above solution is valid as long as the centroid of the opening coincides with the centroid of the section. In the absence of this, eccentricity must be accounted for in the final calculation.

Alternative solution:
The below reference provides means and methods for designing steel and composite beams with web openings:

> *Darwin, D. (1990). Steel and Composite Beams with Web Openings. American Institute of Steel Construction AISC. Steel Design Guide Series 2.*

The following equation found in the abovementioned guide calculates the modified available flexural strength when reinforcements are provided:

$$M_m = M_p \left[1 - \frac{t_w\left(\frac{h_o^2}{4} + h_o e - e^2\right) - A_r h_o}{Z} \right] \leq M_p$$

M_m maximum moment.

A_r is the cross-sectional area of the reinforcement either top or bottom of the section.

h_o height of opening.

e eccentricity - i.e., the distance from the center of the opening to the centroid of the overall section.

t_w web thickness.

Substituting the figures derived in this example into Darwin (1990) equation – taking ***e*** as zero – generates almost similar results compared to the original solution:

$$\phi_b M_m = 2{,}790\ kip.ft \times \left[1 - \frac{1.04\ in \times \left(\frac{(14\ in)^2}{4} + 0 - 0\right) - 2.0\ in^2 \times 14\ in}{744\ in^3}\right]$$

$$= 2{,}703.9\ kip.ft$$

The slight difference between the two methods is attributed to Darwin (1990) not accounting for the thickness of reinforced plates which results in a slightly conservative strength.

SOLUTION 4.21

Using LRFD method:
From Table 7.1 of the AISC *Steel Construction Manual*:

$F_{nv} = 27\ ksi$ – nominal shear strength
$A_{bolt} = 0.442\ in^2$
$\emptyset = 0.75$
$r_{uv} = \emptyset r_n = 8.97\ kip$

Table 7.8 of the same manual with an angle of zero:

$$e = 10\ in + 2.5\ in + \frac{5.5\ in}{2} = 15.25\ in$$

$$n = 4\ rows$$

$$s = 4\ in$$

Using interpolation with $c = 2.4$:

$$\rightarrow R_u = \emptyset R_n = C \times \emptyset\, r_n$$

$$= 2.4 \times 8.97\ kip = 21.5\ kip$$

Using ASD method:
From Table 7.1 of the AISC *Steel Construction Manual*:

$F_{nv} = 27\ ksi$ – nominal shear strength
$A_{bolt} = 0.442\ in^2$
$\Omega = 2.00$
$r_{av} = r_n/\Omega = 5.97\ kip$

Table 7.8 of the same manual with an angle of zero:

$$e = 10\ in + 2.5\ in + \frac{5.5\ in}{2} = 15.25\ in$$

$$n = 4\ rows$$

$$s = 4\ in$$

Using interpolation $c = 2.4$:

$$\rightarrow R_a = \frac{R_n}{\Omega}$$

$$= C \times \frac{r_n}{\Omega}$$

$$= 2.4 \times 5.97\ in$$

$$= 14.3\ kip$$

Correct Answer is (D)

SOLUTION 4.22

From Table 8.3 of the AISC *Steel Construction Manual,* the $E60$ electrode strength coefficient is $C1 = 0.857$.

Table 1.7 of the same manual identifies the centroid of an $L3 \times 3 \times ½$ as 0.929 in from any of its legs, which makes tension loads eccentric to the weld group by:

$$e = \frac{3.0\ in}{2} - 0.929\ in = 0.571\ in$$

Using LRFD method:
From Table 8.4 of the AISC *Steel Construction Manual*, assume $a = 0$ and C is constant for this assumption for all ***k*** values at '3.71':

$$l_{min} = \frac{p_u}{\emptyset\, C\, C_1 D}$$
$$= \frac{90}{0.75 \times 3.71 \times 0.857 \times 5}$$
$$= 7.6\ in$$

Verify assumption:
$l = 7.6\ in, kl = 3\ in \rightarrow k = 0.4$

With $e = al$, where $e = 0.571\ in$ and $l = 7.6\ in \rightarrow a = 0.08$

Using Table 8.4 with $a = 0.08$ and $k = 0.4$ $\rightarrow C = 3.7$

→ *OK*

Using ASD method:
From Table 8.4 of the AISC *Steel Construction Manual*, assume $a = 0$ and C is constant for this assumption for all ***k*** values at '3.71':

$$l_{min} = \frac{\Omega\, p_a}{C\, C_1 D}$$
$$= \frac{2 \times 60}{3.71 \times 0.857 \times 5}$$
$$= 7.6\ in$$

Verify assumption:
$l = 7.6\ in, kl = 3\ in, hence\ k = 0.4$

With $e = al$, where $e = 0.571\ in$ and $l = 7.6\ in \rightarrow a = 0.08$

Using Table 8.4 with $a = 0.08$ and $k = 0.4$ $\rightarrow C = 3.7$

→ *OK*

Hence $l = 7.6\ in \approx 8\ in$ using both LRFD and ASD methods.

Correct Answer is (B)

SOLUTION 4.23

From Table 1.8 of the AISC *Steel Construction Manual*, $WT22 \times 145$ section properties are as follow:

$I_y = 521\ in^4$
$J = 25.4\ in^4$
$S_x = 111\ in^3$
$Z_x = 196\ in^3$
$h = 21.8\ in$

Also:
$d = 0.865\ in$ – d is defined in Section F9-2b of the AISC specification as the width of web when in compression, in which case it is equivalent to t_w of the section.

Using the *AISC Specification for Structural Steel Buildings,* Section F9 of the specification addresses flexural strength for Ts and Double Angles.

When Tee stems are in compression:

$$M_p = M_y = F_y S_x$$
$$= 50\ ksi \times 111\ in^3 \times \left(\frac{1\ ft}{12\ in}\right)$$
$$= 462.5\ kip.ft$$

Assess lateral torsional buckling for stems in compression:
$M_n = M_{cr} \le M_y$

$$M_{cr} = \frac{1.95E}{L_b}\sqrt{I_y J}\left(B + \sqrt{1 + B^2}\right)$$

$$B = -2.3\left(\frac{d}{L_b}\right)\sqrt{\frac{I_y}{J}}$$
$$= -2.3\left(\frac{0.865\ in}{50\ ft \times 12\frac{in}{ft}}\right)\sqrt{\frac{521\ in^4}{25.4\ in^4}}$$
$$= -0.015$$

PART II Structural Depth

$$M_{cr} = \frac{1.95 \times 29{,}000\ ksi}{50\ ft \times 12 \frac{in}{ft}} \times \sqrt{521\ in^4 \times 25.4\ in^4} \times \left(-0.015 + \sqrt{1 + (-0.015)^2}\right)$$

$$= 10{,}680\ kip.in\ (890\ kip.ft) > M_y$$

$$\rightarrow M_n = 462.5\ kip.ft$$

Using ASD method:
Maximum moment occurs when P_a is in the middle of the main beam, while maximum shear occurs when P_a is at the support.

Given the question is asking to assess P_a at flexure, the maximum moment is calculated as follows:

$$M = \frac{w_{DL}L^2}{8} + \frac{P_a L}{4}$$

$$= \frac{0.145\ kip/ft \times (50\ ft)^2}{8} + \frac{P_a \times 50 ft}{4}$$

$$= 45.3 + 12.5 P_a$$

$$M_{allowable} = \frac{M_n}{\Omega_b}$$

$$= \frac{462.5\ kip.ft}{1.67}$$

$$= 277\ kip.ft$$

$$277\ kip.ft = 45.3 + 12.5 P_a$$

$$P_a = 18.5\ kip$$

Correct Answer is (A)

Section 4 Structural Design

SOLUTION 4.24
From Table 1.8 of the AISC *Steel Construction Manual*, $WT22 \times 145$ section properties are as follow:

$I_y = 521\ in^4$
$J = 25.4\ in^4$
$S_x = 111\ in^3$
$Z_x = 196\ in^3$
$r_y = 3.49\ in$

Also:
$d = 0.865\ in$ – d is defined in Section F9-2a of AISC specification as the width of web when in tension, in which case this is equivalent to t_w of the section.

Using the *AISC Specification for Structural Steel Buildings,* Section F9 of the specification addresses flexural strength for Ts and Double Angles.

Checking moment capacity of the section:
When Tee stems are in tension the following applies:

$$M_p = F_y Z_x \leq 1.6 M_y$$

$$= 50\ ksi \times 196\ in^3 \times \left(\frac{1\ ft}{12\ in}\right)$$

$$\leq 1.6 F_y S_x$$

$$= 816.7\ kip.ft \leq 740\ kip.ft$$

$$\rightarrow M_p = 740\ kip.ft$$

With $1.6\ M_y = 1.6 \times 50\ ksi \times 111\ in^3 \times \left(\frac{1\ ft}{12\ in}\right)$

$$= 740\ kip.ft$$

Assess lateral torsional buckling for stems in tension:

$$L_p = 1.76\, r_y \sqrt{\frac{E}{F_y}}$$

$$= 1.76 \times 3.46\ in \sqrt{\frac{29{,}000\ ksi}{50\ ksi}}$$

$$= 146.7\ in\ (12.2\ ft)$$

$$L_r = 1.95 \left(\frac{E}{F_y}\right) \frac{\sqrt{I_y J}}{S_x} \sqrt{2.36 \left(\frac{F_y}{E}\right) \frac{d\, S_x}{J} + 1}$$

$$= 1.95 \left(\frac{29{,}000\ ksi}{50\ ksi}\right) \times \frac{\sqrt{521\ in^4 \times 25.4\ in^4}}{111\ in^3}$$

$$\times$$

$$\sqrt{2.36 \left(\frac{50\ ksi}{29{,}000\ ksi}\right) \frac{0.865\ in \times 111\ in^3}{25.4\ in^4} + 1}$$

$$= 1{,}181.1\ in\ (98.4\ ft)$$

Based on the above:

$L_p(12.2\ ft) < L_b(50\ ft) < L_r(98.4\ ft)$

Hence use the following equation for M_n:

$$M_n = M_p - (M_p - M_y)\left(\frac{L_b - L_p}{L_r - L_p}\right)$$

$$= 740 - (740 - 462.5)\left(\frac{50 - 12.3}{98.4 - 12.3}\right)$$

$$= 618\ kip.ft$$

Using ASD method:

$$M_{allowable} = \frac{w_{DL}L^2}{8} + \frac{P_a L}{4}$$

$$= \frac{0.145\ kip/ft \times (50ft)^2}{8} + \frac{P_a \times 50ft}{4}$$

$$= 45.3 + 12.5P_a$$

$$M_{allowable} = \frac{M_n}{\Omega_b}$$

$$= \frac{618\ kip.ft}{1.67}$$

$$= 370\ kip.ft$$

$$370\ kip.ft = 45.3 + 12.5P_a$$

$$\boldsymbol{P_{a-flexure} = 26\ kip}$$

Using LRFD method:

$$M_{ultimate} = 1.2 \times \frac{w_{DL}L^2}{8} + \frac{P_u L}{4}$$

$$= 1.2 \times \frac{0.145\ kip/ft \times (50ft)^2}{8} + \frac{P_u \times 50\ ft}{4}$$

$$= 54.4 + 12.5P_u$$

$$M_{ultimate} = \emptyset_b M_n$$

$$= 0.9 \times 618\ kip.ft$$

$$= 556.2\ kip.ft$$

$$556.2\ kip.ft = 54.4 + 12.5P_u$$

$$\boldsymbol{P_{u-flexure} = 40\ kip}$$

Checking shear capacity for the section:

Using Chapter G of the AISC *Specification for Structural Steel Buildings*:

$$V_n = 0.6F_y btC_{v2}$$

Given that b is the depth of the web (21.8 in) deducted from it the thickness of the top flange (1.58 in).

The web shear buckling coefficient C_{v2} is determined per Section G.2.2 of the same chapter:

$$\frac{h}{t_w} = \frac{21.8\ in}{0.865\ in} = 25.2$$

$$1.1\sqrt{\frac{k_v E}{F_y}} = 1.1 \times \sqrt{5.34 \times \frac{29{,}000\ ksi}{50\ ksi}}$$

$$= 61.2$$

Were $\boldsymbol{k_v}$ is '5.34' per Section G.2.1.

Given that $\frac{h}{t_w} < \sqrt{k_v E/F_y} \rightarrow C_{v2} = 1.0$

$$V_n = 0.6 \times 50\ ksi \times (21.8 - 1.58)\ in \times 0.865\ in \times 1.0$$

$$= 524.7\ kip$$

Using ASD method:

$$V_{allowable} = \frac{w_{DL}L}{2} + \frac{P_a}{2}$$

$$= \frac{0.145\ kip/ft \times 50\ ft}{2} + \frac{P_a}{2}$$

$$= 3.63 + 0.5P_a$$

$$V_{allowable} = \frac{V_n}{\Omega_v}$$
$$= \frac{524.7\ kip}{1.5}$$
$$= 349.8\ kip$$

$$349.8\ kip = 3.63 + 0.5P_a$$

$$\boldsymbol{P_{a-shear} = 692.3\ kip} > P_{a-flexure}$$

<u>Using LRFD method:</u>

$$V_{ultimate} = 1.2 \times \frac{w_{DL}L}{2} + \frac{P_u}{2}$$
$$= 1.2 \times \frac{0.145\ kip/ft \times 50\ ft}{2} + \frac{P_u}{2}$$
$$= 4.35 + 0.5P_u$$

$$V_{ultimate} = \emptyset_v V_n$$
$$= 1.0 \times 524.7\ kip$$
$$= 524.7\ kip$$

$$524.7\ kip = 4.35 + 0.5P_a$$

$$\boldsymbol{P_{u-shear} = 1{,}040.7\ kip} > P_{u-flexure}$$

$$\rightarrow \boldsymbol{P_a = \min\ (P_{a-shear}\ or\ P_{a-flexure})}$$
$$\boldsymbol{= \min(692.3\ kip\ or\ 26\ kip)}$$
$$\boldsymbol{= 26\ kip}$$

$$\rightarrow \boldsymbol{P_u = \min\ (P_{u-shear}\ or\ P_{u-flexure})}$$
$$\boldsymbol{= \min(1{,}040.7\ kip\ or\ 40\ kip)}$$
$$\boldsymbol{= 40\ kip}$$

Correct Answer is (B)

SOLUTION 4.25

Given the loads in the question are service loads, the ASD method is used with reference to the AISC *Specification for Structural Steel Buildings* Chapter 6:

<u>For Section $W36 \times 330$</u>

$$p = 0.459 \times 10^{-3}\ kip^{-1}$$
$$b_x = 0.274 \times 10^{-3}\ (kip.ft)^{-1}$$
$$pP_r = (0.459 \times 10^{-3}\ kip^{-1}) \times 300\ kip$$
$$= 0.14$$

Since $pP_r < 0.2$, the following equation should be satisfied:

$$\frac{1}{2}pP_r + \frac{9}{8}(b_x M_{rx}) \leq 1.0$$

$$\frac{1}{2} \times 0.14 + \frac{9}{8} \times \begin{bmatrix} 0.274 \times 10^{-3}\ (kip.ft)^{-1} \\ \times \\ 250\ kip.ft \end{bmatrix}$$

$0.07 + 0.08 = 0.15 < 1.0$ →OK

<u>For Section $W30 \times 116$</u>

$$p = 2.35 \times 10^{-3}\ kip^{-1}$$
$$b_x = 1.39 \times 10^{-3}\ (kip.ft)^{-1}$$
$$pP_r = (2.35 \times 10^{-3}\ kip^{-1}) \times 300\ kip$$
$$= 0.71$$

Since $pP_r \geq 0.2$, the following equation should be satisfied:

$$pP_r + b_x M_{rx} \leq 1.0$$
$$0.71 + 1.39 \times 10^{-3}\ (kip.ft)^{-1} \times 250\ kip.ft$$

$0.71 + 0.34 = 1.05 > 1.0$ → NOT OK

<u>For Section $W24 \times 207$</u>

$$p = 0.858 \times 10^{-3}\ kip^{-1}$$
$$b_x = 0.664 \times 10^{-3}\ (kip.ft)^{-1}$$
$$pP_r = (0.858 \times 10^{-3}\ kip^{-1}) \times 300\ kip$$
$$= 0.26$$

Since $pP_r \geq 0.2$, the following equation should be satisfied:

$$pP_r + b_x M_{rx} \leq 1.0$$

$0.26 + 0.66 \times 10^{-3}(kip.ft)^{-1} \times 250\ kip.ft$

$0.26 + 0.17 = 0.43 < 1.0$ → O.K.

For Section $W27 \times 129$

$p = 2.09 \times 10^{-3}\ kip^{-1}$

$b_x = 1.27 \times 10^{-3}\ (kip.ft)^{-1}$

$pP_r = (2.09 \times 10^{-3}\ kip^{-1}) \times 300\ kip$

$= 0.63$

Since $pP_r \geq 0.2$, the following equation should be satisfied:

$pP_r + b_x M_{rx} \leq 1.0$

$0.63 + 1.27 \times 10^{-3}\ (kip.ft)^{-1} \times 250\ kip.ft$

$0.63 + 0.32 = 0.95 < 1.0$ → OK

For Section $W18 \times 119$

$p = 1.7 \times 10^{-3}\ kip^{-1}$

$b_x = 1.62 \times 10^{-3}\ (kip.ft)^{-1}$

$pP_r = (1.7 \times 10^{-3}\ kip^{-1}) \times 300\ kip$

$= 0.51$

Since $pP_r \geq 0.2$, the following equation should be satisfied:

$pP_r + b_x M_{rx} \leq 1.0$

$0.51 + 1.62 \times 10^{-3}\ (kip.ft)^{-1} \times 250\ kip.ft$

$0.51 + 0.41 = 0.92 < 1.0$ → OK

For Section $W14 \times 120$

$p = 1.28 \times 10^{-3}\ kip^{-1}$

$b_x = 1.8 \times 10^{-3}\ (kip.ft)^{-1}$

$pP_r = (1.28 \times 10^{-3}\ kip^{-1}) \times 300\ kip$

$= 0.38$

Since $pP_r \geq 0.2$, the following equation should be satisfied:

$pP_r + b_x M_{rx} \leq 1.0$

$0.38 + 1.8 \times 10^{-3}\ (kip.ft)^{-1} \times 250\ kip.ft$

$0.38 + 0.45 = 0.83 < 1.0$ → OK

For Section $W24 \times 76$

$p = 4.64 \times 10^{-3}\ kip^{-1}$

$b_x = 3.92 \times 10^{-3}\ (kip.ft)^{-1}$

$pP_r = (4.64 \times 10^{-3}\ kip^{-1}) \times 300\ kip$

$= 1.4$

Since $pP_r \geq 1.0$ → NOT OK

Based on the above, the following sections are satisfactory:

1- $W36 \times 330$
2- $W24 \times 207$
3- $W27 \times 129$
4- $W18 \times 119$
5- $W14 \times 120$

The three most economical sections of those are the lightest in weight, and those are:

1- $W27 \times 129$
2- $W18 \times 119$
3- $W14 \times 120$

Correct Answers are (D, E & F)

SOLUTION 4.26

The steel plate shall be assessed based on its Gross Area A_g multiplied by the yielding strength F_y, followed by a check on its Net Area A_n (removing holes and using different paths for possible failures) multiplied by ultimate yielding strength F_u. And finally, a check on block shear failure is carried out.

Gross Area and Net Area are determined using Section B4 Clause 3 of the *AISC Specification.*

Reference is made to Section J of the *ASIC Specification.* Design of Connections, Section J4, Affected Elements of Members and Connecting Elements, Clause 1 Strength of Elements in Tension as follows:

Tensile Yielding of Connecting Elements:

$$A_g = 0.5\ in \times 16\ in$$
$$= 8\ in^2$$

$$R_n = F_y\ A_g$$
$$= 50\ ksi \times 8\ in^2$$
$$= 400\ kip$$

$$R_u = 0.9 \times 400\ kip$$
$$= 360\ kip$$

Tensile Rupture of Connecting Elements

Diameter of holes are determined from Table J3.3 for ½ in bolts as $9/16\ in$. Add to that $1/16\ in$ per Section B4 for use in Net Areas → $d_h = 10/16\ in\ (0.625\ in)$.

Using Failure Path '2 – 4 – 6':

$$A_n = 0.5 \times (16 - 3 \times 0.625)$$
$$= 7.06\ in^2$$

Using Failure Path '2 – 3 – 6':

Use Section B4 of the *AISC Specifications* to calculate inclined sections with pitch $s = 4.5\ in$ and gauge $g = 4\ in$ using the equation $\left(\frac{s^2}{4g}\right)$.

$$A_n = 0.5 \times \left(16 - 3 \times 0.625 + 2 \times \frac{4.5^2}{4\times4}\right)$$
$$= 8.3\ in^2$$

Using Failure Path '2 – 3 – 5':

$$A_n = 0.5 \times \left(16 - 3 \times 0.625 + 1 \times \frac{4.5^2}{4\times4}\right)$$
$$= 7.7\ in^2$$

Effective area is therefore calculated as follow:

$$A_e = A_n U$$

Where A_n is the smallest of the previous A_n calculations taken as $7.06\ in^2$ and U is the shear lag factor taken from Table D3.1 for case 1 as '1.0'

$$A_e = 1.0 \times 7.06$$
$$= 7.06\ in^2$$

$$R_n = F_u\ A_e$$
$$= 65\ ksi \times 7.06\ in^2$$
$$= 458.9\ kip$$

$$R_u = 0.75 \times 458.9\ kip$$
$$= 344.2\ kip$$

Block Shear

The following figure represents how block shear can subject this section to failure:

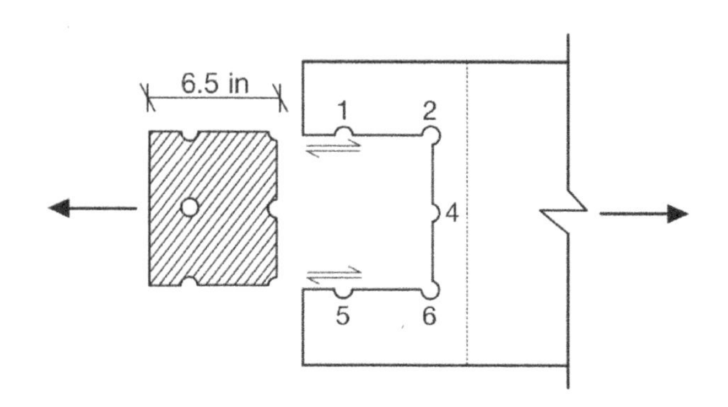

$$R_n = 0.60 F_u A_{nv} + U_{bs} F_u A_{nt}$$
$$\leq 0.60 F_y A_{gv} + U_{bs} F_u A_{nt}$$

Where A_{nv} is Net Area in shear, A_{nt} is Net Area in tension, A_{gv} is Gross Area in shear and U_{bs} is taken as '1.0' identifying that the tension stress is uniform.

$$A_{nv} = 2 \times 0.5 \times \left(6.5 - 0.625 - \frac{0.625}{2}\right)$$
$$= 5.56\ in^2$$

$$A_{gv} = 2 \times 0.5 \times 6.5$$
$$= 6.5\ in^2$$

$$A_{nt} = 0.5 \times (4 + 4 - 2 \times 0.625)$$
$$= 3.38\ in^2$$

$$R_n = 0.60 F_u A_{nv} + U_{bs} F_u A_{nt}$$
$$= 0.60 \times 65 \times 5.56 + 1 \times 65 \times 3.38$$
$$= 436.5\ kip$$
$$\leq 0.60 F_y A_{gv} + U_{bs} F_u A_{nt}$$
$$\leq 0.60 \times 50 \times 6.5 + 1 \times 65 \times 3.38$$
$\leq 414.7\ kip$ use this value as R_n

$$R_u = 0.75 \times 414.7\ kip$$
$$= 311.0\ kip$$

The final ultimate tensile strength is the smallest of the three above calculated values:

$$R_u = min(360, 344.2, 311.0)$$
$$= 311\ kip$$

Correct Answer is (A)

SOLUTION 4.27

Using the *National Design Specification for Wood Construction.*

Reference design values for structural glued laminated timber are dependent on their orientation as explained in Section 5.2.2 of the NDS Specification.
In this question bending and shear occurs around the y-y axis.

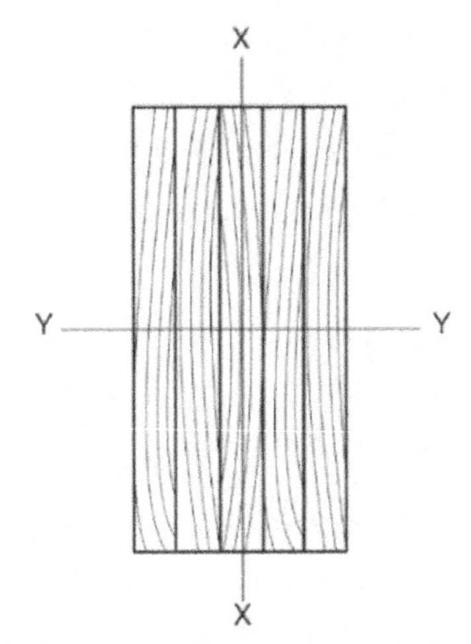

Using Table 5B from the NDS *National Design Specification Supplement*, southern pine glulam grade N2M12 has the following design references:

$F_{by} = 1{,}750\ psi$ for four layers or more.
$F_{vy} = 260\ psi$ for shear parallel to grains.
$E_{min} = 0.74 \times 10^6\ psi$ for stability calcs.

Bending F_{by} and shear F_{vy} should be multiplied by the following adjustment factors:

Load Duration Factor $C_D = 0.9$ taken from Table 2.3.2 of the NDS Specification given that occupancy is for a permanent dead load.

Wet Service Factor $C_{M,bending} = 0.8$ for bending and $C_{M,shear} = 0.875$ for shear are both taken from Table 5B of the NDS Supplement *Adjusted Factors* given that moisture is greater than 16%.

Temperature Factor $C_t = 1.0$ given that section is not subjected to elevated temperatures.

Stability Factor C_L with $d > b$ is calculated per Section 3.3.3.3 as follows:

$$\frac{\ell_u}{d} = \frac{60\ ft}{24.75\ in \times 1\ ft/12\ in}$$

$= 29.1 > 7$

$\ell_e = 1.63\ \ell_u + 3d$ for uniform load

$= 1.63 \times 60\ ft + 3 \times 24.75\ in \times \frac{1\ ft}{12\ in}$

$= 104\ ft$

Slenderness Ratio R_B:

$$R_B = \sqrt{\frac{\ell_e d}{b^2}}$$

$$= \sqrt{\frac{104\ ft \times \frac{12\ in}{1\ ft} \times 24.75\ in}{(10.5\ in)^2}}$$

$= 16.7 < 50 \rightarrow$ OK

$C_L =$

$$\frac{1+(F_{BE}/F_B^*)}{1.9} - \sqrt{\left[\frac{1+(F_{BE}/F_B^*)}{1.9}\right]^2 - \frac{F_{BE}/F_B^*}{0.95}}$$

$F_b^* = F_b \times C_D \times C_M \times C_t$

$= 1{,}750\ psi \times 0.9 \times 0.8 \times 1.0$

$= 1{,}260\ psi$

$E'_{min} = E_{min} \times C_M \times C_t$

$= 0.74 \times 10^6\ psi \times 0.8 \times 1.0$

$= 0.59 \times 10^6\ psi$

$$F_{BE} = \frac{1.2E'_{min}}{R_B{}^2}$$

$$= \frac{1.2 \times 0.59 \times 10^6\ psi}{16.7^2}$$

$= 2{,}538.6\ psi$

$$\frac{1+(F_{BE}/F_B^*)}{1.9} = \frac{1+2{,}538.6/1{,}260}{1.9} = 1.59$$

$$\rightarrow C_L = 1.59 - \sqrt{1.59^2 - \frac{2{,}538.6/1{,}260}{0.95}}$$

$= 0.95$

Volume Factor C_V is not considered given the section is loaded about its y-axis.

Flat Use Factor C_{fu} is not considered, although section is loaded about its y-axis, its depth exceeds 12 in.

Shear Reduction Factor C_{vr} is not applicable here provided none of the conditions mentioned in Section 5.3.10 apply.

$F'_{by} = F_{by} \times C_D \times C_{M,bending} \times C_t \times C_L$

$= 1{,}750\ psi\ \times 0.9 \times 0.8 \times 1.0 \times 0.95$

$= 1{,}197\ psi$

$F'_{vy} = F_{vy} \times C_D \times C_{M,shear} \times C_t$

$= 260\ psi\ \times 0.9 \times 0.875 \times 1.0$

$= 204.75\ psi$

<u>Determine w using moment stresses:</u>

$$M = \frac{w\ell_u{}^2}{8}$$

$$w = \frac{8 \times M}{\ell_u{}^2}$$

$M = F'_{by}\ S_y$

$$= 1{,}197\ psi \times 1{,}072\ in^3 \times \frac{1\ ft}{12\ in} \times \frac{1\ kip}{1{,}000\ lb}$$

$= 106.9\ kip.ft$

$$w = \frac{8 \times 106.9\ kip.ft}{(60\ ft)^2} \times \frac{1{,}000\ lb}{1\ kip}$$

$= 237.6\ lb/ft$

<u>Determine w using shear stresses:</u>

$$V = \frac{w\ell_u}{2}$$

$$w = \frac{2V}{\ell_u}$$

$$V = \frac{2F'_{vy}bd}{3}$$
$$= \frac{2\times204.75\ psi\times10.5\ in\times24.75\ in}{3}$$
$$= 35{,}473\ lb$$

$$w = \frac{2\times35{,}473\ lb}{60\ ft} = 1{,}182.4\ lb/ft$$

The lowest w controls:

$$w = 237.6\ lb/ft$$

Correct Answer is (B)

SOLUTION 4.28

Using the *National Design Specification and Supplement for Wood Construction,* the following reference data can be used for Lumber Fir-Larch No.3:

$F_c = 775\ psi$
$E_{min} = 510{,}000\ psi$
$E = 1{,}400{,}000\ psi$

Column Stability Factor is calculated per Section 3.7.1 as follows:

$$C_p = \frac{1+(F_{cE}/F_c^*)}{2c} - \sqrt{\left[\frac{1+(F_{cE}/F_c^*)}{2c}\right]^2 - \frac{F_{cE}/F_c^*}{c}}$$

$F_c^* = F_c = 775\ psi$ given all adjustment factors equal '1.0'.

$$E'_{min} = E_{min}\times C_M\times C_t = 510{,}000\ psi$$

$$F_{cE} = \frac{0.822\ E'_{min}}{\left(\frac{\ell_e}{d}\right)^2}$$

$$= \frac{0.822\times510{,}000\ psi}{(10ft\times12\frac{in}{ft}\ /\ 2\ in)^2} = 116.45\ psi$$

$c = 0.8$ for sawn lumber

$$\frac{1+(F_{cE}/F_c^*)}{2c} = \frac{1+116.45/775}{2\times0.8} = 0.72$$

$$C_p = 0.72 - \sqrt{0.72^2 - \frac{116.45/775}{0.8}}$$
$$= 0.145$$

$$F'_c = F_c\times C_p$$
$$= 775\ psi\times0.145$$
$$= 112.4\ psi$$

$$A = 1.5\ in\times3.5\ in = 5.25\ in^2$$

$$P = A\times F'_c$$
$$= 5.25\ in^2\times112.4\ psi$$
$$= 590\ lb$$

Correct Answer is (A)

SOLUTION 4.29

Using the *National Design Specification and Supplement for Wood Construction.*

Table 16.2.1A – also equation presented in 16.2.1 – provides the *effective char rate* for $\beta_n = 1.5\ in/hr$ as $\beta_{eff} = 1.67\ in/hr$ for $1.5\ hours$ of endurance with an effective char depth of $a_{char} = 2.5\ in.$

The new dimensions for a $14\ in\times24\ in$ member exposed to fire for 1½ *hours* are as follows:

$$b_{new} = 13.5 - 2.5 - 2.5 = 8.5\ in$$
$$d_{new} = 23.5 - 2.5 - 2.5 = 18.5\ in$$

$$I_{new} = \frac{b_{new}\times d_{new}^3}{12}$$
$$= \frac{8.5\ in\times(18.5\ in)^3}{12}$$
$$= 4{,}484.9\ in^4$$

$$I_{original} = \frac{13.5\ in\times(23.5\ in)^3}{12}$$
$$= 14{,}600.1\ in^4$$

$\Delta I = 14{,}600.1\ in^4 - 4{,}484.9\ in^4$

$= 10{,}115.2\ in^4$

Correct Answer is (B)

SOLUTION 4.30

Using the *National Design Specification for Wood Construction*, Chapter 5, Section 5.3.6:

$$C_v = \left(\frac{21}{L}\right)^{1/x}\left(\frac{12}{d}\right)^{1/x}\left(\frac{5.125}{b}\right)^{1/x} \leq 1.0$$

The above equation is applicable as long as the loading is perpendicular to the wide faces of laminations. Also, 'x' is taken as '10' for *other species* of Pine other than Southern Pine. Moreover, dimensions for d and b in the above equation are in *inches* and L is in *feet*.

$$= \left(\frac{21}{25\ ft}\right)^{1/10}\left(\frac{12}{12.25\ in}\right)^{1/10}\left(\frac{5.125}{19.5\ in}\right)^{1/10}$$

$= 0.86$

Correct Answer is (C)

SOLUTION 4.31

Using the *National Design Specification and Supplement for Wood Construction.*

Table 12B of NDS specifies the lateral design value $Z_{\parallel}$ for Sawn Lumber Douglas Fir-Larch when parallel to grains, for single shear connections, with $t_m = 3.5\ in$ and $t_s = 0.25\ in$, with bolt diameters of $D = 0.75\ in$ as $Z_{\parallel} = 1{,}670\ lb$.

<u>$Z_{\parallel}$ to be multiplied by the relevant adjustment factors found in Table 11.3.1 as follows:</u>

Load Duration Factor $C_D = 1.0$ taken from Table 2.3.2 of the NDS Specification for live load normal occupancy.

Wet Service Factor $C_M = 1.0$ for moisture less than 19% per Table 11.3.3.

Temperature Factor $C_t = 1.0$ for application in normal temperatures.

End Grain Factor C_{eg}, Diaphragm Factor C_{di}, and Toe-Nail Factor C_{tn} are not applicable in this question.

Group Action Factor C_g for bolts < 1 in is calculated as follows:

$$C_g = \left[\frac{m(1-m^{2n})}{n\left[(1+R_{EA}\,m^n)(1+m)-1+m^{2n}\right]}\right] \times \left[\frac{1+R_{EA}}{1-m}\right]$$

$n = 4,\ s = 3.5\ in$

$$R_{EA} = \min\left(\frac{E_sA_s}{E_mA_m}\ \ or\ \ \frac{E_mA_m}{E_sA_s}\right)$$

$E_sA_s = 29{,}000\ ksi \times (8 \times 0.25)in^2$

$= 58{,}000\ kip$

$E_mA_m = 1{,}400\ ksi \times (8 \times 3.5)in^2$

$= 39{,}200\ kip$

$$R_{EA} = \frac{E_mA_m}{E_sA_s} = \frac{39{,}200\ kip}{58{,}000\ kip} = 0.676$$

$$u = 1 + \gamma\frac{s}{2}\left[\frac{1}{E_mA_m} + \frac{1}{E_sA_s}\right]$$

$$= 1 + (270{,}000)(0.75^{1.5}) \times \frac{3.5\ in}{2} \times \left[\frac{1}{39.2 \times 10^6} + \frac{1}{58.0 \times 10^6}\right]$$

$= 1.013$

$m = u - \sqrt{u^2 - 1}$

$= 1.013 - \sqrt{1.013^2 - 1}$

$= 0.85$

$$C_g = \left[\frac{0.85\times(1-0.85^{2\times4})}{4\times[(1+0.676\times 0.85^4)(1+0.85)-1+0.85^{2\times4}]}\right] \times \left[\frac{1+0.676}{1-0.85}\right]$$

$= 0.97$

Geometry Factor C_Δ for bolts with $D \geq 1/4\ in$ is calculated per Section 12.5 of the NDS Specification as follows:

Per Table 12.5.1A, for tension members parallel to grains, and for softwood – Sawn Lumber Douglas Fir-Larch is considered Softwood – minimum end distance (which equals to 4.0 in as given in the question) sits between $3.5D$ (= 2.63 in) and $7D$ (= 5.25 in):

$$C_\Delta = \frac{actual\ end\ distance}{min\ end\ distance\ for\ C_\Delta=1.0} = \frac{4.0\ in}{5.25\ in} = 0.76$$

Per Table 12.5.1B, spacing for fasteners in a row (3.5 in as given in the question), and parallel to grains, exceed both $3D$ (= 2.25 in) and $4D$ (= 3.0 in). In this case $C_\Delta = 1.0$

Given the above two conditions, C_Δ should be the smallest of both per 12.5.1.2, in which case $C_\Delta = 0.76$.

<u>Based on all the above adjustment factors, the finally adjusted service load P is calculated as follows:</u>

$$P = \#bolts \times Z_{\parallel} \times C_D \times C_M \times C_t \times C_g \times C_\Delta$$

$= 8 \times 1.67\ kip \times 1 \times 1 \times 1 \times 0.97 \times 0.76$

$= 9.8\ kip$

Correct Answer is (C)

SOLUTION 4.32

Using the *National Design Specification for Wood Construction.*

Using Section 16.2.1.3 for CLT, the effective char depth a_{char} is calculated as follows:

$$a_{char} = 1.2\left[n_{lam}\,h_{lam} + \beta_n\left(t-(n_{lam}\,t_{gi})\right)^{0.813}\right]$$

$$t_{gi} = \left(\frac{h_{lam}}{\beta_n}\right)^{1.23} = \left(\frac{1.25}{1.5}\right)^{1.23} = 0.8\ hr$$

$$n_{lam} = \frac{t}{t_{gi}} = \frac{2.5\ hrs}{0.8\ hr} = 3.125 \qquad \rightarrow use\ 3.0$$

$$a_{char} = 1.2\left[3.0 \times 1.25 + 1.5 \times (2.5-(3.0\times 0.8))^{0.813}\right]$$

$= 4.8\ in$

Correct Answer is (A)

An alternative method for solving this question is through using Table 16.2.1B, with extrapolation, can provide an indicative result as well.

SECTION 5

Codes and Specifications

Problems & Solutions

PROBLEM 5.1 ***Inspection of Construction Activities***

The following construction activities require continuous special inspections except for:

(A) Collecting concrete specimens for strength testing, slump, temperature, and air content, all prior to concrete placement.

(B) Confirming all sizes and lengths of material for driven deep foundations.

(C) Confirming that the substrate under shallow footings can achieve the bearing capacity that footings were designed for.

(D) Inspecting tension adhesion anchors installed in finished concrete elements that are either horizontal or vertical.

PROBLEM 5.2 ***Reinforced Concrete Basement Wall***

The below is a $7.5\,in$ thick, $8\,ft$ deep, basement reinforced concrete wall and all building floors are made of reinforcement concrete as well.

This wall supports clayey gravel that is poorly graded with clay mixes within its gradation. Reinforced concrete properties are as follows:

$f'_c = 3{,}500\,psi$

$f_y = 40\,ksi$

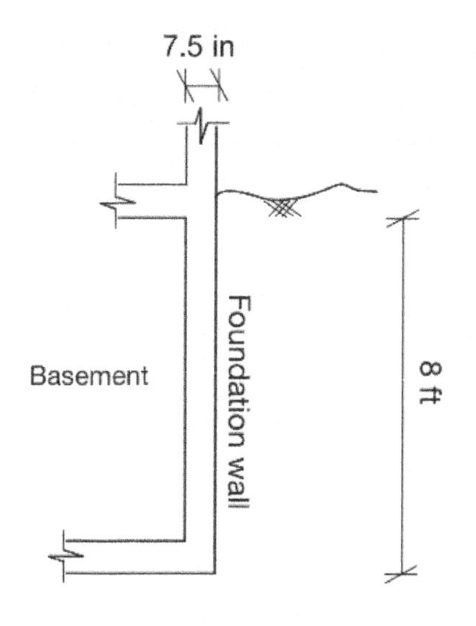

The minimum size and spacing of reinforcements required for this wall are:

(A) $\#\,6\ at\ 21\ in$

(B) $\#\,6\ at\ 32\ in$

(C) $\#\,6\ at\ 43\ in$

(D) $\#\,6\ at\ 29\ in$

PROBLEM 5.3 ***Concrete Mix Design***

The maximum allowable percentage of fly ash that can be used for concrete subjected to extreme freezing and thawing cycles with the frequent use of deicing chemicals is:

(A) 10

(B) 25

(C) 35

(D) 50

PROBLEM 5.4 ***Concrete mix Design***
The below is a section of a waffle slab rib that shall be casted in situ. Specified concrete cover is per the ACI code.

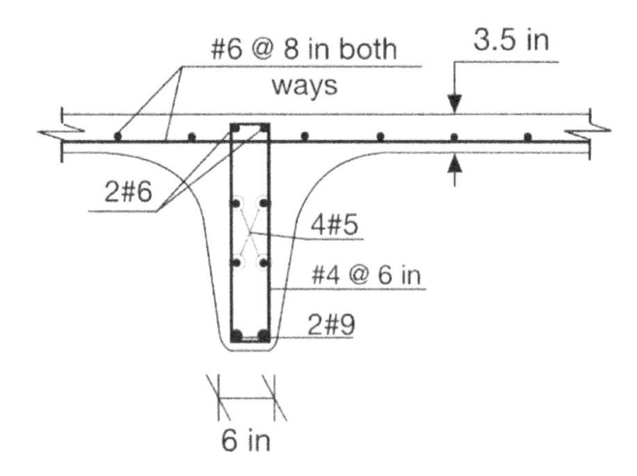

The maximum nominal coarse aggregate size that can be used for the concrete mix is as follows:

(A) ½ *in*

(B) ¾ *in*

(C) 1 *in*

(D) 1 ½ *in*

PROBLEM 5.5 ***Construction Safety***
The type of pedestrian protection required for a $500\,ft \times 500\,ft$ plot with a $150\,ft$ tall under construction building that occupies nearly $400\,ft \times 400\,ft$ of the plot is as follows:

(A) None

(B) Construction railings

(C) Barrier

(D) Barrier and covered walkway

PROBLEM 5.6 ***Concrete Mix Design***
A reinforced concrete deep basement wall that extends above grade in an area that has been known for snow accumulation and the frequent use of de-icing chemicals, has its compressive strength designed at $f_c' = 5{,}500\,psi$, maximum nominal aggregate size used in the mix of $1.5\,in$ and its water cement content < 0.4.

The minimum permitted percentage of air content for this wall's concrete mix is:

(A) 4.5

(B) 5.5

(C) 4.0

(D) 5.0

PROBLEM 5.7 ***Masonry Basement Wall***
The below figure is for a cross section of a masonry basement wall. The flooring system of the building is that of wood and steel as shown. The wall supports well graded clean gravel. Masonry and reinforcement properties are as follows:

$f'_m = 1{,}700\,psi$

$f_y = 60\,ksi$

??
Steel beam and wooden joists floor system
Basement
Masonry wall
8 ft

The minimum design requirements for this masonry wall are:

(A) 12 *in* thick solid masonry units placed in running bond.

(B) 10 *in* thick solid masonry units placed in running bond.

(C) Reinforced 10 *in* thick masonry blocks with # 7 @ 48 *in*.

(D) Reinforced 8 *in* thick masonry blocks with # 6 @ 48 *in*.

PROBLEM 5.8 ***Concrete Strength in Bridge Design***

The minimum allowed specified concrete compressive strength f_c' that can be used in bridge decks and other bridge prestressed applications is:

(A) 3.5 *ksi*

(B) 4.0 *ksi*

(C) 4.5 *ksi*

(D) 10.0 *ksi*

PROBLEM 5.9 ***Abutment Design***

The below is a section taken at the abutment of a straight 125 *ft* bridge located at a Seismic Zone 1 with soil type B. The joint at the bearing is not restrained.

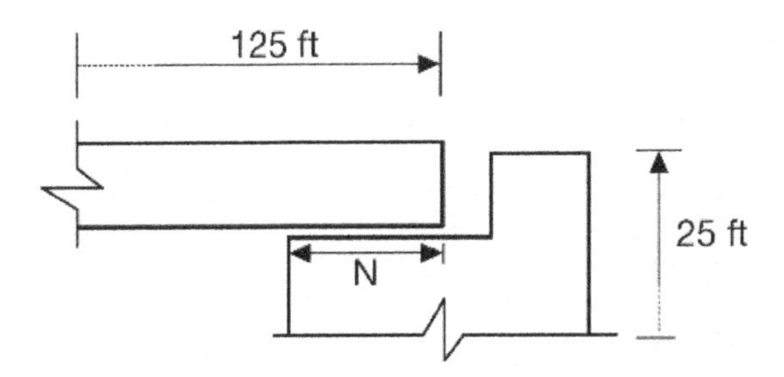

Considering the Peak Ground Acceleration at this site is $PGA = 0.2$, the support length N should not be less than:

(A) 11 *in*

(B) 8 *in*

(C) 15 *in*

(D) 20 *in*

PROBLEM 5.10 ***Concrete testing***

Columns of an existing concrete building are being tested using cut cores. The following statement is correct when it comes to accepting the structural adequacy of those columns:

(A) The average of three cores is at least 85% of the target/design f_c' with no core below 75% of the compressive strength.

(B) The average of three cores is at least 95% of the target/design f_c' with no core below 85% of the compressive strength.

(C) The average of five cores is at least 85% of the target/design f_c' with no core below a 75% of the compressive strength.

(D) The average of five cores is 95% of the target/design f_c' with no core below 85% of the compressive strength.

PROBLEM 5.11 ***Highway Bridge Column Design***

The maximum longitudinal reinforcements for columns of highway bridges located in Seismic Zone 4 is:

(A) $0.04\ of\ A_g$

(B) $0.05\ of\ A_g$

(C) $0.06\ of\ A_g$

(D) $0.01\ of\ A_g$

PROBLEM 5.12 ***Seismology Retaining Wall Safety***

The minimum overturning safety factor for retaining walls subjected to earthquake loading is:

(A) 3.0

(B) 2.0

(C) 1.5

(D) 1.1

PROBLEM 5.13 ***Masonry Wall Openings***

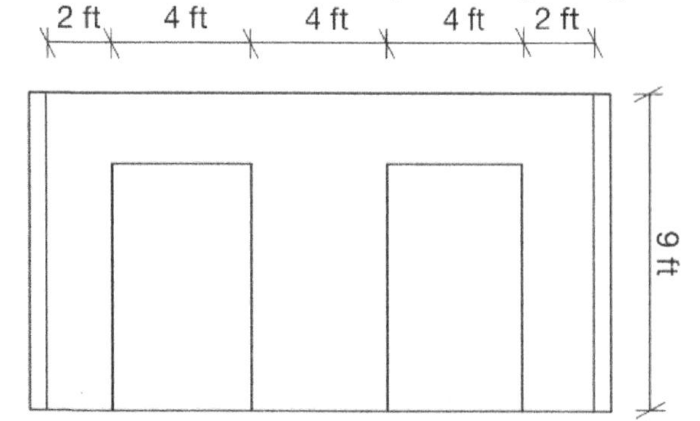

The minimum thickness for the above hollow units' masonry partition with two openings using Portland cement mortar type N and subjected to lateral load of 5 psf is:

(A) 3.2 in

(B) 4.5 in

(C) 4.0 in

(D) 5.0 in

PROBLEM 5.14 ***Masonry Wall Reinforcement Req's***

Minimum cover for masonry joint reinforcing wires when masonry is exposed to earth is:

(A) ½ in

(B) $5/8\ in$

(C) $7/8\ in$

(D) 1 ½ in

PROBLEM 5.15 ***Minimum Distance from Slopes***

The below building is to be constructed next to a slope as shown. The building foundations are 5 ft below surface, edge of its exterior footings is 2 ft away from the face of the building.

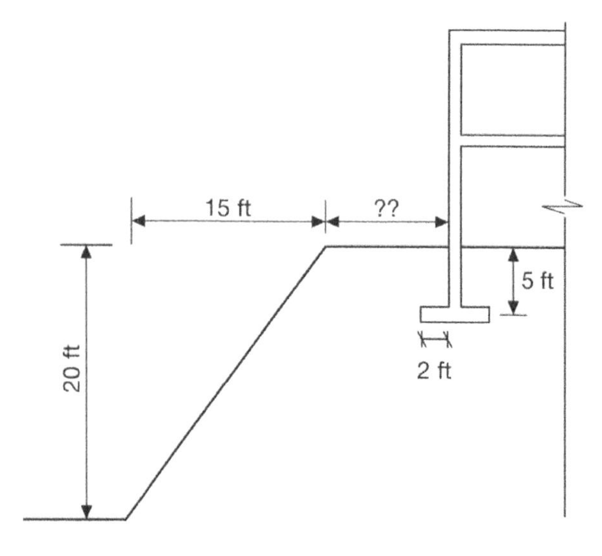

The minimum distance from the exterior face of this building to the top of the slope (the crest) that provides sufficient vertical and horizontal support for its foundations is:

(A) 6 ft

(B) 7 ft

(C) 9 ft

(D) 11 ft

PROBLEM 5.16 M*asonry Mix Design*
The upper limit of carbon black pigments that can be added to pigmented masonry cement mortar is:

(A) 1% of cement weight

(B) 2% of cement weight

(C) 3% of cement weight

(D) 4% of cement weight

PROBLEM 5.17 *Masonry Construction Inspection*
The following constitutes the minimum continuous special inspections for masonry structures assigned to risk category IV:

(A) Checking the size of reinforcements along with prestressing steel, bolts, and anchorage.

(B) Observing the preparation of grout and mortar specimens.

(C) Inspecting the building of masonry units and mortar joints.

(D) The inspection of the location and the sizing of structural elements.

PROBLEM 5.18 *Steel Members Strength during Fire*
A critical component of a structural steel building where fuel will be stored is to be designed for high temperatures in an anticipation of fire and heat exchange.

Members of with $F_y = 50\,ksi$ can still be considered in the design process while using the following new F_y assuming they will be heated up to $900^o\,F$ in worst case conditions:

(A) 50 ksi

(B) 40 ksi

(C) 33 ksi

(D) 0 ksi

PROBLEM 5.19 *Structural Steel Weld Inspection*
The prime responsibility for checking steel fabricated joints prior to welding falls under the:

(A) Contractor's appointed quality control inspector.

(B) Consultant's appointed quality control inspector.

(C) Owner's appointed quality control inspector.

(D) Fabricator's appointed quality control inspector.

PROBLEM 5.20 *Shotcrete Mix Design*
The maximum size of coarse aggregate that can be used in shotcrete, if needed only, is:

(A) ½ in

(B) ¾ in

(C) 1 in

(D) 1 ½ in

PROBLEM 5.21 *Masonry Wall Minimum Requirements*
The minimum wall thickness of a 10 ft high load bearing fully grouted masonry wall is:

(A) 4 in

(B) 5 in

(C) 6 in

(D) 7 in

SOLUTION 5.1

Check IBC Chapter 17 *Special Inspections and Tests*, Section 1705 *Required Special Inspections and Tests*, and the following tables:

- Table 1705.3 *Concrete Construction.*
- Table 1705.6 *Tests of Soils.*
- Table 1705.7 *Driven Deep Foundation Elements.*

The activities mentioned in this question were paraphrased compared to the ones mentioned in the IBC code.

The following items from the abovementioned tables correspond to the statements in the question and are the only ones requiring continuous special inspections:

Item	Continuous Special Inspection
1705.3 Item 4.a	Inspect anchors post-installed in hardened concrete members: (a) Adhesive anchors installed in horizontally or upwardly inclined orientations to resist sustained tension loads. (b) Mechanical anchors and adhesive anchors not defined in 4.a.
1705.3 Item 6	Prior to concrete placement, fabricate specimens for strength tests, perform slump and air content tests, and determine the temperature of the concrete.
1705.7 Item 1	Verify element materials, sizes and lengths comply with the requirements.

The only activity from the statements mentioned in the body of the question that does not require continuous special inspections is activity (C).

The following activity requires periodic special inspection only:

Item	Periodic Special Inspection
1705.6 Item 1	Verify materials below shallow foundations are adequate to achieve the design bearing capacity.

Correct Answer is (C)

SOLUTION 5.2

Reference is made to the IBC code Section 1610 which specifies the minimum load for lateral backfills, and Section 1807.1.6.2 which provides the minimum thickness and minimum vertical reinforcement requirements for foundation walls.

IBC 1610, Table 1610.1 specifies a lateral design load of 60 psf for *at-rest* pressure and for clayey gravels, poorly graded gravel-and-clay mixes (GC).

An *at-rest* pressure is based on the type of supported diaphragm. The code permits the use of *at-rest* pressure for rigid diaphragms as horizontal movement will be restricted, and the use of active pressures for flexible diaphragms as horizontal movement will not.

The supported diaphragm in this case (reinforced concrete slab) is a rigid diaphragm as defined in ASCE 7 Section 12.3.1.2.

The question did not specify whether the soil is expansive or undrained, and hence the assumption would be that it is neither.

IBC Section 1807.1.6.2 specifies that for concrete foundation walls which are 7.5 in thick, with a height and unbalanced backfills of 8 ft a # 6 @ 32 in when $f_y = 60\ ksi$ steel is used.

For 40 ksi steel, the provided spacing shall be multiplied by $'0.67'$, which makes the minimum vertical reinforcements for this wall # 6 @ 21 in.

Correct Answer is (A)

SOLUTION 5.3

ACI 318-14, Chapter 19, Table R19.3.1 assigns an exposure class of F3 for concrete subjected to extreme cases of freezing and thawing with the frequent application of deicing chemicals.

ACI 318-14, Chapter 26, Table 26.4.2.2(b) limits the use of fly ash for concrete assigned to Exposure Class F3 to 25%.

Correct Answer is (B)

SOLUTION 5.4

ACI 318-14, Section 26.4.2 specifies the nominal maximum size of coarse aggregate as the minimum of the following:

(i) One fifth of the narrowest dimension between the sides of formwork. In this case 6 in which is the web thickness.
→ $6/5 = 1.2\ in$

(ii) One third the depth of the slab:
→ $3.5/3 \cong 1.2\ in$

(iii) Three-fourths the minimum clear spacing between reinforcing bars. See below sketch:

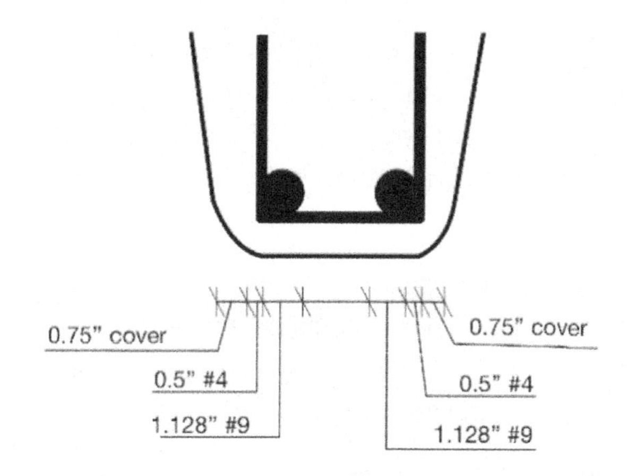

The specified clear cover for cast-in-place, non-prestressed, concrete per ACI 318-14 Table 20.6.1.3.1. Slabs, joists, and walls clear cover is ¾ in for bars $< \#11$.

The remaining clear distance between the two #9 bars is as follows:

$6 - 2 \times (3/4 + 1/2 + 1.128) = 1.244\ in$

$3/4 \times 1.244\ in = 0.933\ in$

Correct Answer is (C)

SOLUTION 5.5

Chapter 33 of the IBC code, Section 3306, and Table 3306.1 specifies that a barrier should be provided to protect pedestrians when the height of construction $> 8\ ft$ and the distance from lot line to construction ($50\ ft$) is between one-fourth ($37.5\ ft$) and half ($75\ ft$) the height of construction.

Correct Answer is (C)

SOLUTION 5.6

ACI 318-14, Chapter 19, Table R19.3.1 assigns an exposure class of F3 for concrete basement walls that are frequently subjected to accumulation of snow and the use of de-icing chemicals.

Table 19.3.2.1 states that concrete strength requirements and minimum w/c for this exposure class should be $> 5{,}000\ psi$ and < 0.4 respectively, which were given in the body of the question anyway.

Table 19.3.3.1 specifies the target air content for concrete exposure classes and states that the target air content with a maximum nominal aggregate size of 1.5 *in* is 5.5%.

Section 19.3.3.3 permits the reduction of air content specified in Table 19.3.3.1 by 1% when $f_c' > 5{,}000\ psi$, in which case it is $5{,}500\ psi$.

This makes the permitted minimum air content in this case 4.5%.

Correct Answer is (A)

SOLUTION 5.7

Reference is made to the IBC code Section 1610 which specifies the minimum load for lateral backfills, and Section 1807.1.6.3 which stipulates the requirements for masonry foundation walls.

IBC 1610, Table 1610.1 specifies a lateral design load of 30 psf for *active* pressure and for well-graded, clean gravel and gravel sand mixes (GW).

An *active* pressure is based on the type of supported diaphragm. The code permits the use of *active* pressure for flexible diaphragms as horizontal movement will not be restricted, and the use of *at-rest* pressures for rigid diaphragms as horizontal movement will be restricted.

The supported diaphragm in this case (Steel and wooden joists) is of the flexible type as defined in ASCE 7 Section 12.3.1.1.

IBC Section 1807.1.6.3 specifies for masonry concrete foundation walls with height, and height of backfill, of 8 ft, **10 *in* solid units**. And Section 1807.1.6.3-7 specifies that masonry **walls in this case should be in running bond**.

Correct Answer is (B)

SOLUTION 5.8

AASTHO LRFD Bridge Design Specification specify in Section 5.4.2.1 that the minimum compressive strength for prestressed concrete and decks is 4.0 ksi.

Correct Answer is (B)

SOLUTION 5.9

Using the *AASTHO LRFD Bridge Design Specification*, Section 4.7.4.4 specifies the minimum support length for Seismic Zone 1 as follows:

$$N = (8 + 0.02 \times L + 0.08 \times H) \times (1 + 0.000125 \times S^2)$$

N is in *inches*. The length of the bridge ***L*** and height of abutment ***H*** are in $feet$. Skew angle ***S*** is zero.

$$N = (8 + 0.02 \times 125 + 0.08 \times 25)(1 + 0) = 10.5\ in$$

Acceleration Coefficient ***As*** to be calculated to determine what percentage to be used from *N* as defined in Table 4.7.4.4-1.

Section 3.10.4.2 along with Table 3.10.3.2.1 of the specification defines ***As*** per Equation 3.10.4.2-2 as follows:

$$A_s = F_{PGA} PGA = 1.0 \times 0.2 > 0.05 \rightarrow use\ 100\%\ N\ of\ 10.5\ in$$

Correct Answer is (A)

SOLUTION 5.10

ACI 318-14 Section 26.12.4.1(d) specifies that core strengths are considered structurally adequate if the average of three cores is equal to at least 85% of f_c' with no single core below 75% of f_c'.

Correct Answer is (A)

SOLUTION 5.11

AASTHO LRFD Bridge Design Specification, Section 5.10.11.4, specifies the upper limit for longitudinal column reinforcements in Seismic Zone 4 as 0.04 of A_g. This limit has been set as such to avoid congestion, extensive shrinkage cracking, and, to permit anchorage of rebars.

Correct Answer is (A)

SOLUTION 5.12

Per IBC code Section 1807.2.3, in the exception part of this section, safety factor for when earthquake loads are included in the design and analysis of retaining walls has been specified as a minimum of 1.1.

Correct Answer is (D)

SOLUTION 5.13

Reference is made to the TMS 402 *Building Code Requirements and Specifications for Masonry Structures*, Chapter 14, Section 14.3.2, openings in partition walls, and Table 14.3.1 (5) which specifies the maximum height over thickness h/t for 5 psf lateral loads for walls spanning vertically.

In this case, h/t for hollow units, with Portland cement mortar type N, is 24. This makes wall thickness as follows:

$$t = \frac{9\,ft \times 12\,\frac{in}{ft}}{24} = 4.5\,in$$

Due to openings in the wall, this value should be divided by the following ratio:

$$\sqrt{W_T/W_S}$$

$\boldsymbol{W_S}$ for laterally supported walls is measured from the edge of the opening to the wall's support, while $\boldsymbol{W_T}$ is measured form the center of the opening to the end of $\boldsymbol{W_S}$ as shown below:

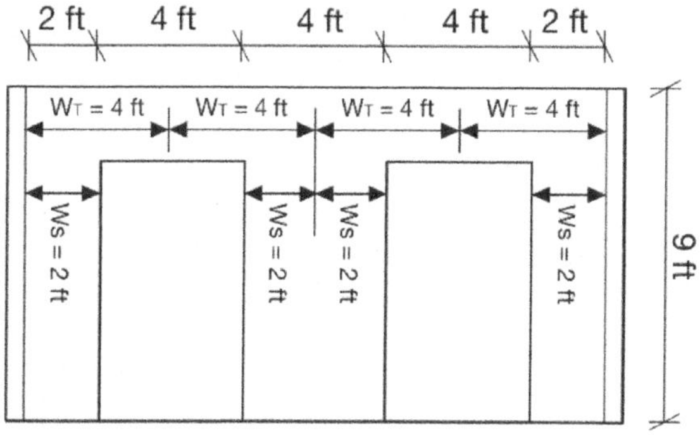

2 ft 4 ft 4 ft 4 ft 2 ft
WT = 4 ft WT = 4 ft WT = 4 ft WT = 4 ft
Ws = 2 ft Ws = 2 ft Ws = 2 ft Ws = 2 ft
9 ft

$$t' = \frac{t}{\sqrt{W_T/W_S}} = \frac{4.5\,in}{\sqrt{4/2}} = 3.2\,in$$

However, the minimum thickness of a partition is 4.0 in per Section 14.2.2.1 of the specification.

Correct Answer is (C)

SOLUTION 5.14

Section 6.1.4.2 of the TMS 402 *Building Code Requirements and Specifications for Masonry Structures* specifies that a minimum cover of 5/8 in is required for joint wire reinforcements for masonry exposed to earth or weather.

Correct Answer is (B)

PART II Structural Depth

Section 5 Codes & Specs

SOLUTION 5.15
IBC code Section 1808.7.2 states that foundations adjacent to descending slopes steeper than 1:1 shall have an imaginary 45° slope constructed from its toe for the purpose of measuring setbacks. The setback from this imaginary slope to the foundation should be the least of:

$$\frac{H}{3} = \frac{20\,ft}{3} \approx 7\,ft$$

Or

$$40\,ft$$

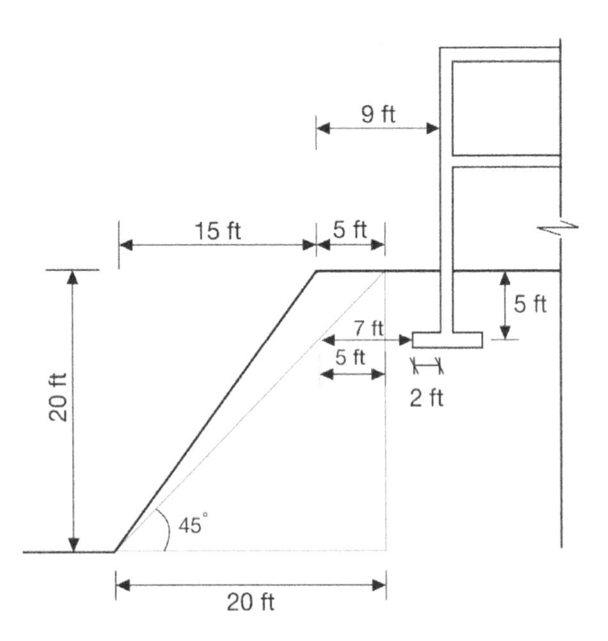

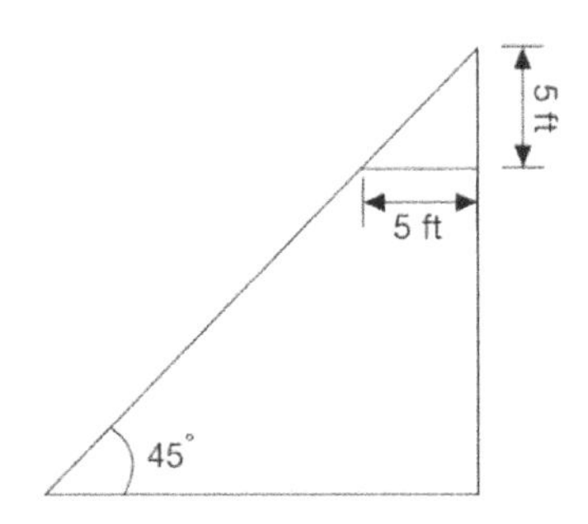

Using trigonometry as shown above, the requested minimum distance is measured as $9\,ft$.

Correct Answer is (C)

SOLUTION 5.16
TMS 602 *Specification for Masonry Structures*, Section 2.6 *Mixing* states that the maximum carbon black pigment percentage by weight of cement that can be used in pigmented masonry cement mortar is 1%.

Correct Answer is (A)

SOLUTION 5.17
Referring to the TMS 402 *Building Code Requirements and Specifications for Masonry Structures*, Chapter 3, Section 3.1.3 assigns a level C quality assurance for buildings assigned to risk category IV.

Table 3.1.3 of the same chapter identifies the minimum special inspections (periodic or continuous) required for such a quality assurance program.

When all options in the question belong to the periodic requirements of inspection of this program, only option B which specifies observing grout specimens' preparation was mentioned under the continuous inspection requirement.

Correct Answer is (B)

SOLUTION 5.18
Reference is made to the AISC *Specification for Structural Steel Buildings*, Appendix 4, Section 4.2.3 *Mechanical Properties at Elevated Temperatures.* Mechanical properties for steel shall be multiplied by

ratios provided in Table A-4.2.1 during elevated temperatures.

Prorating the figures in this table, at temperature $900^0\ F$, F_y shall be reduced by a ratio of $(0.94 + 0.66)/2 = 0.8$

The new F_y will therefore be:

$50 \times 0.8 = 40\ ksi$

Correct Answer is (B)

SOLUTION 5.19

Chapter N, Section N5-1 of the AISC *Specification for Structural Steel Buildings* states that QC inspection tasks shall be performed by the fabricator or the erector Quality Control Inspector QCI. Table N5.4-1 of the same chapter specifies what activities to be observed by this QCI prior to welding.

Correct Answer is (D)

SOLUTION 5.20

Chapter 19 of the IBC code, Section 1908.3 specifies that the maximum size of coarse aggregate if used in shotcrete shall not exceed ¾ *in.*

Correct Answer is (B)

SOLUTION 5.21

TMS 402 *Building Code Requirements and Specifications for Masonry Structures*, Part 5/Appendix A *Empirical Design of Masonry*, Section A.5, Table A.5.1 sets the limit on h/t for load bearing fully grouted walls to 20.

$m_{min} = 10/20 \times 12 = 6\ in$

Correct Answer is (C)

References

Code	Title	Institution
IBC	International Building Code, 2015 edition (without supplements)	International Code Council
ACI 318	Building Code Requirements for Structural Concrete and Commentary, 2014	American Concrete Institute
TMS 402/602	Building Code Requirements and Specifications for Masonry Structures (and companion commentary), 2013	The Masonry Society
PCI	PCI Design Handbook: Precast and Prestressed Concrete 7th edition 2010	Precast/Prestressed Concrete Institute
AISC	Steel Construction Manual, 14th edition, 2011	American Institute of Steel
AWC NDS	National Wood Design Specification for Wood Construction National Design Specification Supplement, Design Values for Wood Construction Special Design Provisions for Wind and Seismic with Commentary	American Wood Council
ASCE 7	Minimum Design Loads for Buildings and Other structures 2010	American Society of Civil Engineers
AASHTO	LRFD Bridge Design Specification, 7th edition, 2014 with 2015 and 2016 interim Revisions	American Association of State Highway & Transportation Officials

Moreover, the following references are essential additions, and it would be advantageous for you to acquaint yourself with them:

CFR TITLE 29
PART 1910
U.S. Department of Labor, Washington, D.C., July 2020. Occupational Safety and Health Standards:

- Subpart I, Personal Fall Protection Systems, 1910.140
- Subpart D, Walking-Working Surfaces, 1910.28-1910.30
- Subpart F, Powered Platforms, Manlifts, and Vehicle-Mounted Work Platforms, 1910.66-1910.68, with Appendix D to 1910.66

PART 1926

Safety and Health Regulations for Construction

- Subpart E, Personal Protective and Life Saving Equipment, 1926.104
- Subpart L, Scaffolding Specifications, Appendix A
- Subpart M, Fall Protection, 1926.500-1926.503, Appendix B-Appendix D
- Subpart Q, Concrete and Masonry Construction, 1926.703-1926.706 with Appendix A
- Subpart R, Steel Erection, 1926.752 & 1926.754 – 1926.758

As indicated in previous sections, all these references can either be purchased, borrowed, or borrowed as an electronic version. Older versions for the above can be downloaded either for free or for a fraction of the original cost, speaking of which, you need to assess the risks and possible variations of an old version compared to a new one.

In addition to the above you need to familiarize yourself and download – free of charge – a copy of the *NCEES PE Civil Reference Handbook the latest version*.

References' Key Chapters

General

The following is a compilation of crucial chapters and their corresponding key points for each reference, it is recommended to use this list as a review aid to ensure comprehensive coverage of the material. However, it is advisable to thoroughly examine all references and their chapters to adequately prepare for the exam.

Once you have thoroughly studied, practiced, and became proficient in the chapters listed below, and have ensured that you are entirely confident with the material covered and have practiced few examples as well, whether from this book, or examples that you can re-write yourself from this book or elsewhere, you may proceed to review other chapters in those references at your own pace.

Finally, it is important to note that the tables below may not correspond with the version of the reference material you are using, as recent versions may have altered chapters or sections' names or numbers or titles. An equation may have been added or omitted for instance. Therefore, it is important to be mindful of the version of the reference you are studying from as well as the version that will be used for the CBT exam during your examination intake.

Chapters per Reference

Code	TITLE	No. of Chapters
IBC	International Building Code, 2015 edition	35
ACI 318	Building Code Requirements for Structural Concrete and Commentary, 2014	27
TMS 402/602	Building Code Requirements and Specifications for Masonry, 2013	14
PCI	PCI Design Handbook: Precast and Prestressed Concrete	11
AISC	Steel Construction Manual, 14th edition, 2011	17
AWC NDS	National Wood Design Specification for Wood Construction	16
	National Design Specification Supplement, Design Values for Wood Construction	NA
ASCE 7	Minimum Design Loads for Buildings and Other structures, 2010	31
AASHTO	LRFD Bridge Design Specification, 7th edition	15

The tables below describe key highlights per chapter, they do not describe the entire chapter:

IBC International Building Code
This code consists of 35 chapters and 17 appendices as presented in its 2015 edition. All these chapters and appendices are important. I recommend that you familiarize yourself with the below chapters first.

Chapters	Title and Description
Chapter 16	***Structural Design*** This chapter discusses all loads and load combinations, most of this information is available with more details in the ASCE 7 code as well, so feel free to use both this chapter, and the relevant chapters in ASCE 7 interchangeably. This chapter also contains the earthquake related MCE diagrams.
Chapter 17	***Special Inspections and Tests*** This chapter defines the requirements for continuous special inspections and periodic special inspections. It specifies when inspections are to take place for steel structures, concrete, soils, foundations, and deep foundations.
Chapter 18	***Soils and Foundations*** This chapter talks about excavations and backfills, soil bearing values and soil lateral resistance. It also specifies requirements for foundation walls made of masonry or concrete. Requirements and safety adjacent to slopes, shallow foundation requirements and deep foundations as well. You will also find requirements for helical piles, micro-piles, pile caps and the likes.
Chapter 19	***Concrete*** This chapter talks about and specifies requirements for reinforced concrete and plain concrete, minimum slab thicknesses and requirements of shotcrete as well.
Chapter 21	***Masonry*** This chapter covers masonry mix design, veneer, design methods and special inspections.
Chapter 23	***Wood*** Although most of this information is covered in the NDS specification, this chapter elaborates on the maximums and minimums when it comes to span length, thicknesses, deflection, wind load on panels, etc.
Chapter 31 to 33	Those chapters address operations during construction, be it the erection of temporary structures, encroachments to the public, or protecting the public during construction activities.

ACI 318 Building Code for Structural Concrete

This code consists of 27 chapters as presented in its 2014 edition. A significant proportion of these chapters hold substantial importance, thereby indicating a high likelihood of a question being posed on any of them. I recommend however that you familiarize yourself with the below chapters first.

Chapter	Title and Description
Chapter 5	***Loads*** This chapter discusses loads, load factors and load combinations.
Chapter 6 **This is a key chapter**	***Structural Analysis*** This chapter talks about how to arrange live load to maximize its effect, simplified method to analyze one-way slabs and beams, effective length for columns and moment magnifications, cracked and uncracked moment of inertia.
Chapter 7	***One-Way Slabs*** Calculation of thickness for one-way slabs, the calculation of flexural and shear forces/stresses and their critical location in the slab. The chapter also specifies minimum and maximum reinforcement requirements for one-way slabs.
Chapter 8	***Two-Way Slabs*** The calculation of minimum thickness for flat slabs, calculation of moments and shears using the direct design method and equivalent frame method, also the calculation of two-way shear (punching shear) and its critical location for flat slabs, along with reinforcement detailing.
Chapter 9	***Beams*** The calculation of beams' depths, T beams effective width, minimum and maximum reinforcement requirements for beams.
Chapter 10	***Columns*** Reinforcement limits and splices.
Chapter 11	***Walls*** Minimum thickness of walls, effective length of the wall and distribution of load on the wall. Walls nominal strength and reinforcement limits.
Chapter 13	***Foundations*** Distribution of reinforcements in footings.
Chapter 14	***Plain Concrete*** Specifies minimum thickness required for bearing for different members along with nominal strengths for plain concrete.
Chapter 19	***Design and Durability Requirements*** Provides concrete properties such as modulus of elasticity, exposure categories and classes, requirements for air content and other additives.
Chapter 20	***Steel Reinforcement Properties, Durability, and Embedments*** Specifies concrete cover to steel reinforcements. Provides limits on prestressing. Loss of

	prestressing, and other relevant topics.
Chapter 21	***Strength Reduction Factors*** Specifies strength reduction factors Ø for structural concrete members and connections.
Chapter 22 **This is a key chapter**	***Sectional Strength*** This chapter specifies how to calculate concrete sectional strength for: flexure, axial, one-way shear, two-way shear (punching shear), torsion, bearing and shear friction.
Chapter 24	***Serviceability Requirements*** This chapter specifies allowable deflection limits, reinforcement spacing at the tension face to reduce cracking, shrinkage reinforcements, permissible prestressing forces and classification of prestressed concrete sections.
Chapter 25	***Reinforcement Details*** Minimum spacing between reinforcements, reinforcements' development length, lap length of spiral reinforcements. Anchorage and anchorage region.

TMS 402/602 Building Code Requirements and Specifications for Masonry Structures

This code consists of 14 chapters as presented in its 2013 edition. All these chapters are important, it is recommended to familiarize yourself with the below chapters first.

Chapter	Title and Description
Chapter 3	***Quality and Construction*** This chapter talks about inspections and quality assurance program.
Chapter 4	***General Analysis and Design Consideration*** This chapter provides material properties, such as modulus of elasticity for masonry, compressive strength, and several other factors and coefficients. It also discusses bearing areas and connections.
Chapter 5	***Structural Elements*** This chapter explains how to distribute concentrated loads on a wall and calculate effective length. Beams, moment of inertia and deflections, general column design and corbels.
Chapter 8	***Allowable Stress Design for Masonry*** Using ASD method, this chapter discusses how to design reinforced and unreinforced elements of masonry with emphasis on section 8.2 onwards.
Chapter 9	***Strength Design of Masonry*** It is important to remember that NCEES are expecting ASD

	method to be used for all masonry design questions – which may be subject to change anytime – apart from Section 9.5.3 for *out of plane walls*.
Chapter 14	***Masonry Partition Walls*** This chapter specifies the limitations to wall thicknesses, lateral supports and opening requirements.

Moreover, all section properties for concrete masonry walls are provided in the National Concrete Masonry Association document *NCMA TEK 14-1B*. Remember that most blocks are hollow, or part hollow, and you are not expected to long calculate a moment of inertia or a radius of gyration for a partly solid section to proceed with the solution of a question on design.

PCI Design Handbook
This handbook consists of 11 chapters as presented in its 2010 edition. All these chapters are important, it is recommended to familiarize yourself with the below chapters first.

Chapter	Title and Description
Chapter 1	***Applications and Materials*** This chapter provides information on material properties such as modulus of elasticity, shear strength, Poisson's ratio, etc.
Chapter 2	***Preliminary Design*** This chapter contains design tables for numerous prestressed/precast sections.
Chapter 3	***Analysis and Design*** This chapter repeats information found in the IBC and the ASCE 7 codes on loads, load combinations, and analysis. It also discusses seismic loads and diaphragm design. What distinguishes this chapter from IBC and ASCE 7 is the provision of solved examples.
Chapter 4	***Design*** This chapter contains several worked design examples for various sections.
Chapter 11	***General Design Information*** Design aids, shear, deflection, and moment diagrams for various loading scenarios and other information on accessories such as bolts and the likes.

AISC Steel Construction Manual

This manual consists of 17 parts as presented in its 2011 edition. All these parts are important, it is recommended to familiarize yourself with the below parts first.

Part	Title and Description
Part 1	***Dimensions and Properties*** This part documents all steel sections' properties.
Part 2	***General Design Considerations*** This part provides load combinations for LRFD and ASD methods along with some other serviceability requirements.
Part 3	***Design of Flexural Members*** This part provides equations that help in designing for flexure along with flexure design tables for all steel sections. It also provides shear, moment, and deflection equations and diagrams for beams similar to the ones provided in the NCEES PE handbook.
Part 4	***Design of Compression Members*** This part provides equations that help in designing for compression along with compression design tables for all steel sections.
Part 5	***Design of Tension Members*** This part provides tension design tables for all steel sections.
Part 6	***Design of Members Subject to Combined Forces*** This part provides equations and tables to easily identify and choose sections to withstand combined forces.
Part 7	***Design Consideration for Bolts*** Along with the equations this part provides tables with strength of bolt groups.
Part 8	***Design Consideration for Welds*** Along with the equations this part provides tables with strength of weld groups.
Part 10	***Design of Simple Shear Connections*** This part provides various design tables to help solve problems similar to problem 4.17 in this guide with various bolt and beam/column arrangements.
Part 16	***Specifications and Codes*** This is a large section which contains all the equations and specifications that support the entire design tables in this manual. You can refer to it when needed only.

NDS National Wood Design Specifications

This specification consists of 16 chapters and 14 appendices as presented in its 2015 edition. All these sections are important, it is recommended to familiarize yourself with the below first.

Chapter	Title and Description
Chapter 2	***Design Values for Structural Members*** This chapter provides wood properties that need to be looked at along with some wood adjustment factors applicable to all wood types.
Chapter 3	***Design Provisions and Equations*** This chapter provides design equations for flexure, shear, compression, deflection, and combined loading. It explains as well how to calculate the stability factor C_L which is key to column and beam design.
Chapter 4 to 10	These chapter present different wood types, one type per chapter, and each of those chapters provides the relevant adjustment factors per type.
Chapter 11	***Mechanical Connections*** This chapter provides the adjustment factors that should be applied to lateral loads on fasteners, split rings, shear plates, timber rivets and spike grids, along with withdrawal loads as well. All these loads are available in chapters 12 to 14. This chapter also explains how to calculate some important adjustment factors as well such as the group action factor.
Chapter 12	***Dowel Type Fasteners*** This chapter consists of tables that provide lateral and withdrawal loads for various fastener arrangements – those loads should have adjustment factors of chapter 11 applied on them. Chapter 13 and 14 are similar to Chapter 12 but used mainly for split rings, shear plates and timber rivets.
Chapter 16	***Fire Design of Wood Members*** This chapter provides methods on how to calculate effective char rates and depths.

It is important to note that all design values – e.g., moduli of elasticity, shear and bending allowable stresses, etc. – as well as all section properties such as length, width, and section modulus, should all be taken from the *NDS National Wood Design Supplement*.

Also, it is important to remember that NCEES are expecting ASD method to be used for all wood design questions – which may be subject to change anytime.

ASCE 7 Minimum Design Loads for Buildings and Other structures

This specification consists of 31 chapters as presented in its 2010 edition. All these chapters are important, it is recommended to familiarize yourself with the below chapters first.

Chapter	Title and Description
Chapter 1	***General*** This chapter talks about risk categories, minimum lateral loads, and importance factors.
Chapter 2	***Combination of Loads*** This chapter talks about load combinations in a similar fashion to Part 16 of the IBC code.
Chapter 3	***Dead Loads, Soil Loads and Hydrostatic Pressures*** This chapter specifies design lateral loads for different backfills.
Chapter 4	***Live Loads*** This chapter provides numerous properties and details relevant to live loads, such as reduction and maximization. It also defines live loads for various occupancies.
Chapter 7	***Snow Loads*** This chapter explains how to calculate the snow loads for various roof slopes, types, and various areas as well.
Chapter 8	***Rain Loads*** This chapter provides and explains the equations used to calculate rain loads with.
Chapter 11	***Seismic Design Criteria*** This chapter teaches you how to calculate seismic site coefficients such as the S_{DS} and the S_{D1}. It also shows you how to determine the Seismic Risk Category which is a prerequisite to the remainder of the seismic design procedure.
Chapter 12	***Seismic Design Requirements for Building Structures*** This chapter provides numerous valuable information on seismic design coefficients and procedures for buildings. *The following are few important highlights:* Section 12.2 is used for seismic system selection, it also specifies important factors and coefficients to be taken forward for the calculation of base shear. Section 12.3 defines diaphragm flexibility along with horizonal and vertical irregularities. Section 12.5 talks about the direction of seismic loading. Section 12.6 is a *key section*, it points you to which analytical procedure to use as those procedures differ with the Seismic Design Category. Section 12.8 explains the use of the Equivalent Lateral Force procedure. It explains how the seismic base shear is calculated, how to distribute forces vertically and horizontally.

	Section 12.14 discusses the simplified alternative design procedure for bearing/shear walls.
Chapter 20	***Soil Classification Procedure for Seismic Design*** Site values taken from this chapter can be used in equations mentioned in Chapter 12 and other relevant chapters.
Chapter 22	***Long Period Transition and Risk Coefficient Maps*** This chapter provides maps for different locations in the United States detailing information on spectral coefficients. This information to be used in chapter 12 and other relevant chapters.
Chapter 26	***Wind Loads: General Requirements*** This chapter provides most of the basic wind loads parameters and shall *be frequently referred to* as a starting point for any problem concerning wind loads.
Chapter 27	***Wind Loads on Buildings – MWFRS (Directional Procedure)*** This chapter explains the directional procedure, a slightly lengthier wind loads calculation procedure that applies to all building shapes and heights.
Chapter 28	***Wind Loads on Buildings – MWFRS (Envelope Procedure)*** This chapter refers to the envelope procedure for low rise $\leq 60\,ft$ buildings, and its methods are summarized in two parts: Part I contains a slightly more detailed procedure. Part II presents the simplified procedure with conditions.
Chapter 29	***Wind Load on Other Structures and Building Appurtenances*** This chapter takes care of any attachments to buildings, such as roof top equipment, parapets, or sign boards.

AASHTO LRFD Bridge Design Specification

This specification consists of 15 sections as presented in its 7th edition. All these sections are important. I recommend that you familiarize yourself with the below sections first.

Section	Title and Description
Section 3	***Loads and Load Factors*** This section discusses loads that should be used for bridge design and their corresponding load factors. Loads discussed include: dead loads, earth loads, vehicular live loads and the design trucks, fatigue loads, pedestrian loads, wind loads, water loads, etc. Some of the factors and allowances discussed in this section are the multiple presence factor, dynamic load allowance for normal, buried and wood components, collision and break forces, friction forces, etc.
Section 5	***Concrete Structures*** This section discusses material properties and requirements of concrete, reinforcing steel and prestressing steel used in bridge construction. It also discusses methods for the design of axial, torsion, shear, and flexure in different regions. This section goes further and talks about seismic design requirements in different seismic zones. It also covers the design of beams, girders, footings, piles, arches, and diaphragms. Lastly it talks about durability, corrosion resistance, freeze thaw resistance, etc.

Made in the USA
Las Vegas, NV
14 December 2023

82785395R00142